T0181944

Springer Collected Works in Mathematics

More information about this series at http://www.springer.com/series/11104

Andrei N. Kolmogorov

Selected Works III

Information Theory
and the Theory of Algorithms

Editor
Albert N. Shiryaev

Reprint of the 1993 Edition

 Springer

Author
Andrei N. Kolmogorov (1903 – 1987)
Moscow State University
Moscow, Russia

Editor
Albert N. Shiryaev
Russian Academy of Sciences
Moscow, Russia

ISSN 2194-9875
Springer Collected Works in Mathematics
ISBN 978-94-024-1710-4

This Springer imprint is published by the registered company Springer Nature B.V.
The registered company address is: Van Godewijckstraat 30, 3311 GX Dordrecht, The Netherlands

Selected Works of A. N. Kolmogorov

Volume III
Information Theory and the Theory of Algorithms

edited by

A. N. Shiryayev

Translated from the Russian by A. B. Sossinsky

KLUWER ACADEMIC PUBLISHERS

DORDRECHT / BOSTON / LONDON

ISBN 90-277-2798-8 (Vol. III)
ISBN 90-277-2795-3 (Set)

Published by Kluwer Academic Publishers,
P.O. Box 17, 3300 AA Dordrecht, The Netherlands.

Kluwer Academic Publishers incorporates
the publishing programmes of
D. Reidel, Martinus Nijhoff, Dr W. Junk and MTP Press.

Sold and distributed in the U.S.A. and Canada
by Kluwer Academic Publishers,
101 Philip Drive, Norwell, MA 02061, U.S.A.

In all other countries, sold and distributed
by Kluwer Academic Publishers Group,
P.O. Box 322, 3300 AH Dordrecht, The Netherlands.

Printed on acid-free paper

This is an annotated translation from the original work
ТЕОРИЯ ИНФОРМАЦИИ И ТЕОРИЯ АЛГОРИТМОВ

Published by Nauka, Moscow, © 1987

'Et moi, ..., si j'avait su comment en revenir, je
n'y serais point allé.'

Jules Verne

The series is divergent; therefore we may be
able to do something with it.

O. Heaviside

One service mathematics has rendered the
human race. It has put common sense back
where it belongs, on the topmost shelf next to
the dusty canister labelled 'discarded nonsense'.

Eric T. Bell

Mathematics is a tool for thought. A highly necessary tool in a world where both feedback and nonlineari-
ties abound. Similarly, all kinds of parts of mathematics serve as tools for other parts and for other sciences.

Applying a simple rewriting rule to the quote on the right above one finds such statements as: 'One service topology has rendered mathematical physics ...'; 'One service logic has rendered computer science
...'; 'One service category theory has rendered mathematics ...'. All arguably true. And all statements
obtainable this way form part of the raison d'être of this series.

This series, *Mathematics and Its Applications*, started in 1977. Now that over one hundred volumes have
appeared it seems opportune to reexamine its scope. At the time I wrote

"Growing specialization and diversification have brought a host of monographs and textbooks
on increasingly specialized topics. However, the 'tree' of knowledge of mathematics and
related fields does not grow only by putting forth new branches. It also happens, quite often in
fact, that branches which were thought to be completely disparate are suddenly seen to be
related. Further, the kind and level of sophistication of mathematics applied in various sciences has changed drastically in recent years: measure theory is used (non-trivially) in
regional and theoretical economics; algebraic geometry interacts with physics; the Minkowsky
lemma, coding theory and the structure of water meet one another in packing and covering
theory; quantum fields, crystal defects and mathematical programming profit from homotopy
theory; Lie algebras are relevant to filtering; and prediction and electrical engineering can use
Stein spaces. And in addition to this there are such new emerging subdisciplines as 'experimental mathematics', 'CFD', 'completely integrable systems', 'chaos, synergetics and large-
scale order', which are almost impossible to fit into the existing classification schemes. They
draw upon widely different sections of mathematics."

By and large, all this still applies today. It is still true that at first sight mathematics seems rather frag-
mented and that to find, see, and exploit the deeper underlying interrelations more effort is needed and so
are books that can help mathematicians and scientists do so. Accordingly MIA will continue to try to make
such books available.

If anything, the description I gave in 1977 is now an understatement. To the examples of interaction
areas one should add string theory where Riemann surfaces, algebraic geometry, modular functions, knots,
quantum field theory, Kac-Moody algebras, monstrous moonshine (and more) all come together. And to
the examples of things which can be usefully applied let me add the topic 'finite geometry'; a combination
of words which sounds like it might not even exist, let alone be applicable. And yet it is being applied: to
statistics via designs, to radar/sonar detection arrays (via finite projective planes), and to bus connections
of VLSI chips (via difference sets). There seems to be no part of (so-called pure) mathematics that is not
in immediate danger of being applied. And, accordingly, the applied mathematician needs to be aware of
much more. Besides analysis and numerics, the traditional workhorses, he may need all kinds of combina-
torics, algebra, probability, and so on.

In addition, the applied scientist needs to cope increasingly with the nonlinear world and the extra mathematical sophistication that this requires. For that is where the rewards are. Linear models are honest and a bit sad and depressing: proportional efforts and results. It is in the nonlinear world that infinitesimal inputs may result in macroscopic outputs (or vice versa). To appreciate what I am hinting at: if electronics were linear we would have no fun with transistors and computers; we would have no TV; in fact you would not be reading these lines.

There is also no safety in ignoring such outlandish things as nonstandard analysis, superspace and anticommuting integration, p-adic and ultrametric space. All three have applications in both electrical engineering and physics. Once, complex numbers were equally outlandish, but they frequently proved the shortest path between 'real' results. Similarly, the first two topics named have already provided a number of 'wormhole' paths. There is no telling where all this is leading - fortunately.

Thus the original scope of the series, which for various (sound) reasons now comprises five subseries: white (Japan), yellow (China), red (USSR), blue (Eastern Europe), and green (everything else), still applies. It has been enlarged a bit to include books treating of the tools from one subdiscipline which are used in others. Thus the series still aims at books dealing with:

- a central concept which plays an important role in several different mathematical and/or scientific specialization areas;
- new applications of the results and ideas from one area of scientific endeavour into another;
- influences which the results, problems and concepts of one field of enquiry have, and have had, on the development of another.

The roots of much that is now possible using mathematics, the stock it grows on, much of that goes back to A.N. Kolmogorov, quite possibly the finest mathematician of this century. He solved outstanding problems in established fields, and created whole new ones; the word 'specialism' did not exist for him.

A main driving idea behind this series is the deep interconnectedness of all things mathematical (of which much remains to be discovered). Such interconnectedness can be found in specially written monographs, and in selected proceedings. It can also be found in the work of a single scientist, especially one like A.N. Kolmogorov in whose mind the dividing lines between specialisms did not even exist.

The present volume is the third of a three volume collection of selected scientific papers of A.N. Kolmogorov with added commentary by the author himself, and additional surveys by others on the many developments started by Kolmogorov. His papers are scattered far and wide over many different journals and they are in several languages; many have not been available in English before. If you can, as Abel recommended, read and study the masters themselves; this collection makes that possible in the case of one of the masters, A.N. Kolmogorov.

The shortest path between two truths in the real domain passes through the complex domain.

J. Hadamard

La physique ne nous donne pas seulement l'occasion de résoudre des problèmes ... elle nous fait pressentir la solution.

H. Poincaré

Never lend books, for no one ever returns them; the only books I have in my library are books that other folk have lent me.

Anatole France

The function of an expert is not to be more right than other people, but to be wrong for more sophisticated reasons.

David Butler

Amsterdam, September 1992

Michiel Hazewinkel

CONTENTS

Papers by A. N. Kolmogorov

Comments and addenda

From A. N. Kolmogorov's recollections

GREETINGS TO A. N. KOLMOGOROV
FROM THE MOSCOW MATHEMATICAL SOCIETY

Dear Andrei Nikolayevich!

The Managing Board of the Moscow Mathematical Society, members of the Society heartily congratulate you on the occasion of your 80th anniversary. Your entire scientific life has been most closely connected with the Moscow Mathematical Society. You first spoke at a session of the Society on October 8, 1922; your talk then, on "An Example of a Fourier-Lebesgue Series Diverging Almost Everywhere", made you, a 19-year-old student, renowned throughout the mathematical world. Beginning with that moment, you have given important reports on very varied topics at sessions of the Society no less than 98 times.

The list of your reports strikes by the wide range of topics considered. It suffices to enumerate only some of the titles of your reports:

"On Fourier series diverging everywhere" (11.16.1926);

"About a general scheme for the theory of probability" (03.20.1928);

"A new interpretation of intuitionist logic" (12.18.1928);

"On the geometric ideas of Plücker and Klein" (12.11.1932);

"Markov chains and the reversibility of the laws of nature" (01.05.1935);

"Statistical theory of the crystallization of solidified metals" (02.16.1937);

"Current questions of point set geometry" (12.22.1938);

"Stationary sequences of elements of Hilbert space" (06.04.1939);

"On two types of axiomatic methods" (03.18 and 04.01.1941);

"On measures invariant with respect to transformation groups" (04.16.1941);

"Unitary representations of unitary groups" (02.02.1944);

"The mathematical theory of turbulence" (10.03.1944);

"The structure of complete metric Boolean algebras" (12.09.1947);

"Best approximations of complex functions" (02.24.1948);

"On certain mathematical problems related to production control" (01.09.1951);

"Two-valued functions of two-valued variables and their applications to switching circuits" (11.27.1951);

"On the spectra of dynamical systems on the torus" (09.30.1952);

"On almost periodic motion of a solid about a fixed point" (05.26.1953);

"Estimates of the minimal number of elements in ε-nets in various functional classes and their applications to the represention of functions of several variables as a superposition of functions of a lesser number of variables" (04.27.1954);

'"On certain asymptotic characteristics of completely bounded metric space" (06.05.1956);

"Uniform limiting theorems for sums of independent summands" (12.18.1956);

"Small denominators in problems of mechanics and analysis" (01.13.1956);

"What is information" (06.04.1961);

"Self-constructing devices" (11.21.1961);

"Computable functions and the foundations of information theory and probability theory" (19.11.1963);

"Experiments and mathematical theory in the study of turbulence" (05.18.1965);

"Statistical hydrodynamics of the ocean" (02.24.1970);

"Complexity of definition and complexity of construction for mathematical objects" (11.23.1971);

"On statistical solutions of the Navier-Stokes equations" (01.18.1978).

You have delivered many survey lectures which served as the source of ideas for further research:

"Measures and probability distributions in function spaces" (11.30.1948);

"Solved and unsolved problems related to the 13th Hilbert problem" (05.17.1960);

"Mathematical methods in the study of Russian verse" (12.27.1960);

"On uniform limiting theorems for sums of independent variables" (03.26.1963);

"The complexity of algorithms and an objective definitions of randomness" (04.16.1974).

Andrei Nikolayevich, you have been the initiator and lecturer at numerous sessions of the Society devoted to discussions on projects of articles prepared for the *Great Soviet Encyclopaedia*. In 1938 (on March 10) you participated in the session devoted to a discussion of your famous article for the encyclopaedia called "Mathematics". Several other times you made reports on your articles for the *Great Soviet Encyclopaedia*: "Axioms", "Infinitely Small Magnitudes" (18th and 19th of October, 1949) and others.

Of special interest to the society were your reports on the following topics:

"The development in the USSR of mathematical methods for the study of nature" (11.10.1937);

"On Ostrogradsky's criticism of the works of Lobachevsky" (09.29.1948);

"The development of mathematics in the Soviet period" (12.23.1949);

"Elements of Mathematics by Nicholas Bourbaki" (05.29.1956);

"On certain traits of the current stage in the development of mathematics" (04.21.1959);

"From my experience" (04.25.1963);

"On the International Congress of Mathematicians in Amsterdam" (09.28.1954);

"On my scientific trip to France, GDR and Poland" (02.28.1956).

You have always given great importance to discussions with the mathematical community of questions related to secondary school education.

On November 22 and 28, 1937, a session of the Society was devoted to discussing a new outline of a textbook in elementary algebra prepared by yourself and P.S. Alexandrov. You also made the following reports:

"On a project of a syllabus in mathematics for secondary schools" (03.03.1946);

"On a project of a syllabus for secondary schools" (02.17.1948);

"On the optional study of mathematics in secondary schools in the school year of 1967/1968" (02.14.1967).

Dear Andrei Nikolayevich, you are member of the Mathematical Society since February 1, 1930, an Honorary Member since April 28, 1953, now you are the President of our Society and one can say with assurance that, together with P. S. Alexandrov, you have determined the essence of today's Moscow Mathematical Society.

Your anniversary is a joyous occasion for all of us. Please accept, dear Andrei Nikolayevich, our feelings of deepest respect and admiration.

We wish you the best of health and happiness!

The Managing Board of the Moscow Mathematical Society

ANDREI NIKOLAEVICH KOLMOGOROV[*]

(on the occasion of his 80th anniversary)

N. N. Bogolyubov, B. V. Gnedenko, S. L. Sobolev

April 25, 1983, was the 80th birthday of Andrei Nikolaevich Kolmogorov. In the greetings from the Mathematics Section of the USSR Academy of Sciences, the Moscow Mathematical Society and the Editorial Board of the journal Uspekhi Matematicheskykh Nauk on the occasion of Andrei Nikolaevich's 70th birthday, it was stated: "You have made an outstanding contribution to the development of contemporary mathematics and its applications. Your fundamental research has determined the entire outlook of many branches of 20th century mathematics. The theory of trigonometric series, measure theory, set theory, integration theory, constructive logic, topology, approximation theory, probability theory, the theory of random processes, information theory, mathematical statistics, dynamical systems, finite automata, the theory of algorithms, mathematical linguistics, turbulence theory, celestial mechanics, differential equations, Hilbert's 13th problem, the theory of artillery fire, applications of mathematics to problems of biology, geology, cristallization of metals is but an incomplete list of branches of mathematics and its applications that you have enriched with your profound ideas. You have always attached great importance to the applications of mathematics and its relationship with other sciences. The extraordinary breadth of your research interests, your pedagogical talent have always attracted talented young scientists. You have created a world famous scientific school. Representatives of this school work in the most varied areas of mathematics and other sciences"[1].

In the greetings on the occasion of Andrei Nikolaevich's 70th aniversary from the Presidium of the USSR Academy of Sciences, it is stated, in particular: "You have made an important contribution to the dissemination of mathematical knowledge by heading the Mathematical Section of the Great Soviet Encyclopaedia. You have created a large scientific school, from which numerous outstanding scientists have emerged. You play an important role in the activities of the Mathematics Section of the USSR Academy of Sciences. Your contribution to the development of higher education in our country is widely known..."[2].

---

[*]Uspekhi Matem. Nauk, 1983, vol. 38, vyp. 4, pp. 11-23

[1]Uspekhi Mat.Nauk, 1978, vol. 33, vyp. 2, 212-213.

[2]Uspekhi Mat.Nauk, 1973, vol. 28, vyp. 5, p.3.

A. N. Kolmogorov was born in Tambov. His father was an agronomist. Andrei Nikolaevich's mother died during his birth. Her sister, Vera Yakovlevna, a very independent woman of high social ideals, took upon herself the child's upbringing. She succeeded in conveying these ideals to her nephew, taught him the feelings of responsibility, independence of judgement, intolerance of laziness and of a poorly done job, the desire to understand and not only memorize. He began to work early, and before entering Moscow university actually served as a railroad ticket collector for a certain time.

Andrei Nikolaevich entered Moscow University in the fall of 1920, but did not decide to be a mathematician from the outset. For a long period of time his main interest was Russian history, and he took an active part in the seminar headed by professor S. V. Bakhrushin. He carried out a serious research study of 15-16th century cadasters in connection with land appropriation in ancient Novgorod. It is known that Andrei Nikolaevich put forward a conjecture in the 20ies about the settling of the upper Pinega valley, and this Kolmogorov conjecture was later substantiated by an expedition to that area headed by P. S. Kuznetsov.

A. N. Kolmogorov's interest in the humanities reasserted itself with new force and new possibilities in the 60ies, when Andrei Nikolaevich gave much time and attention to questions of versification and to the application of mathematical methods to the study of poetry; he wrote several articles, gathered several younger researchers, participated in scientific conferences, made reports on these topics. Systematic work began in 1960, when A. N. Kolmogorov, finally realizing his long-standing aspiration, began research work on mathematical linguistics in the framework of the chair of probability theory and his newly-created problem laboratory of mathematical statistics. This work immediately was met with great interest and without mistrust by experts in linguistics and literature. Already in 1961, at the 4th Mathematical Congress in Leningrad, and then at a conference in Gorki on the application of mathematical methods to the study of the language of literary works, Andrei Nikolaevich's young collaborators reported the first results on probabilistic methods in the study of versification and on the determination of the entropy of Russian speech by means of probabilistic experiments.

But let us return to the early years of Andrei Nikolaevich's life. His development as a mathematician was significantly influenced by V. V. Stepanov's seminar on trigonometric series. Participants of the seminar were presented with unsolved problems whose answers seemed of immediate interest.

In 1922, after A. N. Kolmogorov wrote his first independent work (on the values of Fourier coefficients) he became the pupil of N. N. Luzin.

Participation in V. V. Stepanov's seminar and subsequent work under the guidance of N. N. Luzin marks the first creative period in the life of Kolmogorov the mathematician. This period is characterized by the fact that the problems considered and Kolmogorov's approach to them remained in the mainstream of the basic interests, at that time, of the Moscow mathematical school.

In particular, under the influence of work by M. Ya. Suslin, N. N. Luzin and P. S. Alexandrov, A. N. Kolmogorov aspired to develop a general theory of operations on sets. Already in the spring of 1922 A. N. Kolmogorov concluded a large research work on the theory of operation on sets. He introduced a very wide class of operation on sets, namely δs-operations. For reasons independent of the author, this work was only published in 1928. It attracted the attention of researchers and the ideas contained in it were later developed by a several scientists, L. V. Kantorovich, A. A. Lyapunov and others.

In June 1922, A. N. Kolmogorov constructed an example of a Fourier-Lebesgue series diverging almost everywhere and soon after that a Fourier-Lebesgue series diverging at every point. Both of these examples were completely unexpected for the experts and produced a huge impression. An important researcher had entered the theory of functions of a real variable. Interest in the theory of Fourier series and the theory of orthogonal functions remained with Andrei Nikolaevich for the rest of his life and from time to time he returned to problems from this branch of study, setting new problems before younger researchers.

In the theory of functions, A. N. Kolmogorov was interested in all the basic problems: questions of derivation, integration, measure theory, the theory of approximation of functions, etc. In each of the questions that he studied he introduced an element of essential renewal.

It is in 1924 that A. N. Kolmogorov became interested in probability theory, the branch of science where his authority is especially important. The first step in this new branch of research he carried out jointly with A. Ya. Khinchin. They succeeded in finding necessary and sufficient conditions for the convergence of series whose terms are mutually independent random variables. The importance of this paper is not limited to the complete solution of an important problem, it laid down the foundations of a method that was used afterwards numerous times and with great success for the solution of very varied questions. In 1928, A. N. Kolmogorov succeeded in indicating necessary and sufficient conditions for the law of large numbers, a problem to whose solution important contributions were made by P. L. Chebyshev and A. A. Markov (senior). A year later, a large paper appeared in which the law of repeated logarithms for sums of independent random variables was proved under very wide conditions imposed on the summands. A few years before that, the law of repeated logarithm had been discovered by A. Ya. Khinchin first for the Bernoulli scheme and later carried over to the Poisson scheme. In the same year an outstanding result was obtained: wide conditions for the applicability of the strengthened law of large numbers. A while later these conditions were used as the basis by A. N. Kolmogorov for finding necessary and sufficient condition for a strengthened law of large numbers in the case of identically distributed independent summands. This condition is remarkably simple: it is the existence of a finite expectation for each of the summands.

Once again we should stress that in all the papers mentioned above it is not only the final results that are important but also the methods, which later became part of the arsenal used contemporary probability theory. In particular, the famous "Kolmogorov inequality", generalizing the classical "Chebyshev inequality" was proved. Special mention should be made of the small note by Andrei Nikolaevich called *The general theory of measure and the calculus of probabilities*, which first presented an original sketch of the axioms of probability theory based on measure theory and the theory of functions of a real variable. By itself this approach was not entirely new (since it had been outlined by E. Borel and A. Lomnitsky), but it is precisely in the hands of A. N. Kolmogorov that it obtained its conclusive form, as simple and clear mathematical statements that were to gain universal acceptance. E. Borel as early as 1909 expressed general considerations on the importance of measure theory for the construction of the foundations of probability theory. In 1923 the general ideas obtained a development in a large article by A. Lomnitsky and in 1929 became the object of study of A. N. Kolmogorov. Four years later, in 1933, E. Borel's original idea acquired its final form in the classical monograph by A. N. Kolmogorov *Main notions of probability theory* published in German by Springer-Verlag. Four years later it appeared in Russian translation and soon became a bibliographic rarity (in 1974 the second Russian edition was published). This monograph not only determined a new stage in the development of probability theory as a mathematical science, but gave the necessary foundations for the construction of the theory of random processes.

In 1931 A. N. Kolmogorov's remarkable paper *Analytical methods in probability theory* appeared. It contained the foundations of the contemporary theory of Markov random processes and uncovered the deep relationships existing between probability theory and the theory of differential equations, ordinary differential equation as well as second order partial differential equations. Some disconnected results relating the problems of the theory of Markov chains and Markov random walks on the line with equations of parabolic type had been obtained a long time before that by Laplace and later by Focker and Planck. But these were only examples. The true theory of Markov processes or, as A. N. Kolmogorov would say, "processes without afteraction", were founded by Andrei Nikolaevich in this paper. In it he obtained direct as well as inverse equations. A new branch of mathematics with numerous openings to the applications appeared. It was immediately picked up by people working in physics and biology, chemistry and engineering, and soon became one of the most powerful mathematical tools of contemporary science.

A lasting impression was created by the article *Mathematics* written by A. N. Kolmogorov for the second edition of the *Great Soviet Encyclopaedia*. In this article A. N. Kolmogorov (in condensed form and on a new basis and principle) follows the historical development of mathematics, indicates the key periods of this development, and proposes a new original outline of periodization. Unfortunately, this article was never published in our country as a separate book (probably it is not too late to do this; in general A. N. Kolmogorov's articles for the GSE warrant special attention and edition). But even without this, the article *Mathematics* had an important influence on the direction of study in the domain of history of mathematics, and not only in our country. It should be noted that in the German Democratic Republic a separate edition of this work was published.

A. N. Kolmogorov's mind worked in the most diverse directions, setting forth new problems and solving questions of principle. In the ten years that directly preceded World War II, he published more than 60 papers that were concerned with probability theory, projective geometry, mathematical statistics, the theory of functions of a real variable, topology, mathematical logic, mathematical biology, philosophy and the history of mathematics. Among the works of that decade were some which set the foundation of new directions of mathematical thought in probability theory, in topology, in the theory of approximation of functions.

It is precisely in this period that A. N. Kolmogorov introduced the notion of upper boundary operator or ∇-operator in topology simultaneously with the American mathematician Alexander and independently of the latter. By means of this operatòr, he constructed the theory of cohomology groups or, as they were then called, ∇-groups (first for complexes and then for arbitrary bicompact spaces). The notion of cohomology group turned out to be very convenient and a very strong tool for the study of numerous questions of topology, in particular those related to the study of continuous maps. It is on this basis that he constructed the notion of cohomology ring, one of the important concepts of topology. It is necessary to indicate also the extremely general statement of the duality laws relating to closed sets located in arbitrary locally bicompact completely regular topological spaces satisfying only the conditions of acyclycity in the dimensions that concern the statement of the result itself.

It is then that he constructed an example of an open map of a compact set onto a compact set of higher dimension.

During the same period, in probability theory he wrote the monograph *Fundamental notions of probability theory*, gave a presentation of unboundedly divisible distributions, found the limit distribution for the maximal discrepancy of an empiric distribution from the true distribution (the Kolmogorov criterion), constructed the theory of Markov chains with a countable number of states, found a relationship between the geometry of Hilbert space and a series of problems in the theory of stationary sequences.

At the end of the 30s A. N. Kolmogorov's attention was more and more attracted by the mechanics of turbulence, i.e. by the laws of those flows of liquid and gas which often appear in practice and which are characterized by disorderedly pulsations in velocity pressure and other hydrodynamic expressions. A correct mathematical description of such flows necessarily must be statistical in nature and be based on the general notion of random function, as was already explained in 1938 in the first paper by M. D. Millionshchikov, the first representative of the Kolmogorov school in turbulence theory. A rigorous statistical approach was systematically used by A. N. Kolmogorov himself; it is precisely in the work of A. N. Kolmogorov and his pupils that the theory of turbulence acquired a clear mathematical formulation in the form of an applied chapter of the theory of measure in functional spaces. However, the main papers by Andrei Nikolaev on the mechanics of turbulence that appeared in 1941 had a physical rather than a formal mathematical character. These papers were based on a deep understanding of the essence of the processes themselves, typical of an extremely complicated non-linear physical system with very many degrees of freedom, such as turbulent flow is; it is precisely his deep physical intuition that helped A. N. Kolmogorov display very refined properties of "local self-similitude" in these processes (this notion later played an important role in many problems of theoretical physics) and use it to deduce fundamental quantitative relationships, which were essentially new laws of nature. Among these laws let us note the famous Kolmogorov "law of two thirds" possessing, as is typical of the fundamental laws of nature, the property of extreme simplicity: in any developed turbulent flow the mean square of the difference of velocities at two points located at a distance r (which is not too small and not too large) is proportional to $r^{2/3}$.

By the time Kolmogorov's first papers on this topic first appeared, no experimental data allowing the verification of the quantitative laws that he had put forth were available; later, however, these laws were compared many times with measurement data in natural media (atmosphere, ocean) as well as in laboratory devices, and always turned out to be justified to a high degree of precision. A. N. Kolmogorov also made certain qualitative predictions; thus he predicted the laminary structure of the ocean that was later supported by experiments, the effect now known as the "pancake effect". In 1961-1962 A. N. Kolmogorov once more returned to the results of his own studies in mechanics of turbulence and showed that if one no longer admits one of the assumptions he had used earlier and which had seemed quite natural, it is possible to obtain certain specifications of the laws that he had obtained, which are very difficult to discover in experiments because of their small size, but which nevertheless has been verified by a series of experiments in the very last years.

The return to old topics is typical of the creative process of A. N. Kolmogorov overall: he can only open new and new domains for himself, but he cannot leave them once and for all. This trait is especially noticeable in the postwar period, when Andrei Nikolaevich again returned both to turbulence and to functions of a real variable (in connection with the 13th Hilbert problem and to the logical foundations of mathematics in their appliation to geometry, probability theory and information theory). Here at the same time Andrei Nikolaevich was involved in an extremely wide spectrum of topics, including classical mechanics, ergodic theory, the theory of functions and the theory of information, the theory of algorithms), and this spectrum is continuous, its topics which at first seem far away from each other, turn out to be related in completely unexpected ways discovered by A. N. Kolmogorov. The number of papers in each topic is usually not very large, but all of them are fundamental, not only by the strength of the results but by the influence on the entire further development of the corresponding domain of science.

To illustrate the above it is sufficient to consider the work of Andrei Nikolaevich in the theory of dynamical systems, in which all the originality of his talent appeared most explicitly. These papers are very few in number, but they are important in that they open new facets of the theory which eventually developed into large new directions. The papers fall into two cycles, one of which originates in problems of classical mechanics and the other in problems of information theory.

Works of the first cycle were devoted to the general theory of Hamiltonian systems and were carried out in a short two-year period of 1953-1954. On the conclusions of these papers A. N. Kolmogorov made an important review report at the International mathematical congress in Amsterdam in 1954. The problem on the evolution of orbits in the problem of three or more bodies date back to Newton and Laplace; in the case of small masses of planets, this problem is a particular case of the behaviour of quasiperiodic motions of Hamiltonian systems under small changes of the Hamiltonian function. It is this last problem that Poincaré called the "main problem of dynamics". In the works of A. N. Kolmogorov, this problem was solved for most of the initial conditions in the case of general position. The applications of Kolmogorov theory to various specific problems made it possible, in subsequent years, to solve numberous problems which had awaited their solution for decades. For example, it follows from A. N. Kolmogorov's theorem that the motion of an asteroid in the flat circular problem of three bodies is stable, as well as in the rapid rotation of a heavy nonsymmetric solid. Kolmogorov's theorem is also applied in the study of magnetic power lines, which is of great importance in plasma physics.

The method used to obtain these results itself turned out to be extremely fruitful. Later on it was used in an improved direction by Andrei Nikolaevich's pupil
V. I. Arnold and by Yu. Moser, and is now known generally as KAM theory (i.e.,
the theory of Kolmogorov-Arnold-Moser).

The work of A. N. Kolmogorov and V. I. Arnold on the theory of perturbations
of Hamiltonian systems received the Lenin Prize in 1965.

The second cycle of papers on dynamical systems consisted in the application of
the ideas of information theory to the ergodic theory of such systems.

The interest of Andrei Nikolaevich in information theory dates back to 1955. He
did a great deal in order to disseminate and popularize this theory in our country.
In June 1956 he (jointly with I. M. Gelfand and Ya. M. Yaglom) made a report on
this topic at the 3rd All-Union Mathematical Congress in Moscow and in October
1956 made a report before the general meeting of the Academy of Sciences of the
USSR. Concentrated reflection on the ideas of Shannon's theory of information led
A. N. Kolmogorov to a completely unexpected and courageous synthesis of these
ideas, first with the ideas of approximation theory that he had developed in the 30s
and then with the ideas of the theory of algorithms. Concerning the algorithmic
theory of information, we shall have a lot to say further, on but now let us note
the extremely fruitful introduction of information characteristics (namely entropy
characteristics) in the study both of metric spaces and of dynamical systems.

In his papers of 1955-1956 A. N. Kolmogorov introduces the notion of ε-entropy
of sets in a metric space and thus obtains a means to estimate the "metric mass"
of functional classes and spaces. Using this notion, Andrei Nikolaevich gives an
entropy interpretation of the remarkable results of A. G. Vitushkin on the non-
representability of functions in n variables of smoothness r as the superposition of
function of m variables of smoothness l if $n/r > m/r$. The thing is, the ε-entropy
of any class of functions is, roughly speaking, the amount of information which
allows to indicate a function of this class with precision ε. As A. N. Kolmogorov
showed, ε-entropy of natural classes of r times differentiable functions in n variables
grows as $\varepsilon^{-n/r}$ (the necessary amount of information is greater when the number of
independent variables is greater and is lesser when the smoothness requirements are
higher and it is precisely the ratio of dimension to smoothness which is essential).
This immediately implies the Vitushkin theorem: indeed, the amount of information
necessary to give the superposition cannot be greater (in order of magnitude) than
the amount of information sufficient to give the constituents of the superposition
function of a lesser number of variables).

These studies led A. N. Kolmogorov to attack the 13th Hilbert problem directly;
this problem consists in proving that it is impossible to represent some continuous
function of three variables in the form of a superposition of continuous functions
of two variables. A. N. Kolmogorov proved an extremely unexpected result in
1956: each continuous function of any number of variables can be represented
in the form of a superposition of continuous function of three variables. As to
the Hilbert problem itself, it turned out to be reduced to a certain problem of
representing functions given on universal trees in three-dimensional space. This
last problem was solved by V. A. Arnold in a paper written under the guidance
of A. N. Kolmogorov in 1957, and it was solved by obtaining an answer which

contradicted the Hilbert conjecture: each continuous function of three variables turned out to be representable as the superposition of continuous functions of two variables. That same year A. N. Kolmogorov made the decisive step, proving that a continuous function of any number of variables can be represented in the form of a superposition of continuous functions of one variable and addition.

A great impression was produced by the introduction of entropy characteristics in the theory of dynamical systems. This introduction was effected out in the small note by Andrei Nikolaevich called *A New metric invariant of transitive dynamical systems and automorphisms of Lebesgue spaces*, and note which opened his second cycle of works on dynamical systems and played an extremely important role in the development of ergodic theory of dynamical systems. By then, the existence of metric (i.e. not depending on any structures other than measure) invariants (spectrum, mixing) did not allow to distinguish from the metrical point of view even such systems as the simplest (i.e., with two outcomes) Bernoulli automorphisms with different probabilities p. A. N. Kolmogorov's note solved this classical problem: systems with different p turn out to be non-isomorphic, as was expected. The success was guaranteed by the introduction of an entirely new metric invariant, coming from information theory, the entropy of the dynamical system. This note opened long series of studies in our country and abroad, which were concluded in the last years by Ornstein's theorems, which claim that for a sufficiently rapidly mixing dynamical system the Kolmogorov invariant entirely determines the systems up to metric isomorphism (so that, for example, Bernoulli automorphisms with equal entropies are metrically isomorphic). Another important notion introduced in the same paper by A. N. Kolmogorov was the notion of quasiregular or, as it is now called, K-system ("K" stands for A. N. Kolmogorov). This notion plays an important role in the analysis of classical dynamical systems with strong stochastic properties. The appearance of such properties in dynamical systems is explained by their internal dynamical non-stability and the accepted method of study of these properties consists in discovering that the system under consideration is a K-system. There is no doubt that the widespread character of the evolution of the applications of the ideas of ergodic theory to problems of physics, biology, chemistry has as one of its sources precisely this paper by A. N. Kolmogorov.

In 1958-1959 Andrei Nikolaevich brought together in one seminar studies in the ergodic theory of dynamical systems and in hydrodynamical non-stability. The program of this seminar supposed constructing (at least in simple model cases) the ergodic theory of the notion of flows established after loss of stability by laminary flows. This program for applying the ideas of ergodic theory to phenomena of turbulence type had an important influence on all subsequent papers in this direction.

If in the 50s, A. N. Kolmogorov's efforts were directed to the use of the notions of information theory in other domains of science, in the 60s he undertakes a reconstruction of information theory itself, a reconstruction based on the algorithmic approach. In two fundamental papers published in 1965 and 1969 in the journal *Problems of Information Transmission*, Andrei Nikolaevich creates a new domain of mathematics, the algorithmic theory of information. The central position in this theory is occupied by the notion of complexity of a finite object for a fixed (algorithmic) method of its description. This complexity is defined in a very natural way, as the minimal volume of description. Kolmogorov's theorem establishes that among all possible algorithmic methods of description there exist optimal ones, those for which the complexity of the objects described turns out to be relatively small; although the optimal method is not unique, for two given optimal methods, the corresponding complexities differ no more than by an additive constant. As always in A. N. Kolmogorov's work, the new notions turn out to be at the same time quite natural, unexpected and simple. In the framework of these ideas it turned out to be possible, in particular, to define the notion of individual random sequence (which is impossible in terms of classical probability theory): one must call random any sequence whose complexity (under any optimal method of description of its initial segment) grows sufficiently fast as the length of the segment increases.

Since 1980 Andrei Nikolaevich heads the chair of mathematical logic of the mechanics and mathematics department of Moscow University. Mathematical logic (in its wider sense, including the theory of algorithms and the foundations of mathematics) is an early and late love in his life. Already in 1925 A. N. Kolmogorov published in the *Matematicheskiy Sbornik* a paper on the law of the excluded third, which once and for all became part of the "golden reserve" of papers in mathematical logic. This was the first Russian publiation in mathematical logic containing new mathematical results (which were actually very important) and the first systematic study in the world of intuitionist logic (logic that does not accept the law of *tertio non datur*). In this paper, in particular, the so-called embedding operations used to include one logical calculus in another first appeared. By means of one of these operations (historically the first, now known as the "Kolmogorov operation") the paper under consideration succeded in including classical logic into intuitionist logic, thereby proving that the application of the law of the excluded third by itself cannot lead to a contradiction. Then in 1932 Andrei Nikolaevich published a second paper in intuitionist logic in which he proposed the first semantics for this logic free of the philosophical considerations of intuitionism. It is precisely this paper that gave the possibility to interpret intuitionist logic as constructive logic.

In 1952 A. N. Kolmogorov proposes the most general definition of a constructive object and the most general definition of algorithm. The importance of these definitions is being understood completely only today. In 1954 he states the initial notions of the theory of numerations.

In 1972 Andrei Nikolaevich's initiative leads to the introduction of a course of mathematical logic in the general curriculum at the mechanics and mathematics department of Moscow University. A. N. Kolmogorov creates the syllabus of this course, still in force today and is the first to deliver the course.

And here we come to one of the most typical traits of the biography of A. N. Kolmogorov. All his life Andrei Nikolaevich not only enhanced the development of science, but taught people of the younger generation.

The pedagogical activity of A. N. Kolmogorov began in 1922, when he became a teacher of an experimental school of Narcompros of the Russian Federation. He worked there until the year of 1925. This activity he found very rewarding, and for the rest of his life he remained in spirit a teacher and educator. In 1931 Andrei Nikolaevich became a professor at Moscow University and from 1933 to 1939 was also the director of the Scientific Research Institute in Mathematics at Moscow State University. Many efforts and initiatives were carried out by him in order to make the Mathematics institute of the University one of the leading research centres of the country. Special attention was paid to the education of young scientists , great pains were taken and notable success was achieved in finding talented and dedicated young scientists among students and research fellows. One of the pupils of A. N. Kolmogorov was N. D. Milionshchikov, later to become Vice President of the USSR Academy of Sciences, who first came to his seminar after being only involved in practical work. He then did his graduate work under the guidance of Andrei Nikolaevich and in 1938 was accepted with the latter's recommendation to the Institute of Theoretical Geophysics of the Academy of Sciences of the USSR, headed at the time by O. Yu. Shmidt.

One of the authors recalls the seminar in probability theory in 1934, then headed by A. N. Kolmogorov and A. Ya. Khinchin. The character of the seminar's sessions, where the reports were very sharply discussed, where conjectures were put forward, propositions were made, solutions other than the ones proposed by the reporter were suggested, produced a very strong impression.

Among Andrei Nikolaevich's graduate students at that time, we note A. I. Maltsev, S. M. Nikolsky, B. V. Gnedenko, I. M. Gelfand, G. M. Bavli, I. Ya. Verchenko. Their areas of interest were respectively mathematical logic, functional analysis, probability theory, functional analysis again, probability theory, theory of functions.

Andrei Nikolaevich strived to create a team of graduate students that would constantly be in a state of scientific ferment and intense search. And he succeeded in this. As the director of the Mathematics Research Institute, he systematically met with all the grduate students, whether they were his own pupils or not, learning what they were working on, what problems they were trying to solve. If it turned out that one of them had no problem to think about, he would propose one immediately. Later he would inquire about how the solution was coming along, thus involving the graduate student in active research. For all of Kolmogorov's pupils the years of graduate study were always an unforgettable period of their lives, the time when they entered into scientific research, pondered on the purpose of science, formed their belief in the boundless possibilities of human creative powers. At the same time, Andrei Nikolaevich tried to develop their general cultural interests, above all in art, architecture and literature, as well as in sports (he was himself an accomplished cross-country skier, swimmer and untiring hiker).

Andrei Nikolaevich's unique personality, his broad erudition, the variety of his scientific interests, the depth of his penetration into any topic, all this served his students as a perhaps unattainable but very inspiring role model.

The out of town excursions organized by Andrei Nikolaevich for undergraduate and graduate students, to which he would invite three-five people who interested him, were quite unforgettable. These excursions were filled with talks about problems of mathematics and its applications, about the architectural points of interest along the way, about outstanding events in cultural life. But most important were conversations about specific problems proposed to the attention of his younger colleagues for individual study. Andrei Nikolaevich possessed the art of setting a problem so that it would interest the person for whom it was intended, seem close to him, give rise to strong desire to solve it. It is amazing how Andrei Nikolaevich was always capable of choosing a problem accessible to the given person, yet requiring a complete mobilisation of his forces. His questions about what one has achieved kept people under constant pressure and helped them think in a well-defined direction. It was always embarassing to have to say that nothing had been obtained, that there is nothing new to say. In this individual work Andrei Nikolaevich's immense and unique pedagogical talent as a teacher of creative young people was most obvious. He made people work without forcing them to; taught them to believe in the power of reason and in the necessity of scientific research. During these excursions, each of Andrei Nikolaevich's pupils would talk to him several times, he would change the topics of conversation, always keeping its directions in hand. Many excursions would end in Komarovka, where Andrei Nikolaevich and Pavel Sergeevich Alexandrov would have dinner with all the guests. The participants of the excursion, undergraduate and graduate students, tired, filled with impressions of the conversations and new mathematical ideas, would walk back to the railroad station to return to Moscow. It is impossible to overestimate the educational value of these excursions.

The people listed above were still graduate students, but Andrei Nikolaevich was already looking among undergraduates for talented young students ready to give all their forces to science. Thus appeared G. E. Shilov, M. K. Fage, as well as two graduate students from Saratov: A. M. Obukhov and V. N. Zasukhin (the latter, after having defended his Ph. D. Thesis in 1941, was killed in the war). The active involvement in research work of younger people did not stop during the war: in 1942 A. S. Monin became a graduate student of Moscow University under A. N. Kolmogorov (soon afterwards he left for the front and returned to his graduate work in 1946); in 1943 A. M. Yaglom became a graduate student at the Mathematics Institute of the Academy of Sciences.

After the war Andrei Nikolaevich was surrounded by new young mathematicians whom he introduced to this science. Among them B. A. Sevastyanov, S. Kh. Siradzhinov, M. S. Pinsker, Yu. V. Prokhorov, G. I. Barenblatt, L. N. Bolshev, R. L. Dobrushin, Yu. T. Medvedev, V. S. Mikhalevich, V. A. Uspensky, A. A. Borovkov, V. M. Zolotarev, V. M. Alekseev, Yu. K. Belyaev, L. D. Meshalkin, V. D. Yerokhin, Yu. A. Rozanov, Ya. G. Sinai, V. M. Tikhomirov, A. N. Shiryaev, V. I. Arnold, L. A. Bassalygo, Yu. P. Ofman. Somewhat later, to this list of A. N. Kolmogorov's pupils were added A. B. Prokhorov, M. V. Kozlov, I. G. Zhurbenko, A. M. Abramov, A. V. Bulinsky, as well as a number of foreign mathematicians, among them the well-known Swedish researcher P.Martin-Löv.

Some of Andrei Nikolaevich's pupils were later elected to the Academy of Sciences of the USSR or to the Academy of Sciences of the Union Republics. These are the full members of the USSR Academy of Sciences A. I. Maltsev (algebra, mathematical logic), M. D. Milionshchikov (mechanics, applied physics), S. N. Nikolsky (theory of functions), A. M. Obukhov (atmospheric physics), Yu. V. Prokhorov (probability theory); the corresponding members of the USSR Academy of Sciences L. N. Bolshev (mathematical statistics), A. A. Borovkov (probability theory, mathematical statistics), I. M. Gelfand (functional analysis), A. S. Monin (oceanology), the real members of the Ukrainian Academy of Sciences B. V. Gnedenko (problems of probability theory, history of mathematics), V. C. Mikhalevich (cybernetics), the full member of the Academy of Sciences of the Uzbek SSR S. Kh. Siradzhinov (probability theory)[4]. What an amazing variety of fields of research!

During the last 20 years or so Andrei Nikolaevich has given much energy to problems of secondary education. In this as in all he gives all of himself to a question of paramount importance for society.

Let us begin with the organization in Moscow of boarding school No.18 or, as it is usually called, the Kolmogorov school. In the first years of its existence, he not only lectured and worked in exercise groups with the pupils but also wrote notes of his own lectures for them. Besides classes prescribed by the syllabus, he lectured on music, art, literature.

The fact that the general idea of this boading school was sound was borne out by all its activity. It is enough to say that among the pupils of the school 97-98% of its graduates have been accepted to the leading colleges of the country; approximately 400 of its allumni have successfully completed or are completing their graduate work at Moscow University; over 250 of them have defended their Ph.D. Several of the graduates have also defended Doctor of Mathematics dissertations. And yet the school is not very large. It has only 360 students. It should be added that the students of the school systematically take leading places at All-Union and International mathematical olympiads. Undoubtedly the success of the school is basically due to Andrei Nikolaevich's total dedication to the education of young people and to the development of their creative possibilities.

In 1964 A. N. Kolmogorov heads the mathematical section of the Committee of the Academy of Sciences of the USSR and the Academy of Pedagogical Sciences of the USSR created to determine the syllabus of secondary education. The new syllabus developd by this commission in the years of 1965-1968 for grades 6 to 8 and 9 to 10 was approved by the mathematical section of the Academy of Sciences of the USSR and still is, despite numerous changes, the basis of further improvement of mathematical education and the foundation for writing textbooks. Andrei Nikolaevich himself took part directly in writing new textbooks for the general secondary school. The textbook written under his guidance, called *Algebra and Elements of Analysis 9-10* is still being used in all general schools in the Soviet Union. The textbook *Geometry 6-10* was met with fairly strong criticism. At the present time it is being replaced by other textbooks. But in the creation of these new textbooks, their content in ideas was greatly influenced by the concepts contained in the textbooks written under the guidance of A. N. Kolmogorov.

Soon after he was elected in 1939 to full membership of the Academy of Sciences of the USSR, Andrei Nikolaevich was elected academician-secretary of the section of physics and mathematics. He devoted himself to problems related to the section with the passion that is so characteristic of him. A great amount of work was carried out by Andrei Nikolaevich when he held the post of chairman of the publishing house *Foreign Literature Publishers* and when he was the editor of the mathematical section of the *Great Soviet Encyclopaedia*. The articles about mathematics that he wrote or that were published under his editorship for the encyclopaedia will for many years remain a model for encyclopaedias for other authors.

---

[4]In 1984 I. M. Gelfand and V. S. Mikhalevich was elected to full membership of the Academy of Sciences of the USSR and V. I. Arnold and B. A. Sevastyanov became corresponding members of the Academy of Sciences of the USSR. V. I. Arnold became full member in 1990 (*Editorial note*).

From December 1, 1964 to December 13, 1966 and from November 13, 1973, A. N. Kolmogorov is the president of the Moscow Mathematical Society; from 1946 to 1954 and from 1983 the editor-in-chief of *Uspekhi Matematicheskith Nauk*.

But the centre of A. N. Kolmogorov's activities was always Moscow University, where he works continuously from 1925 after graduating from the department of physics and mathematics. At the university he headed: from 1938 to 1966 the Chair of probability theory (that he founded), from 1966 to 1976 the Interdepartmental laboratory of statistical methods (which he founded), from 1976 to 1980 the Chair of mathematical statistics (also founded by him), from 1980 to the present day the Chair of Mathematical logic. From 1951 to 1953 A. N. Kolmogorov is the director of the Institute of mathematics and mechanics of Moscow University, from 1954 to 1956 and from 1987 to the present time, A. N. Kolmogorov heads the section of mathematics of the mechanics and mathematics department of Moscow University; occupying this post as before (when he was the head of the Research institute in mathematics at the University and the Institute of mechanics in mathematics at Moscow University), he has done a great deal to improve the training of graduate students in mathematical specialties. From 1954 to 1958 A. N. Kolmogorov was the dean of the mechanics and mathematics department of Moscow University.

Special notice should be paid to the activities of Andrei Nikolaevich in the organization at Moscow University of a small but very necessary subdivision: the Statistics laboratory. The idea of creating a statistics laboratory in the framework of the Chair of probability theory came to Andrei Nikolaevich at the beginning of 1960. He began implementing this idea by organizing a workshop for people working at the chair in the solution of applied statistical problems, where he himself delivered several introductory lectures with a demonstration of examples. Within the laboratory, it turned out to be possible to create a sector of computational work provided with a computer. Without a well-organized library, which would contain the leading journals and would be systematically refilled by new books, it would have been impossible to fruitfully carry out scientific research. In order to organize a specialized library in statistics, Andrei Nikolaevich not only managed to obtain the necessary funds, but also made a gift of an important part of the international Bolzano prize that he had received to buy the necessary literature.

The laboratory statistics is not the only subdivision of Moscow University (and not only of the University) that was created on Andrei Nikolaevich's initiative. Thus he was the one who proposed in 1946 to create a small laboratory of atmospheric turbulence at the Institute of theoretical geophysics of the Academy of Sciences of the USSR, which eventually developed into an important part of the Institute of Physics of the Atmosphere.

Concern about libraries, about the publication of mathematical literature, about the training of specialists at all levels, about the creation of new chairs and laboratories and many other things, all these were the manifestations of Andrei Nikolaevich's feeling of personal responsibility for the future of science and education. This feeling led him to take an active part in the formation of syllabuses and programs at all levels of mathematical education. This same feeling led Andrei Nikolaevich to decisively support, in its time, the budding science of cybernetics. This support was essential in gaining acceptance to the scientific status of cybernetics not only in our country but also abroad. In the greetings to the 80th anniversary of A. N. Kolmogorov from the V. M. Glushkov Institute of Cybernetics of the Ukrainian Academy of Sciences, it is stated: "In our institute we all always remember your outstanding contribution to the establishment of Soviet cybernetics. Your theories and scientific ideas have been implemented in many cybernetic developments. Many of the Soviet cybernetists are your pupils or pupils of your pupils." It is well known that cybernetics had to establish itself in a struggle against those who would not accept it. Such was also the fate of genetics, and here also A. N. Kolmogorov courageously upheld and supported scientific truth. In an article published in 1940 in the section of "Genetics" of the *Doklady* of the USSR Academy of Sciences, he showed that the materials gathered by the followers of Academician T. D. Lysenko turned out to be, despite the belief of their authors, a new ground for supporting the Mendel laws.

The belief in personal responsibility is one of the components of A. N. Kolmogorov's ethical stand in science. Other components of his position are his objectivity, the absence of exaggeration in the assessment of his pupils' worth, the readiness to support any notable achievements, modesty in questions of priority and authorship (Andrei Nikolaevich is the *de facto* coauthor of many of his pupils' papers), constant readiness to help.

The scientific achievements of A. N. Kolmogorov are highly recognized both in our country and abroad. He has received seven orders of Lenin, has the title of Hero of Socialist Labor, is the laureate of Lenin and State prizes.

Over twenty scientific organizations have elected Andrei Nikolaevich to membership. He is, in particular, member of the Royal Academy of Sciences of the Netherlands (1963), of the London Royal Society (1964), the National Academy of Sciences of the USA (1967), the Paris Academy of Sciences (1968), the Rumanian Academy of Sciences (1965), the "Leopoldina" Academy of Natural Sciences (1959), the American Academy of Arts and Sciences in Boston (1959). He is doctor *honoris causa* of the universities of Paris, Stockholm and Warsaw, honorary member of the Mathematical Societies of Moscow, India, Calcutta, of the London Royal Statistical Society, the International Statistical Institute, the American Society of Meteorology. In 1963 he was awarded the International Bolzano Prize. (Simultaneously, this same prize was awarded for other activities to the Pope Johann XXIII, the historian S. Morison, the biologist K. Frisch, the composer P. Hindemith).

Andrei Nikolaevich Kolmogorov occupies a unique position in modern mathematics, in fact more generally in world science as well. By the breadth and diversity of his scientific interests, he brings to mind the classics of the natural sciences of previous centuries.

In each of the domains of sciences where Andrei Nikolaevich worked, he reached the summit. His works may justly be called classical; their influence as time goes by not only does not decrease, but constantly grows.

The scientific image of A. N. Kolmogorov is characterized by a rare combination of the traits of an abstract mathematician and of a natural scientist, of a theoretician and a practitioner; from the axiomatic foundations of probability theory he moves on to statistical quality control in industry, from theoretical hydromechanics to personal participation in oceanographic expeditions.

A. N. Kolmogorov's entire life is an exemplary feat in the name of science. A. N. Kolmogorov is one of the all-time greats of Russian science.

ON THE NOTION OF ALGORITHM*)

We start from the following intuitive considerations about algorithms:

1) An algorithm Γ applied to any "condition" ("initial state") A from some set $\mathfrak{G}(\Gamma)$ ("domain of applicability" of the algorithm Γ) gives a "solution" ("concluding state") B.

2) The algorithmic process may be subdivided into separate steps of apriori bounded complexity; each step consists of an "immediate processing" of the state S (that occurs at this step) into the state $S^* = \Omega_\Gamma(S)$.

3) The processing of $A^0 = A$ into $A^1 = \Omega_\Gamma(A^0)$, A^1 into $A^2 = \Omega_\Gamma(A^1)$, A^2 into $A^3 = \Omega_\Gamma(A^2)$, etc., continues until either a nonresultative stop occurs (if the operator Ω_Γ is not defined for the state that just appeared) or until the signal saying that the "solution" has been obtained occurs. It is not excluded that the process will continue indefinitely (if the signal for the solution's appearance never occurs).

4) Immediate processing of S into $S^* = \Omega_\Gamma(S)$ is carried out only on the basis of information on the form of an apriori limited "active part" of the state S and involves only this active part.

A rigorous definition reflecting all these intuitive considerations in strict mathematical terms is possible.

The states are represented by topological complexes of special form. The active part of the state is formed from all elements of the complex whose distance from some initial vertex is no greater than a fixed number. The operator Ω_Γ is given by a finite number of rules. Each of these rules is of the form $U_i \to W_i$; its application consists in the following: if the active part of the complex S is U_i, then in order to obtain $S^* = \Omega_\Gamma(S)$ it is necessary to replace U_i by W_i without changing $S \setminus U_i$.

The equivalence between the notion of algorithm (in the sense specified by this definition) and the notion of recursive function is established.

---

*)Uspekhi Mat. Nauk, 1953, vol. 8,no. 4, pp.175-176.

ON THE GENERAL DEFINITION OF
THE QUANTITY OF INFORMATION*)

Jointly with I. M. Gelfand and A. M. Yaglom

The definitions (2), (4) and the properties I-IV of the expression $I(\xi, \eta)$ presented below were given in A. N. Kolmogorov's report at the meeting on probability theory and mathematical statistics (Leningrad, June 1954). Theorem 4 and Theorem 5 on the semicontinuity of $I(\xi, \eta)$ under weak convergence of distributions were found by I. M. Gelfand and A. M. Yaglom. After that, A. N. Kolmogorov proposed the final version of this article, which stresses that basically the passage from the finite case to the general one, and the computation and estimates of the amount of information obtained by taking limits are absolutely trivial if the exposition is carried out in terms of normed Boolean algebras. It would not be difficult, of course, to reformulate more general principles of the passage to the limit: not as $n \to \infty$, but by using some partial ordering.

§1. In this section the system $\mathfrak{G}$ of all "random events" $A, B, C, \ldots$ is assumed to be a Boolean algebra with the operations of taking the complement A', the union $A \cup B$, and the product AB, containing the unit element ε and the zero element N. The probability distributions $P(A)$ considered are non-negative functions defined on $\mathfrak{G}$ satisfying the additivity condition $P(A \cup B) = P(A) + P(B)$, if $AB = N$, and the normalization condition $P(\varepsilon) = 1$.

The classical notion of "trial" can be naturally identified with the notion of subalgebra of the algebra $\mathfrak{G}$ (in intuitive terms: the subalgebra $\mathfrak{U}$ consists of all the events whose result becomes known only after the given trial is carried out). If the subalgebras $\mathfrak{U}$ and $\mathfrak{B}$ are finite, then "the amount of information contained in the results of the trial $\mathfrak{U}$ with respect to the results of the trial $\mathfrak{B}$" will be defined by means of the Shannon formula

$$I(\mathfrak{U}, \mathfrak{B}) = \sum_{i,j} P(A_i B_j) \log \frac{P(A_i B_j)}{P(A_i) P(B_j)}. \tag{1}$$

In the general case, it is natural to put

$$I(\mathfrak{U}, \mathfrak{B}) = \sup_{\mathfrak{U}_1 \subset \mathfrak{U}, \mathfrak{B}_1 \subset \mathfrak{B}} I(\mathfrak{U}_1, \mathfrak{B}_1), \tag{2}$$

*)Doklady Akad. Nauk SSSR, 1956, vol. 111, no. 4, pp. 745-748.

where the supremum is taken over all finite subalgebras $\mathfrak{U}_1 \subset \mathfrak{U}$ and $\mathfrak{B}_1 \subset \mathfrak{B}$. Symbolically we can write the definition (2) in the form

$$I(\mathfrak{U}, \mathfrak{B}) = \int_{\mathfrak{U}} \int_{\mathfrak{B}} P(d\mathfrak{U} d\mathfrak{B}) \log \frac{P(d\mathfrak{U} d\mathfrak{B})}{P(d\mathfrak{U}) P(d\mathfrak{B})}. \tag{3}$$

In the case of finite $\mathfrak{U}$ and $\mathfrak{B}$, the amount of information $I(\mathfrak{U}, \mathfrak{B})$ is a real non-negative number. In the general case, besides real non-negative values, I can only assume the value $I = +\infty$. The following properties of I, well-known in the finite case, remain valid here ($[\mathfrak{M}]$ denotes the minimal subalgebra of the algebra $\mathfrak{G}$ containing the set $\mathfrak{M}$):

1. $I(\mathfrak{U}, \mathfrak{B}) = I(\mathfrak{B}, \mathfrak{U})$.
2. $I(\mathfrak{U}, \mathfrak{B}) = 0$ if and only if the systems of events $\mathfrak{U}$ and $\mathfrak{B}$ are independent.
3. If $[\mathfrak{U}_1 \cup \mathfrak{B}_1]$ and $[\mathfrak{U}_2 \cup \mathfrak{B}_2]$ are independent, then

$$I([\mathfrak{U}_1 \cup \mathfrak{U}_2], [\mathfrak{B}_1 \cup \mathfrak{B}_2]) = I(\mathfrak{U}_1, \mathfrak{B}_1) + I(\mathfrak{U}_2, \mathfrak{B}_2).$$

4. If $\mathfrak{U}_i \subset \mathfrak{U}$, then $I(\mathfrak{U}_1, \mathfrak{B}) \leqslant I(\mathfrak{U}, \mathfrak{B})$.

The following theorems are almost obvious, but when applied to concrete cases they yield useful methods for the computation and estimate of the amount of information by passing to the limit.

Theorem 1. *If the algebra $\mathfrak{U}_1 \subset \mathfrak{U}$ is everywhere dense in the algebra $\mathfrak{U}$ in the sense of the metric*

$$\rho(A, B) = P(AB' \cup A'B)$$

then

$$I(\mathfrak{U}_1, \mathfrak{U}) = I(\mathfrak{U}, \mathfrak{B}).$$

Theorem 2. *If*

$$\mathfrak{U}_1 \subset \mathfrak{U}_2 \subset \cdots \subset \mathfrak{U}_n \subset \ldots,$$

then for the algebra $\mathfrak{U} = \bigcup_n \mathfrak{U}_n$ we have the relation

$$I(\mathfrak{U}, \mathfrak{B}) = \lim_{n \to \infty} I(\mathfrak{U}_n, \mathfrak{B}).$$

Theorem 3. *If the sequence of distributions $P^{(n)}$ converges on $[\mathfrak{U} \cup \mathfrak{B}]$ to the distribution P, i.e. if*

$$\lim_{n \to \infty} P^{(n)}(C) = P(C) \quad \text{for} \quad C \in [\mathfrak{U} \cup \mathfrak{B}],$$

then for the amounts of information defined by means of (1) and (2) using the distributions $P^{(n)}$ and P of the amount of information, we have the inequality

$$\lim_{n \to \infty} \inf I^{(n)}(\mathfrak{U}, \mathfrak{B}) \geqslant I(\mathfrak{U}, \mathfrak{B}).$$

§2. We shall now assume that the main Boolean algebra $\mathfrak{G}$ is a σ-algebra and all the distributions $P(A)$ under consideration will be assumed σ-additive. We shall denote by the same letter X the "measurable space", i.e. the set X and the σ-algebras S_X of certain of its subsets with the unit X (this is a deviation from a terminology of [1], where S_X can be a σ-ring). By a "random element" ξ of the space X we mean a homomorphic map

$$\xi^*(A) = B$$

of the Boolean algebra S_X into the Boolean algebra $\mathfrak{G}$. Intuitively $\xi^*(A)$ denotes the "event consisting in the fact that $\xi \in A$".

The map ξ^* takes the algebra S_X to a subalgebra of the algebra $\mathfrak{G}$ that will be denoted by

$$\mathfrak{G}_\xi = \xi^*(S_X).$$

It is natural to put by definition

$$I(\xi, \eta) = I(\mathfrak{G}_\xi, \mathfrak{G}_\eta). \tag{4}$$

In the usual way, along the lines of the exposition in [1], let us define the measurable space $X \times Y$. The condition

$$(\xi, \eta)^*(A \times B) = \xi^*(A)\eta^*(B)$$

uniquely determines the homomorphism $(\xi, \eta)^*$ of the Boolean σ-algebra $S_{X \times Y}$ into $\mathfrak{G}$. This homomorphism is used as the definition of the pair (ξ, η) of random objects ξ and η, yielding a new random object: a random element of the space $X \times Y$. The formula

$$\mathcal{P}_\xi(A) = P(\xi^*(A))$$

determines a measure in the space X, called the distribution of the random object ξ. Finally, in accordance to [1], let us define the measure in the space $X \times Y$:

$$\Pi = \mathcal{P}_\xi \times \mathcal{P}_\eta.$$

Applying the Radon-Nikodim theorem to the two measures Π and $\mathfrak{P}(\xi, \eta)$ given on $X \times Y$, we can write the relation

$$\mathcal{P}_{\xi, \eta}(C) = \iint_C a(x, y) d\mathcal{P}_\xi d\mathcal{P}_\eta + S(C),$$

where the measure S is singular with respect to the measure Π.

Theorem 4. *The amount of information $I(\xi, \eta)$ can be finite only in the case $S(C) = 0$. In this case*

$$I(\xi, \eta) = \int_X \int_Y a(x, y) \log a(x, y) d\mathcal{P}_\xi d\mathcal{P}_\eta =$$

$$= \int_{X \times Y} \log a(x, y) d\mathcal{P}(\xi, \eta). \tag{5}$$

Let us also note that for any random element ξ of the space X and any Borel measurable map $y = f(x)$ of the space X into the space Y, the random element $\eta = f(\xi)$ of the space Y is defined by the map

$$\eta^*(A) = \xi^* f^{-1}(A)$$

of the algebra S_Y into the algebra $\mathfrak{G}$. Define the notion of mathematical expectation $\mathbf{E}$ in the ordinary way; then we can write formula (5) in the form

$$I(\xi, \eta) = \mathbf{E} \log a(\xi, \eta). \tag{6}$$

From the properties 1-4 of the expression $I(\mathfrak{U}, \mathfrak{B})$, we get the following properties of the expression $I(\xi, \eta)$:

I. $I(\xi, \eta) = I(\eta, \xi)$.

II. $I(\xi, \eta) = 0$ if and only if ξ and η are independent.

III. If the pairs (ξ_1, η_1) and (ξ_2, η_2) are independent, then

$$I((\xi_1, \xi_2), (\eta_1, \eta_2)) = I(\xi_1, \eta_1) + I(\xi_2, \eta_2).$$

IV. If ξ_1 is a function of ξ (in the sense indicated above), then

$$I(\xi_1, \eta) \leqslant I(\xi, \eta).$$

Similarly to pairs (ξ, η), it is easy to define sequences

$$\xi = (\xi_1, \xi_2, \ldots, \xi_n, \ldots)$$

as random elements of the space $X = \times_{n=1}^{\infty} X_n$. It follows obviously from Theorems 1 and 2 that in that case we shall always have

$$I(\xi, \eta) = \lim_{n \to \infty} I((\xi_1, \xi_2, \ldots, \xi_n), \eta).$$

In conclusion let us indicate one more application of Theorem 3. Suppose X and Y are complete metric spaces. If for S_X and S_Y we take the algebras of their Borel sets, then the spaces X and Y become measurable spaces at the same time. In this situation we have the following

Theorem 5. *If for the random elements $\xi \in X$ and $\eta \in Y$ the distributions $\mathcal{P}^{(n)}(\xi, \eta)$ weakly converge to the distribution $\mathcal{P}(\xi, \eta)$, then for the corresponding amounts of information we have the inequality*

$$\liminf_{n \to \infty} I^{(n)}(\xi, \eta) \geqslant I(\xi, \eta).$$

For the proof it suffices to apply Theorem 3 to the algebras $\mathfrak{U}$ and $\mathfrak{B}$ consisting respectively of events $\xi \in A$, $\eta \in B$, where A and B are the sets of continuous distributions of $\mathcal{P}_\xi$ and respectively $\mathcal{P}_\eta$ (i.e. sets whose boundaries are respectively $\mathcal{P}_\xi = 0$ and $\mathcal{P}_\eta = 0$) and note that the algebra $[\mathfrak{U} \cup \mathfrak{B}]$ is everywhere dense on $\mathfrak{G}_{X \times Y}$ in the metric defined by the limiting distribution P.

October 8, 1956.

References

[H1] P. H. Halmos, *Measure theory*, Van Nostrand, 1966.

3
THE THEORY OF TRANSMISSION
OF INFORMATION*)

I. The appearance of the theory and its contents.

Speaking for the second time to the General Assembly of our Academy, I should like to begin by remarking that the topic of today's communication is intimately connected with that of my report "Statistical Theory of Oscillations with Continuous Spectrum" which I delivered at the General Assembly of the Academy in 1947.[1]

Indeed, the branch of information theory that is richest in content and most stimulating from the purely mathematical point of view, namely the theory of stationarily operating transmission channels that process continuous messages, could not have been created without the preliminary development of the theory of stationary random processes and, in particular, without the spectral theory of such processes.

Interest in various problems of transmission and storage of information has an extensive history. The question of estimating the "amount" of information actually arose a long time ago. The possibility of introducing a universal numerical measure for the amount of information is especially important in cases when it is necessary to process information of one kind into information which is qualitatively of a different kind. A typical problem in the processing of information is the problem of tabulating continuous functions of continuously changing arguments. If, for example, a function $f(x, y)$ of two arguments

$$0 \leqslant x \leqslant 1, \; 0 \leqslant y \leqslant 1$$

---

*)In the book The Session of the Academy of Sciences of the USSR on Scientific Problems of Industrial Automation, October 15-20, 1956, Plenary Session. Moscow Academy of Sciences USSR Publishers, 1957, pp. 66-99. A. N. Kolmogorov's intervention during the discussion of the report is presented at the end of the article.

[1]See [1]. A more detailed exposition of the mathematical aspects of the questions to which my 1947 report was devoted can be found in the article [2]. For the present report, a similar role is played by part II of the present paper which reproduces with small changes the report that B. V. Gnedenko made in my name at the Symposium in Information Theory of the American Institute of Radio Engineers (Massachussetts Institute of Technology, September 1956) that was based in many of its parts on the report presented jointly with I. M. Gelfand and A. M. Yaglom at the All-Union Mathematical Congress in June 1956.

must be given with precision ε, and it is known that its increment Δf is no greater in absolute value than ε when

$$|\Delta x| \leqslant \varepsilon_x, \ |\Delta y| \leqslant \varepsilon_y,$$

then it is certainly sufficient to tabulate the function f with step ε_x with respect to x and with step ε_y with respect to y, i.e. to give approximately

$$N \sim 1/\varepsilon_x \varepsilon_y$$

values of this function. If we assume that $|f| \leqslant 1$, then in order to fix one value of the function, it is sufficient to use approximately

$$K \sim \log_{10}(1/\varepsilon)$$

decimal places. Thus if we follow the outline presented above, we must introduce approximately

$$NK \sim (1/\varepsilon_x \varepsilon_y) \log_{10}(1/\varepsilon)$$

decimal places in our table (not counting the notation for the values of the argument x and y which is standard for all functions f satisfying the above assumptions). It is well known that such a primitive approach to the problem, especially in the case of functions of several variable, would lead to tables of huge volume and is practically impossible to carry out. Actually in the tabulation of sufficiently smooth functions, one chooses a step in the table with respect to the independent variable which is considerably larger, and to find the values of the functions for intermediate values of the argument one uses an interpolation procedure involving differences of certain order (in the case of tables for functions of several varriables, often up to the fourth or sixth order). Further, often the first decimal places of $f(x,y)$ which repeat each other for close values of the arguments are not written out in each space of the table corresponding to the given pair of values (x, y); instead they are written out, say, only in the beginning of the line consisting of many such spaces.

Despite the fact that the methods for constructing tables described above are well known and of ancient origin, the corresponding general theoretical studies aimed at finding the minimal amount of information necessary for fixing, with a given precision ε, an arbitrary function f satisfying only certain general conditions are still in the embryonal stage. For example, only recently in my note [3] the following explicit formula[2]

$$H_\varepsilon(F_{p,\alpha}^n) \asymp (1/\varepsilon)^{n/(p+\alpha)}, \qquad (1)$$

indicating the order of growth as $\varepsilon \to 0$ of the amount of information necessary to fix a function of n variables defined in a bounded domain with bounded derivatives of the first p orders and derivatives of order p satisfying Lipschitz condition of degree α was presented.

---

[2]This formula may well be called the A. G. Vitushkin formula, since in his paper [4] the role of the exponent $n/(p + \alpha)$ is clarified in another context; this paper [4] essentially contains half of the proof of formula (1), namely the proof of the fact that the order $H_\varepsilon(F_{p,\alpha}^n)$ cannot be less than the one indicated by formula (1).

It is quite clear that questions related to the processing and storage of information have a decisive importance in the design of modern computers and methods of using them. For computers the "volume of memory" is characterized by the number of binary digits which this memory can hold. The method of measuring the amount of information by comparing this information with that presented by a certain number of binary digits has now become standard in information theory.

Passing from problems in the general theory of "connecting channels" to those related to the functioning of telegraph or telephone channels or similar devices, let us begin with the very simple problem of transmitting a decimal sequence

$$a_1, a_2, \ldots, a_n, \ldots \tag{2}$$

by means of a binary sequence

$$b_1, b_2, \ldots, b_n, \ldots \tag{3}$$

Assume that the sequence (2) is generated in time one digit per unit of time. We are asked at what rate must we transmit binary symbols of the sequence (3) if we desire to recover the original sequence (2) from the sequence (3) without giving rise to unboundedly increasing delay. If we represent each digit a_n separately by binary digits, then for each digit a_n we must use up four binary symbols (since $2^3 = 8 < 10$ and only $2^4 = 16 \geqslant 10$). If we transmit each pair (a_{2n-1}, a_{2n}) of digits of the sequence (2) by a group of binary symbols, then we shall use up seven binary symbols (since $2^6 = 64 < 100$ and only $2^7 = 128 \geqslant 100$), while for the transmission of three decimal symbols it suffices to use up ten binary symbols (since $2^9 < 1000$ but $2^{10} = 1024 \geqslant 1000$) etc. To these systems of transmission correspond different rates of the formation of symbols from the sequence (3), i.e. the corresponding number of binary symbols that we must transmit per unit of time on the average are equal to

$$\nu_1 = 4; \quad \nu_2 = 7/2 = 3,5; \quad \nu_3 = 10/3 = 3,33\ldots$$

It is easy to prove that the series of numbers arising in this way converges to the limit

$$\nu = \log_2 10 = 3,32\ldots$$

It is precisely this ν which is the greatest lower bound of those rates of transmission for the sequence (3) allowing the transmission of the sequence (2) without any systematically increasing delay in time.

This especially chosen naive example already displays many typical phenomena discovered by contemporary information theory in its applications to many considerably more general types of transmission channels.

1. The rate of information transmission, despite its nonuniform qualitative character, admits a reasonable quantitative measurement. In particular, the measure of the rate of information transmission when an average of ν symbols, each of which can assume k values, is transmitted per unit of time, can reasonably be taken equal to

$$\bar{H} = \nu \log_2 k.$$

(In our case

$$\bar{H} = \log_2 10 = 3,32\ldots$$

for the sequence (2) and

$$\bar{H}' = \nu \log_2 2 = \nu$$

for the sequence (3) in the transmission of ν symbols per unit time.)

For the sequel let us note that in the case of exact (errorless) transmission channels the speed of the information they give out can naturally be called their channel capacity. In our case the channel capacity of the sequence (3), viewed as a transmission channel, is equal to $\bar{C} = \bar{H}' = \nu$. In the general case the channel capacity is defined in a slightly different way that will be explained further on. Now let us state the second general principle.

2. To make possible the transmission of information arising at the rate $\bar{H}$ through a transmission channel of channel capacity $\bar{C}$ without systematically increasing delay, the following condition is necessary

$$\bar{H} \leqslant \bar{C}.$$

It is the third principle, however, which is the most important. Its content as formulated here is still very unprecise, but it is this principle that contains the most important and essentially new contribution to the theory of information transmission by connecting channels that arose in the work of Shannon.

3. Even in the case when we are dealing with qualitatively nonhomogeneous information, if the information to be transmitted, as well as that which the transmission channel is capable of accepting, satisfies

$$\bar{H} < \bar{C},$$

it is in principle possible to carry out the transmission without systematically increasing delay if the information to be transmitted is coded before it enters the transmission channel in an appropriate way and decoded at the output. However, in general, when $\bar{H}$ approaches $\bar{C}$, the methods for coding and decoding necessarily become more complicated and the delay in transmission (as $\bar{H} \to \bar{C}$) becomes greater and greater.

Note that the delay mentioned in the third principle increases as $\bar{H} \to \bar{C}$, but for a fixed source of information with $\bar{H} < \bar{C}$ can be estimated by means of a certain magnitude τ, which remains the same no matter how long the transmission channel works.

For the next example, let us consider the transmission of a text consisting of letters of the cyrillic alphabet. Since there are 33 letters in this alphabet, one can formally constitute

$$N = 33^n$$

various "texts" of length n (i.e. sequences of n letters). The amount of information contained in the choice of one such specific text is equal, by the above, to

$$I = n \log 33$$

(here and further on we always use logarithms to the base 2, without mentioning this explicitly). The existence of stenography shows, however, that real language texts may be transmitted by much shorter means. This is quite understandable: the number N^* of "meaningful" texts of n letters is clearly much less than N. Hence, in principle, it is possible to develop a system of notation (the "ideal" stenography) that would require for the choice of any of these N^* meaningful texts only

$$I^* \sim \log N^*$$

binary symbols. The possibility of "compressing text", of course, will never be used fully in any kind of stenographic systems for coding texts according to sufficiently simple formal rules: rules for constructing real language texts will hardly ever be completely formalized.

We shall direct our attention now to one particular trait of language texts: various letters appear in them with various frequencies. If we fix the numbers

$$n_1, n_2, \ldots, n_k$$

of occurences in a text of length n of each of the letters

$$a_1, a_2, \ldots, a_k$$

(naturally we then have

$$n_1 + n_2 + \cdots + n_k = n),$$

then the number of texts of length n will decrease to

$$N' = \frac{n!}{n_1! n_2! \ldots n_k!}.$$

Using the so-called Stirling formula, needed here only in its weak form

$$\log(n!) \sim n \log n,$$

it is easy to calculate that for large n_i we have

$$I' = \log N' \sim -n \sum_i p_i \log p_i,$$

where

$$p_i = n_i/n$$

are the frequencies of occurence of individual letters. The result obtained may be expressed in the following way: when using individual letters with frequencies p_i, the amount of information transmitted by "one letter of the text" is equal to

$$H = -\sum_i p_i \log p_i. \tag{4}$$

In the case of equal frequencies

$$p_1 = p_2 = \cdots = p_k = 1/k$$

we again obtain

$$H = \log k;$$

for any other frequencies p_i

$$H < \log k.$$

It is easy to indicate methods of coding that allow the transmission of arbitrary texts in which the letters

$$a_1, a_2, \ldots, a_k$$

occur with frequencies

$$p_1, p_2, \ldots, p_k$$

by using a number of binary symbols (in the mean, when transmitting sufficiently long texts) for each letter so that this number differs as little as we wish from H, where H is determined by formula (4). This is only another expression of the third principle stated above. We shall now leave the consideration of elementary examples, however, and pass to elements of the general theory.

Formula (4) admits also, besides a purely statistical interpretation (in which the p_i are frequencies), a probabilistic interpretations. Suppose, in the most general way, that we are given the probability distribution

$$\mathbf{P}\{\xi = x_i\} = p_i \tag{5}$$

of a random object ξ. If only this distribution is given, then the answer to the question: which of the possible values

$$x_1, x_2, \ldots, x_k$$

is assumed in fact by ξ remains somewhat indetermined. If we say that ξ is equal to some definite x_i, we remove this indeterminacy, i.e. communicate some additional information about the object ξ. The measure of the indeterminacy of the distribution (5) is its entropy, expressed by formula (4) which we already know. This same formula also expresses the amount of information necessary to remove the indeterminacy of the value of ξ, if the latter is given only by the distribution (5), i.e. of the information contained in indicating the exact value of ξ.

Now suppose further that we are given the joint probability distribution

$$\mathbf{P}\{\xi = x_i, \eta = y_j\} = p_{ij}$$

of two random objects ξ and η. If $\eta = y_j$ is given, then ξ has the conditional distribution

$$\mathbf{P}\{\xi = x_i | \eta = y_j\} = p_{i|j}.$$

The amount of information contained in indicating the exact value of ξ, if the value $\eta = y_j$ is already known, is equal to

$$H(\xi | \eta = y_j) = -\sum_i p_{i|j} \log p_{i|j},$$

and in the mean it is equal to

$$\mathbf{E}H(\xi|\eta) = -\sum_j \mathbf{P}\{\eta = y_j\} \sum_i p_{i|j} \log p_{i|j}.$$

It is natural to assume that the difference

$$I(\eta, \xi) = H(\xi) - \mathbf{E}H(\xi|\eta) \tag{6}$$

is precisely the amount of information concerning ξ which is already contained in η. It is easy to calculate that $I(\eta, \xi)$ may be written in the form

$$I(\eta, \xi) = \sum_{i,j} p_{ij} \log \frac{p_{ij}}{\mathbf{P}\{\xi = x_i\}\mathbf{P}\{\eta = y_j\}}, \tag{7}$$

where ξ and η play exactly similar roles: the amount of information in η with respect to ξ and the amount of information in ξ with respect to η are numerically equal to each other.

It is easy to check that we always have

$$I(\xi, \xi) = H(\xi). \tag{8}$$

Now it is not difficult to define the speed of transmission of information through a transmission channel working with errors and the channel capacity of such channels (the channel capacity is still defined as the least upper bound of possible information transmission rates). Let me give one more elementary example. Suppose that on the input end of a channel the symbols η, equal to 0 or 1 are being fed, while at the output end the corresponding symbols η' are obtained, where

$$\mathbf{P}\{\eta' = 0\} = 1 - \Delta, \quad \mathbf{P}\{\eta' = 1\} = \Delta \quad \text{for } \eta = 0,$$
$$\mathbf{P}\{\eta' = 0\} = \Delta, \quad \mathbf{P}\{\eta' = 1\} = 1 - \Delta \quad \text{for } \eta = 1.$$

It is easy to calculate the least upper bound

$$C = \sup I(\eta, \eta') = 1 + [\Delta \log \Delta + (1 - \Delta) \log(1 - \Delta)] \tag{9}$$

which is assumed when

$$\mathbf{P}\{\eta = 0\} = \mathbf{P}\{\eta = 1\} = 1/2.$$

According to formula (9), the amount of information $I(\eta, \eta')$ transmitted by our channel for one letter is clearly equal to zero if $\Delta = 1/2$. This is understandable, since from the probabilistic point of view in this case the symbols η and η' are independent. The maximal value of the channel capacity $C = 1$ is obtained when $\Delta = 0$ and $\Delta = 1$. In the case $\Delta = 0$, the channel works without error with probability 1, so we have $\eta = \eta'$. Although in the second case we have $\eta \neq \eta'$ with probability 1, it is clear that η can be easily recovered from η': in this case one must simply read 1 instead of 0 and vice versa. In all other cases

$$0 < C < 1.$$

When Shannon, at the end of the forties, proposed measuring the amount of information by means of formulas (4), (6), (7) and discovered, for transmission channels working with errors, under fairly wide assumptions concerning the source of information and the structure of the channel, that the second and third principles formulated above remain valid, this created the foundations of modern information theory as a discipline capable of systematic development.

In any important discovery there are always unexpected elements. It is by this trait that an important discovery differs from the gradual collection of the results of current research work. I shall now try to describe what I feel are the qualitatively new and unexpected traits appearing in the elementary part of information theory described above.

1. In its original understanding "information" is not a scalar magnitude. Various forms of information may be extremely varied. It could have been expected in advance that various methods for measuring the "amount" of information can be proposed, but it was unclear whether there is a unique method possessing advantages of principle over all other methods and, what is most important, it was completely unclear whether qualitatively different types of data appearing in practice, carrying the same amount of information (in the sense of the chosen measure), are equivalent in the sense of the difficulty involved in their transmission through channels or their storage in various data storing devices.

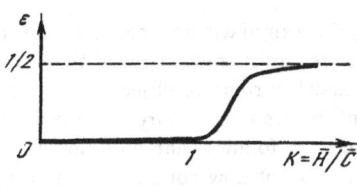

FIGURE 1

It turned out that such a basically "correct" measure of the amount of information exists and allows the complete solution of a wide class of problems about which, apriori, the independence of the solution from more delicate qualitative traits of the data itself was entirely unclear.

2. In the case of transmission channels or data storage devices working with errors, one could fear that it is only possible to achieve transmission of information with a small probability of error ε only by means of a large increase in the rate of transmission. Despite these fears, it turned out that under the condition $\bar{H} < \bar{C}$ it is always possible to transmit information with error probability as small as we wish. To explain this circumstance, let me indicate the following. Suppose we are given a transmission channel with channel capacity C. Consider the problem of transmitting binary symbols, H of which are generated per unit time.

Then the minimum probability of error in the transmission of a separate symbol that can be achieved to any degree of precision is a universal function $\varepsilon(K)$, namely the ratio $K = \bar{H}/\bar{C}$ (shown on Figure 1). When $K \leqslant 1$ it equals zero, while with the further increase of K it begins to grow and as $K \to \infty$ it tends to $1/2$ (analytically, when $K > 1$, the function $\varepsilon(K)$ is determined by the equations

$$1/K = 1 + \varepsilon \log \varepsilon + (1 - \varepsilon) \log(1 - \varepsilon), \ \varepsilon < 1/2).$$

Thus, from values which are exactly equal to zero when $K \leqslant 1$, our function begins to grow drastically when $K > 1$. Actually, as we pointed out above, when $\bar{H}$ is close to $\bar{C}$, it is possible to obtain a small probability of error only by using very compliated rules for coding information, which leads to a large delay in its output from the channel. But it is precisely the great strength of complicated coding methods that eliminate errors which is the basically important conclusion; its applications will be discussed further on. Two approaches may be considered to justify the methods proposed by Shannon for measuring the amount of information, i.e. formulas (4) and (7). First of all one may present various natural axiomatic requirements that the amount of information must satisfy. In their application to the magnitude $H(\xi)$, i.e. in accordance to formula (8), the amount of information contained in determining a random object ξ with respect to the random object ξ itself, this approach to the foundations of information theory has been presented with remarkable clarity in the work of A.Ya.Khinchin [5]. It would be of a certain interest to give just as natural an axiomatic approach to the more general notion of the amount of information $I(\xi, \eta)$ contained in one random object about another random object. The thing is (compare part II below) that usually in the continuous case $H(\xi)$ and $H(\eta)$ taken separately are infinite, so that $I(\xi, \eta)$ cannot be computed in accordance to formula (6).

As the result of studying the axiomatic approach, we can regard as established that no other equally natural scalar characteristic of the information contained in one random object ξ about another random object η other than $I(\xi, \eta)$ can exist. However, since "information" by its own nature necessarily must not be (and in fact is not) a scalar magnitude, it follows that no axiomatic studies of the above approach can answer the question of how completely the magnitude $I(\xi, \eta)$ characterizes the "information" that we are interested in. A certain answer to this question, as could be understood from the previous exposition, is given by theorems establishing the second and third principles of the theory of data transmission stated above. By these theorems, in a very wide class of cases, the quality of the transmission channel or the memorizing device for the transmission or storage of data can be characterized in a sufficiently complete way by a single number C (or, per a unit time, by the number $\bar{C}$) while the adaptability of the data itself to transmission or storage is characterized by a single number H (or per unit time by the number $\bar{H}$). In this case it turns out that the qualitative character of the data presented is not essential. It is difficult to overestimate the importance of results of this type, if one takes into consideration the widespread character of their applications in modern technology, say when we are dealing with the transformation of images transmitted by television into electric oscillations and vice versa (in the terminology of the theory of data transmission, this is a particular case of "coding" and "decoding"), etc. At the same time one must understand that no matter how interesting the ideas of information theory concerning the suppression of the qualitative traits of data are, this is only true in a certain approximate sense or under special conditions. These conditions in well studied cases, roughly speaking, involve the storage of large amounts of uniform data and the possibility of implementing complicated coding methods without important delays in data output. At the same time one should clearly understand that the rigourous mathematical justification of this approach has only been carried out under fairly restrictive assumptions. Shannon's remarkably rich ideas appear in his works in very unclear form. It is only later, in a series of works by pure mathematicians for the case of stationarily working channels transmitting discrete signals, that the "Shannon theorems" were proved rigorously under sufficiently general assumptions. The most finished of the works in this direction is the paper by A. Ya. Khinchin [6].

The Soviet reader may become quite familiar with the development of information theory due to researchers of a more applied direction and preceeding fundamental mathematical studies, by reading the books [7-9] ([7] contains a translation of Shannon's main paper [10]).

Essential elements of information theory for the continuous case, which seems more difficult, appeared before Shannon. The logarithmic measure of the amount of information, similar to Shannon's expression $H(\xi)$, lies at the foundations of asymptotic statistical methods developed by Fischer in 1921-1925 (see [11, Chapters 32-33])[3].

Even closer to the works of Shannon are the results of V. A. Kotelnikov [12] obtained as early as 1933. Here the fundamental idea of using spectral theory for data transmission by continuous signals was stated; this I shall discuss in more detail in Chapter II[4].

II. PRINCIPLES FOR CONSTRUCTING THE THEORY
IN THE CASE OF CONTINUOUS MESSAGES

The role of entropy of a random object ξ which can assume values $x_1, x_2, \ldots, x_n$ with probabilities $p_1, p_2, \ldots, p_n$, given by the formula

$$H(\xi) = -\sum_k p_k \log p_k$$

may be considered sufficiently understood in information theory and in the theory of data transmission by means of discrete signals. Further I insist on the idea that the main notion that generalizes to absolutely arbitrary continuous messages and

---

[3]It is also well known that $H(\xi)$ is formally similar to the expression for entropy in physics. This coincidence I feel is quite sufficient to justify calling the expression $H(\xi)$ entropy in information theory: such mathematical analogues should always be stressed since attention to them helps the progress of science. But it would be an exaggeration to think that physical theories related to the notion of entropy already contained elements of information theory in final form. The first use of expressions of entropy type as a measure of the amount of information that I know appears only in Fischer's paper mentioned above.

[4]Chapter II of the present paper contains more mathematical details than the actual oral report.

signals is not entropy, but the amount of information $I(\xi, \eta)$ in the random object ξ about the object η. In the discrete case, this expression can be correctly computed by the well-known Shannon[5] formula

$$I(\xi, \eta) = H(\eta) - \mathbf{E}H(\eta|\xi).$$

For finite-dimensional distributions possessing a density, the expression $I(\xi, \eta)$ can be defined according to Shannon by a similar formula

$$I(\xi, \eta) = h(\eta) - \mathbf{E}h(\eta|\xi),$$

where $h(\eta)$ is the "differential entropy"

$$h(\eta) = -\int p_\eta(y) \log p_\eta(y) dy,$$

while $h(\eta|\xi)$ is the conditional differential entropy defined in a similar way. It is well known that the expression $h(\xi)$ does not have a straightforward meaningful interpretation and is even noninvariant with respect to transformations of coordinates in the space $x_1, \ldots, x_n$. For infinite-dimensional distributions, there is, generally speaking, no analogue of the expression $h(\xi)$.

In the strict meaning of the word, the entropy of an object ξ with continuous distribution is always infinite. If continuous signals cannot nevertheless be used for the transmission of boundless amount of information, this is only because they can only be observed with limited precision. It is therefore natural, having fixed the precision of the observation ε, to define the corresponding "ε-entropy" $H_\varepsilon(\xi)$ of the object ξ. Precisely this, using the term "rate of message creation" was done by Shannon. Although the choice of new term for this expression does not change the essence of things, I take it upon myself to propose the following change of terminology, which underscores the wider interest of the notion and its deep analogy to ordinary exact entropy. I shall indicate in advance, as pointed out in §3, that the theorem on the extremal role of normal distributions (both in the finite-dimensional and in the infinite-dimensional case) remains valid. Further in §§1,2 I shall give, without pretending to be entirely original, an abstract formulation of the definition and the main properties of the expresion $I(\xi, \eta)$ and a review of the main problems in the theory of message transmission according to Shannon. Sections 3-6 contain certain specific results recently obtained by Soviet mathematicians. I would especially like to stress the interest in the study of the asymptotical behaviour of ε-entropy as $\varepsilon \to 0$. The cases studied previously

$$H_\varepsilon(\xi) \sim n \log(1/\varepsilon), \quad \bar{H}_\varepsilon(\xi) \sim 2W \log(1/\varepsilon),$$

where n is the number of measurements, while W is the width of the strip of the spectrum, are only very particular cases of rules that can be observed here. To understand the vistas that open up before us, perhaps my note [3], although expressed in different terms, may be useful.

---

[5]In a notation which I feel is satisfactory, $H(\eta|\xi)$ is the conditional entropy of η when $\xi = x$, while $\mathbf{E}H(\eta|\xi)$ is the mathematical expectation of this conditional entropy for the variable ξ.

§1. **The amount of information in one random object about another.**
Suppose ξ and η are random objects with domains of possible values X and Y,

$$P_\xi(A) = \mathbf{P}\{\xi \in A\}, \ P_\eta(B) = \mathbf{P}\{\eta \in B\}$$

are the corresponding probability distributions and

$$P_{\xi\eta} = \mathbf{P}\{(\xi, \eta) \in C\}$$

is the joint probability distribution of the objects ξ and η. By definition, the amount of information in the random object ξ about the random object η is given by the formula

$$I(\xi, \eta) = \int_X \int_Y P_{\xi\eta}(dxdy) \log \frac{P_{\xi\eta}(dxdy)}{P_\xi(dx)P_\eta(dy)}. \tag{1}$$

The exact meaning of this formula requires certain explanations, while the general properties of the expression $I(\xi, \eta)$ presented below are valid only under certain set theoretic restrictions on the distributions P_ξ, P_η and $P_{\xi,\eta}$, but here I shall not dwell on these questions. Certainly the general theory may be developed without important difficulties so as to be applicable to random objects ξ and η of very general nature (vectors, functions, distributions etc.). The definition (1) may be assumed to belong to Shannon, although he restricted himself to the case

$$P_\xi(A) = \int_A p_\xi(x)dx, \quad P_\eta(B) = \int_B p_\eta(y)dy,$$
$$P_{\xi\eta}(C) = \iint_C p_{\xi\eta}(x, y)dxdy,$$

when (1) becomes

$$I(\xi, \eta) = \int_X \int_Y p_{\xi\eta}(x, y) \log \frac{p_{\xi\eta}(x, y)}{p_\xi(x)p_\eta(y)} dxdy.$$

Sometimes it is useful to represent the distribution $P_{\xi\eta}$ in the form

$$P_{\xi\eta}(C) = \int \int_C a(x, y)P_\xi(dx)P_\eta(dy) + S(C), \tag{2}$$

where the function $S(C)$ is singular with respect to the product

$$P_\xi \times P_\eta.$$

If the singular component S is absent, then the formula

$$\alpha_{\xi,\eta} = a(\xi, \eta) \tag{3}$$

determines uniquely (up to probability 0) the random magnitude $\alpha_{\xi,\eta}$. Sometimes the following theorem, stated by I. M. Gelfand and Ya. M. Yaglom in [13], is useful.

Theorem. If $S(X \times Y) > 0$, then $I(\xi, \eta) = \infty$. If $S(X \times Y) = 0$, then

$$I(\xi, \eta) = \int_X \int_Y a(x, y) \log a(x, y) P_\xi(dx) P_\eta(dy) =$$
$$= \int_X \int_Y \log a(x, y) P_{\xi\eta}(dx dy) = \mathbf{E} \log \alpha_{\xi, \eta}.$$

(4)

Let us list certain basic properties of the expression $I(\xi, \eta)$.

I. $I(\xi, \eta) = I(\eta, \xi)$.

II. $I(\xi, \eta) \geqslant 0$; $I(\xi, \eta) = 0$ only in the case when ξ and η are independent.

III. If the pairs (ξ_1, η_1) and (ξ_2, η_2) are independent, then

$$I((\xi_1, \xi_2), (\eta_1, \eta_2)) = I(\xi_1, \eta_1) + I(\xi_2, \eta_2).$$

IV. $I((\xi, \eta), \zeta) \geqslant I(\eta, \zeta)$.

V. $I((\xi, \eta), \zeta) = I(\eta, \zeta)$ if and only if ξ, η, ζ form a Markov sequence, i.e. if the conditional distribution of ζ for fixed ξ and η depends only on η.

Concerning property IV, it is useful to note the following. In the case of the entropy

$$H(\xi) = I(\xi, \xi),$$

besides the estimates of the entropy of the pairs (ξ, η) from below

$$H(\xi, \eta) \geqslant H(\xi), \quad H(\xi, \eta) \geqslant H(\eta)$$

which follow from I and IV, we have the following estimate from above

$$H(\xi, \eta) \leqslant H(\xi) + H(\eta).$$

For the amount of information in ζ about the pair (ξ, η), a similar estimate does not exist: from

$$I(\xi, \zeta) = 0, \quad I(\eta, \zeta) = 0$$

it does not follow (as can be shown by elementary examples) that

$$I((\xi, \eta), \zeta) = 0.$$

For the sequel, let us note especially the case when ξ and η are random vectors

$$\xi = (\xi_1, \ldots, \xi_m),$$
$$\eta = (\eta_1, \ldots, \eta_n) = (\xi_{m+1}, \ldots, \xi_{m+n})$$

and the expressions

$$\xi_1, \xi_2, \ldots, \xi_{m+n}$$

are distributed normally with second central moments

$$s_{ij} = \mathbf{E}[(\xi_i - \mathbf{E}\xi_i)(\xi_j - \mathbf{E}\xi_j)].$$

If the determinant

$$C = |s_{ij}|, \ 1 \leqslant i, j \leqslant m + n$$

is non-zero, then as computed by I. M. Gelfand and Ya. M. Yaglom, we have

$$I(\xi, \eta) = \frac{1}{2} \log(AB/C), \tag{5}$$

where

$$A = |s_{ij}|, \ 1 \leqslant i, j \leqslant m,$$
$$B = |s_{ij}|, \ m < i, j \leqslant m + n.$$

It is often more convenient to use a somewhat different approach to the situation, applicable without the restriction $C > 0$. As is well known [14], after an appropriate linear transformation of coordinates in the spaces X and Y, all the second moments s_{ij} other than those for which $i = j$ or $j = m + i$ vanish. For such a choice of coordinates,

$$I(\xi, \eta) = -\frac{1}{2} \log \sum_k [1 - r^2(\xi_k, \eta_k)], \tag{6}$$

where the sum is taken over those

$$k \leqslant \min(m, n)$$

for which the denominator in the expression for the correlation coefficient

$$r(\xi_k, \eta_k) = s_{k,m+k}/\sqrt{s_{k,k} s_{m+k,m+k}}$$

is non-zero.

§2. Abstract exposition of the foundations of Shannon theory.
Shannon considers the transmission of messages according to the scheme

$$\xi \to \eta \to \eta' \to \xi',$$

where the "transmitting device"

$$\eta \to \eta'$$

is characterized by the conditional distribution

$$P_{\eta'|\eta}(B'|y) = \mathbf{P}\{\eta' \in B' | \eta = y\}$$

of the "output signal" η' for a given "input signal" η and a certain restriction

$$P_\eta \in V$$

on the distribution P_η^1 of the input signal. The "coding" operation

$$\xi \to \eta$$

and the "decoding" operation

$$\eta' \to \xi'$$

are characterized by conditional distributions

$$P_{\eta|\xi}(B|x) = \mathbf{P}\{\eta \in B | \xi = x\},$$
$$P_{\xi'|\eta'}(A'|y') = \mathbf{P}\{\xi' \in A' | \eta' = y'\}.$$

The main problem considered by Shannon is the following. The spaces X, X', Y, Y' of possible values of input messages ξ, output messages ξ', input signals η and output signals η' are given, as well as the characteristics of the transmitting device, i.e. the conditional distribution $P_{\eta|\eta'}$ and the class V of admissible distributions P_η of the input signal; finally, the distribution

$$P_\xi(A) = \mathbf{P}\{\xi \in A\}$$

of input messages and the "requirements to the precision of transmission"

$$P_{\xi\xi'} \in W$$

are given, where W is a certain class of joint distributions

$$P_{\xi\xi'}(C) = \mathbf{P}\{(\xi, \xi) \in C\}$$

of input and output messages. One asks if is it possible, and if so, how is it possible, to determine rules of coding and decoding, i.e. conditional distributions $P_{\eta|\xi}$ and $P_{\xi'|\eta'}$, so that in the calculation of the distribution $P_{\xi\xi'}$ from the distributions $P_\xi, P_{\eta|\xi}, P_{\eta'|\eta}, P_{\xi'|\eta'}$ under the assumption that the sequence

$$\xi, \eta, \eta', \xi'$$

is Markov, we obtain

$$P_{\xi\xi'} \in W?$$

Following Shannon, let us define the "channel capacity" of the transmitting device by the formula

$$C = \sup_{P_\eta \in V} I(\eta, \eta')$$

and introduce the expression

$$H_W(\xi) = \inf_{P_{\xi\xi'} \in W} I(\xi, \xi')$$

which, when calculated per unit time, is what Shannon calls "rate of message creation". Then property V from 1 immediately implies a necessary condition for the transmission to be possible

$$H_W(\xi) \leqslant C. \tag{7}$$

As we already noted, an uncomparably deeper idea of Shannon is that condition (7), when applied to a sufficiently long work interval of the "transmission channels" is in a certain sense and under certain very general conditions, also "almost sufficient". From the mathematical point of view, here we are talking about the proof of limiting theorems of the following type. It is assumed that the spaces X, X', Y, Y', the distributions P_ξ and $P_{\eta'|\eta}$, the classes V and W and therefore the expressions C and $H_W(\xi)$ depend on the parameter T (which in the applications is the duration of work of the transmitting devices). It is required to establish, under certain very general assumptions, that the condition

$$\liminf_{T \to \infty} \frac{C^T}{H_W^T(\xi)} > 1 \tag{8}$$

is sufficient for a transmission satisfying the conditions stated above (for sufficiently large T) to be possible. Naturally, in such a setting of the problem is somewhat vague. However, here I have purposely avoided using the terminology of the theory of stationary random processes, since it is possible to obtain certain meaningful results in the direction indicated without assuming stationarity [15].

In the discrete case, the deduction of limiting theorems of the type mentioned have been carried out in many remarkable papers. Among them especially important is the previously quoted paper by A.Ya.Khinchin [6].

For the general continuous case, here practically nothing has been done.

§3. The computation and estimate of ε-entropy in certain particular cases.

If the condition

$$P_{\xi\xi'} \in W$$

is chosen in the form requiring the exact coincidence of ξ and ξ'

$$\mathbf{P}\{\xi = \xi'\} = 1,$$

then

$$H_W(\xi) = H(\xi).$$

Correspondingly, in the general case it would seem natural to call $H_W(\xi)$ "the entropy of the random object ξ under the exact reproduction of W".

Now let us assume that the space X' coincides with X, i.e. we are studying methods of approximate transmission of messages on the position in space of the point $\xi \in X$ by indicating a point ξ' of the same space X under the assumption that a "distance" $\rho(x, x')$ has been introduced in this space and this distance satisfies the usual axioms of "metric spaces". It seems natural to require that

$$\mathbf{P}\{\rho(\xi, \xi') \leqslant \varepsilon\} = 1 \tag{W_ε^0}$$

or that

$$\mathbf{E}\rho^2(\xi, \xi') \leqslant \varepsilon^2. \tag{W_ε}$$

These two forms of "ε-entropy" of the distribution P_ξ will be denoted by

$$H_{W_\varepsilon^0}(\xi) = H_\varepsilon^0(\xi), \ H_{W_\varepsilon}(\xi) = H_\varepsilon(\xi).$$

As to the ε-entropy $H^0_\varepsilon(\xi)$, I would like to point out here certain estimates for

$$H^0_\varepsilon(X) = \sup_{P_\xi} H^0_\varepsilon(\xi),$$

where the upper bound is taken over all probability distributions P_ξ on the space X. When $\varepsilon = 0$, as is well known,

$$H^0_0(X) = \sup_{P_\xi} H(\xi) = \log N_X,$$

where N_X is the number of elements of the set X. When $\varepsilon > 0$,

$$\log N^c_X(2\varepsilon) \leqslant H^0_\varepsilon(X) \leqslant \log N^a_X(\varepsilon),$$

where $N^a_X(\varepsilon)$ and $N^c_X(\varepsilon)$ are the characteristics of the space X which I have introduced in my note [3]. The asymptotic properties as $\varepsilon \to 0$ of the functions $N_X(\varepsilon)$ have been studied for a series of specific spaces X in [3] and are interesting analogues of the asymptotic behaviour of the function $H_\varepsilon(\xi)$ developed below.

Now let us consider the ε-entropy $H_\varepsilon(\xi)$. If X is a n-dimensional Euclidean space and

$$P_\xi(A) = \int_A p_\xi(x)dx_1 dx_2 \ldots dx_n,$$

then (at least in the case of sufficiently smooth function $p_\xi(x)$), we have the well-known formula

$$H_\varepsilon(\xi) = n\log(1/\varepsilon) + [h(\xi) - n\log\sqrt{2\pi e}] + o(1), \tag{9}$$

where

$$h(\xi) = -\int_X p_\xi(x)\log p_\xi(x)dx_1 dx_2 \ldots dx_n$$

is the "differential entropy" introduced in Shannon's first papers. Thus the asymptotic behaviour of $H_\varepsilon(\xi)$ in the case of sufficiently smooth continuous distributions in n-dimensional space is determined first of all by the dimension of the space, and differential entropy $h(\xi)$ appears only in the form of the second term of the expression for $H_\varepsilon(\xi)$.

It is natural to expect that for typical distributions in infinite-dimensional spaces the growth of $H_\varepsilon(\xi)$ as $\varepsilon \to 0$ will be much faster. As a simplest example, consider Wiener's random function $\xi(t)$ determined for $0 \leqslant t \leqslant 1$ by normally distributed independent increments

$$\Delta\xi = \xi(t + \Delta t) - \xi(t)$$

for which

$$\xi(0) = 0, \ \mathbf{E}\Delta\xi = 0, \ \mathbf{E}(\Delta\xi)^2 = \Delta t.$$

In this case Ya. M. Yaglom has found, in the metric of L^2, that

$$H_\varepsilon(\xi) = \frac{4}{\pi}\frac{1}{\varepsilon^2} + o\left(\frac{1}{\varepsilon^2}\right). \tag{10}$$

In a more general way, for Markov process of diffusion type on the segment of time $t_0 \leqslant t \leqslant t_1$, where

$$\mathbf{E}\Delta\xi = A(t, \xi(t))\Delta t + o(\Delta t),$$
$$\mathbf{E}(\Delta\xi)^2 = B(t, \xi(t))\Delta t + o(\Delta t),$$

it is possible to obtain, under certain natural assumptions, the formula

$$H_\varepsilon(\xi) = \frac{4}{\pi}\chi\frac{1}{\varepsilon^2} + o(\frac{1}{\varepsilon^2}), \tag{11}$$

where

$$\chi(\xi) = \int_{t_0}^{t_1} \mathbf{E}B(t, \xi(t))dt.$$

In the case of a normal distribution in n-dimensional Euclidean space or in Hilbert space, the ε-entropy H_ε may be computed exactly; the n-dimensional vector ξ, after an appropriate orthogonal transformation, is of the form

$$\xi = (\xi_1, \xi_2, \ldots, \xi_n),$$

where the coordinates ξ_k are mutually independent and distributed normally. For given ε, the parameter θ can be determined from the equation

$$\varepsilon^2 = \Sigma \min(\theta^2 \mathbf{D}\xi_k)$$

and in the case of a normally distributed ξ, we have

$$H_\varepsilon(\xi) = \frac{1}{2} \sum_{\mathbf{D}\xi_k > \theta^2} \log \frac{\mathbf{D}\xi_k}{\theta^2}. \tag{12}$$

The approximating vector

$$\xi' = (\xi'_1, \xi'_2, \ldots, \xi'_n)$$

must be chosen so as to have

$$\xi'_k = 0,$$

while if we have

$$\mathbf{D}\xi_k \leqslant \theta^2,$$

then

$$\xi_k = \xi'_k + \Delta_k, \ \mathbf{D}\Delta_k = \theta^2, \ \mathbf{D}\xi'_k == \mathbf{D}\xi_k - \theta^2$$

while in the case

$$\mathbf{D}\xi_k > \theta^2$$

the vectors ξ_k and Δ_k are mutually independent. The infinite-dimensional case does not differ from the finite-dimensional one in any way.

Finally, it is very important to note that the maximal value of $H_\varepsilon(\xi)$ for the vector ξ (n-dimensional or infinite-dimensional) for fixed second central moments is achieved in the case of a normal distribution. This result may be obtained directly, or from the following proposition due to M. S. Pinsker (compare [16]):

Theorem. *Suppose we are given a positive definite symmetric matrix with entries* s_{ij}, $0 \leqslant i, j \leqslant m + n$, *and a distribution* P_ξ *of the vector*

$$\xi = (\xi_1, \xi_2, \ldots, \xi_m),$$

whose central second moments are equal to s_{ij} *(when* $0 \leqslant i, j \leqslant m$*). Assume that the condition* W *on the joint distribution* $P_{\xi\xi'}$ *of the vector* ξ *and the vector*

$$\xi' = (\xi_{m+1}, \xi_{m+2}, \ldots, \xi_{m+n})$$

is that the central second moments of the magnitudes

$$\xi_1, \xi_2, \ldots, \xi_{m+n}$$

are equal to s_{ij} *(when* $0 \leqslant i, j \leqslant m + n$*). Then*

$$H_W(\xi) \leqslant \frac{1}{2} \log(AB/C). \tag{13}$$

The notation in formula (13) corresponds to the exposition in §1. If we compare the above to §1, we see that inequality (13) is transformed into an equality in the case of a normally distributed P_ξ.

The principles for the solution of variational problems which appear in the calculation of the "rate of message creation" were indicated by Shannon fairly long ago. In [10] Shannon and Weaver write: "Unfortunately these formal solutions in particular cases are difficult to estimate numerically and therefore their value is not very great." However, in essence, many problems of this type, as can be seen from the above, are sufficiently simple. Possibly the slow tempo of research in this direction is related to the insufficient understanding of the fact that in typical cases the solution of the variational problems often turns out to be degenerate; for example, in the problem of computing $H_\varepsilon(\xi)$ for a normally distributed vector ξ in n-dimensional space, considered above, the vector ξ' often turns out not to be n-dimensional, but only k-dimensional with $k < n$, while in the infinite-dimensional case the vector ξ' always turns out to be finite-dimensional.

§4. The amount of information and the rate of creation of messages in the case of stationary processes.

Consider two stationary and stationarily related processes

$$\xi(t), \eta(t), \quad -\infty < t < \infty.$$

Let us denote by ξ_T and η_t the segments of the processes ξ and η in time $0 < t \leqslant T$ and by ξ_- and η_- the development of the processes ξ and η on the negative semiaxis $-\infty < t \leqslant 0$. To choose a pair (ξ, η) of stationarily related processes ξ and η means to choose a probability distribution $P_{\xi\eta}$ invariant with respect to translations along the t-axis in the space of pairs of functions $\{x(t), y(t)\}$. If we fix ξ, then the distribution $P_{\xi\eta}$ yields the conditional distribution

$$P_{\xi_T, \eta | \xi_-}(C | x_-) = \mathbf{P}\{(\xi_T, \eta) \in C | \xi_- = x_-\}.$$

By means of this distribution, according to §1, it is possible to calculate the conditional amount of information

$$I(\xi_T, \eta | x).$$

If the expectation

$$\mathbf{E}I(\xi_T, \eta | \xi_-)$$

is finite for some $T > 0$, then it is finite as well for all other $T > 0$ and

$$\mathbf{E}I(\xi_T, \eta | \xi_-) = T\vec{I}(\xi, \eta).$$

The expression $\vec{I}(\xi, \eta)$ can naturally be called the "rate of creation of information about the process η while observing the process ξ". If the process ξ can be extrapolated exactly from the past into the future, then

$$\vec{I}(\xi, \eta) = 0.$$

In particular this will be the case if the process ξ has a bounded spectrum. Generally speaking, the equality

$$\vec{I}(\xi, \eta) = \vec{I}(\eta, \xi) \tag{14}$$

does not hold. However, under certain fairly wide conditions of "regularity" of the process[7] ξ, we have the equality

$$\vec{I}(\xi, \eta) = \bar{I}(\xi, \eta),$$

where

$$\bar{I}(\xi, \eta) = \lim_{T \to \infty} \frac{1}{T} I(\xi_T, \eta_T).$$

Since $I(\xi_T, \eta_T) = I(\eta_T, \xi_T)$, we always have

$$\bar{I}(\xi_T, \eta_T) = \bar{I}(\eta_T, \xi_T)$$

and therefore in the case when both equalities $\vec{I}(\xi, \eta) = \bar{I}(\xi, \eta)$ and $\vec{I}(\eta, \xi) = \bar{I}(\eta, \xi)$ are satisfied, we obtain (14). Now suppose W is a certain class of joint distributions $P_{\xi\xi'}$ of two stationary and stationarily related processes ξ and ξ'. It is natural to call the expression

$$\vec{H}_W(\xi) = \inf_{P_{\xi\xi'} \in W} \vec{I}(\xi', \xi)$$

the "speed of creation of messages in the process ξ for the exactness of reproduction W". Under certain assumptions of regularity of the process ξ and for certain natural types of conditions W, it can be proved that

$$\vec{H}_W(\xi) = \bar{H}_W(\xi),$$

where

$$\bar{H}_W = \inf_{P_{\xi\xi'} \in W} \bar{I}(\xi, \xi').$$

---

[7]The regularity of the process here and further means, roughly speaking, that the segments of the process corresponding to two sufficiently distant segments of the t-axis are almost independent. In the case of Gaussian processes, here we can apply the well known definition of regularity introduced in my paper [17].

§5. Computation and estimate of the amount of information and of the rate of message creation in terms of spectra.

In the case when a distribution $P_{\xi\eta}$ is normal and at least one of the processes ξ or η is regular, we have the formula

$$\bar{I}(\xi,\eta) = -\frac{1}{4\pi} \int_{-\infty}^{+\infty} \log[1 - r^2(\lambda)]d\lambda,$$

proposed by M. S. Pinsker [18], where

$$r^2(\lambda) = \frac{|f_{\xi\eta}(\lambda)|^2}{f_{\xi\xi}(\lambda)f_{\eta\eta}(\lambda)}, \tag{15}$$

while $f_{\xi\xi}, f_{\xi\eta}, f_{\eta\eta}$ are spectral densities. In the case of processes with discrete time t, for differential entropy of the normal process per unit time

$$\bar{h}(\xi) = \lim_{T\to\infty} h(\xi_1, \xi_2, \ldots, \xi_T)$$

the following expression

$$\bar{h}(\xi) = \log(2\pi\sqrt{e}) + \frac{1}{2\pi} \int_{-\pi}^{+\pi} \log f_{\xi\xi}(\lambda)\, d\lambda \tag{16}$$

is known. However, in the case of continuous time and unbounded spectrum no analogue of the expression $\bar{h}(\xi)$ exists and the Pinsker formula must be derived independently.

It is natural to characterize the precision of reproduction of the stationary process ξ by means of the stationary (and stationarily related to ξ) process ξ' by using the expression

$$\sigma^2 = \mathbf{E}[\xi(t) - \xi'(t)]^2$$

and in the case of condition W of the form

$$\sigma^2 \leqslant \varepsilon^2$$

to call the expression

$$\bar{H}_\varepsilon(\xi) = \bar{H}_W(\xi)$$

the ε-entropy per unit time of the process ξ, and, under the assumption that

$$\bar{H}_W(\xi) = \bar{H}_W(\xi),$$

call it the speed of creation of messages in the process ξ with mean precision of transmission ε. From the corresponding proposition for finite-dimensional distributions (see §3) it is possible to deduce that for a given spectral density $f_{\xi\xi}(\lambda)$ the expression $\bar{H}_\varepsilon(\xi)$ reaches its maximum in the case of a normal process ξ. In the normal case, the expression $\bar{H}_\varepsilon(\xi)$ may also be easily calculated from the spectral density $f_{\xi\xi}(\lambda)$ by a method quite similar to the one that was explained in §3 as

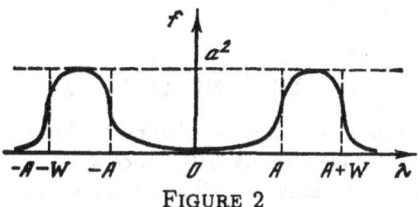

FIGURE 2

applied to the expression $H_\varepsilon(\xi)$ for n-dimensional distributions. The parameter θ is determined from the equation

$$\varepsilon^2 = \int_{-\infty}^{+\infty} \min(\theta^2, f_{\xi\xi}(\lambda))\, d\lambda. \tag{17}$$

By means of this parameter, the value of $\bar{H}_\varepsilon(\xi)$ can be found from the formula

$$\bar{H}_\varepsilon(\xi) = \frac{1}{2} \int_{f_{\xi\xi}(\lambda) > \theta^2} \log \frac{f_{\xi\xi}(\lambda)}{\theta^2}\, d\lambda. \tag{18}$$

It is of practical interest to find the spectral densities of the form shown on Figure 2, which are well approximated by the function

$$\varphi(\lambda) = \begin{cases} a^2 & \text{for } A \leqslant |\lambda| \leqslant A + W, \\ 0 & \text{in other cases.} \end{cases}$$

In this case it is easy to calculate that for ε not too small, we have approximately (for a normal process)

$$\theta^2 \asymp \varepsilon^2/2W, \quad \bar{H}_\varepsilon(\xi) \asymp W \log(2Wa^2/\varepsilon^2). \tag{19}$$

Formula (19) is of course none other than the well-known Shannon formula

$$R = W \log(Q/N). \tag{20}$$

What is essentially new here, however, is that now we see why and within what limits (for not too small ε) this formula may be applied to processes with unbounded spectrum, and such are all practically interesting processes in the theory of data transmission. Writing (19) in the form

$$\bar{H}_\varepsilon(\xi) \approx 2W(\log(1/\varepsilon) + \log(a\sqrt{W})) \tag{21}$$

and comparing it to (9), we see that the double width $2W$ of the frequency strip plays the role of the number of measurements. This idea of the equivalence of the double width of the frequency strip to the number of measurements carried out, in a certain sense, per unit time, was apparently first pointed out by V.A.Kotelnikov (see [12]). In order to establish this idea, Kotelnikov indicated the fact that the function whose spectrum can be placed in a strip of width $2W$ is uniquely determined by the values of the function at the points

$$\ldots, -\frac{2}{2W}, -\frac{1}{2W}, 0, \frac{1}{2W}, \frac{2}{2W}, \ldots, \frac{k}{2W}, \ldots$$

This same argument is also presented by Shannon, who uses the representations obtained in this way to deduce formula (20). Since a function with bounded spectrum is always singular in the sense of [17] and the observation of such a function is not related in general to a stationary input of new information, the meaning of such arguments remains quite unclear, so that the new proof of the approximate formula (21) presented here seems to me to be of interest.

For small ε and for any normally distributed regular random function, the growth of $\bar{H}_\varepsilon(\xi)$ with the decrease of ε takes place much faster than this would be obtained according to formula (21). In particular, if $f_{\xi\xi}(\lambda)$ as $\lambda \to \infty$ is of order $\lambda^{-\beta}$, then $\bar{H}_\varepsilon(\xi)$ has order $\varepsilon^{-2/(\beta-1)}$.

§6. Calculation and estimates of the channel capacity in certain particular cases.

First let us consider the case when the input signal is an m-dimensional vector

$$\eta = (\eta_1, \ldots, \eta_m),$$

and the output signal is an n-dimensional vector

$$\eta' = (\eta'_1, \ldots, \eta'_n).$$

As we mentioned above, the transmitting device $\eta \to \eta'$ is characterized by the conditional distribution of probability $P_{\eta'|\eta}$ and certain restrictions on the input signal η. Let us assume that the dependence of η' on η is of the character of a linear normal correlation, i.e. that

$$\eta' = A\eta + \zeta, \tag{22}$$

where the operator A is linear while the vector ζ is independent of η and satisfies the n-dimensional Gauss distribution. Concerning the conditions on the input signal, we assume that they are of the form

$$\mathbf{E}Q(\eta) \leqslant \varepsilon^2, \tag{23}$$

where Q is a certain positive definite quadratic form in the coordinates $\eta_1, \ldots, \eta_m$. By means of an appropriate choice of linear transformations of coordinates in the spaces Y and Y', the general case can be reduced to the case when

$$Q(\eta) = \Sigma \eta_i^2, \tag{24}$$

while the dependence (22) is written in the form

$$\eta'_i = a_i\eta_i + \zeta_i, \; a_i \neq 0 \quad \text{for } 1 \leqslant i \leqslant k,$$
$$\eta'_i = \zeta_i \quad \text{for } i > k, \tag{25}$$

where $k = \min(m, n)$.

Under the above assumptions the channel capacity, i.e. the least upper bound C of the amount of information $I(\eta, \eta')$ which satisfies the above assumptions can easily be found. This upper bound is achieved if the η_i are mutually independent random Gauss magnitudes

$$
\begin{aligned}
\mathbf{D}\eta_i &= \theta^2 - \mathbf{D}\zeta/a_i^2 \quad \text{if} \quad i \leqslant k \quad \text{and} \quad a_i^2\theta_i^2 > \mathbf{D}\zeta_i, \\
\mathbf{D}\eta_i &= 0 \quad \text{for other } i,
\end{aligned}
\tag{26}
$$

where the constant θ^2, as can be easily seen, is uniquely determined by the requirements (26) and

$$
\sum_i \mathbf{D}\eta_i = \varepsilon^2.
\tag{27}
$$

The corresponding value $C = I(\eta, \eta')$ is (see §3)

$$
C = \frac{1}{2} \sum_{a_i^2\theta^2 > \mathbf{D}\zeta_i} \log \frac{a_i^2\theta^2}{\mathbf{D}\zeta_i}.
\tag{28}
$$

The situation is similar in the case of stationary linear channels with "Gaussian noise", i.e. in the case

$$
\eta'(t) = A\eta(t) + \zeta(t),
\tag{29}
$$

where A is a linear operator and $\zeta(t)$ is a Gauss stationary random function independent of $\eta(t)$. As is well known, the corresponding spectral densities satisfy the relation

$$
f_{\eta'\eta'}(\lambda) = a^2(\lambda)f_{\eta\eta}(\lambda) + f_{\zeta\zeta}(\lambda).
\tag{30}
$$

The condition which restricts the power of the input channel will be chosen in the form

$$
\int_{-\infty}^{+\infty} f_{\eta\eta}(\lambda)d\lambda \leqslant \varepsilon^2.
\tag{31}
$$

The formal solution of the problem is quite similar to the solution of the finite-dimensional problem considered above. The greatest channel capacity of the channel $\bar{C}$ is obtained if

$$
\begin{aligned}
f_{\eta\eta}(\lambda) &= \theta^2 - f_{\zeta\zeta}(\lambda)/a^2(\lambda) \quad \text{for} \quad a^2(\lambda)\theta^2 > f_{\zeta\zeta}(\lambda), \\
f_{\eta\eta}(\lambda) &= 0 \quad \text{for} \quad a^2(\lambda)\theta^2 \leqslant f_{\zeta\zeta}(\lambda).
\end{aligned}
\tag{32}
$$

The corresponding value of $\bar{I}(\eta, \eta')$, according to formula (15), is equal to

$$
\bar{C} = \frac{1}{4\pi} \int_{a^2(\lambda)\theta^2 > f_{\zeta\zeta}(\lambda)} \log \frac{a^2(\lambda)\theta^2}{f_{\zeta\zeta}(\lambda)} \, d\lambda.
\tag{33}
$$

It should immediately be noted that in the cases interesting for practical applications, the spectral density $f_{\eta\eta}(\lambda)$ computed according to formulas (32) vanishes outside a certain bounded interval of the λ-axis and therefore is the spectral density of the singular process. In this case the process $\eta'(t)$ will be (under the natural assumption that $\zeta(t)$ is regular) a mixed process, i.e. will consist of the sum of a singular component $A\eta$ and a regular component ζ. In this case the rate of information creation $\vec{I}(\eta', \eta)$ about the process η when the process η' is observed is no longer expressed by the formula $\vec{I}(\eta', \eta) = \bar{C}$, but is simply equal to zero (see §5). By means of certain supplementary considerations, it can be shown, however, that if we introduce certain regular processes (which in principle can be realized and in fact carry information arising in time) in the form of the input signal $\eta(t)$, it is possible to obtain a rate of information creation $\vec{I}(\eta', \eta) = \vec{I}(\eta, \eta')$ as close as we want to $\bar{C}$. This conclusively justifies formula (33).

REFERENCES

[1] A. N. Kolmogorov, *Statistical oscillation theory with continuous spectrum*, General Session of the Academy of Sciences devoted to the 30th anniversary of the Great October Socialist Revolution, Moscow, Leningrad, Academy of Science Publishers, 1948, pp. 465–472.

[2] A. N. Kolmogorov, *Statistical oscillation theory with continuous spectrum*, Anniversary Collection devoted to the 30th anniversary of the Great October Socialist Revolution, vol. 1, Moscow, Leningrad, Academy of Science Publishers, 1947, pp. 242-252.

[3] A. N. Kolmogorov, *On certain asymptotic characteristics of completely bounded metric spaces*, Dokl. Akad. Nauk SSSR 108,3 (1956), 385-388.

[4] A. G. Vitushkin, *Concerning Hilbert's thirteenth problem*, Dokl. Akad. Nauk SSSR, 95,4 (1954), 701-704.

[5] A. Ya. Khinchin, *The notion of entropy in probability theory*, Uspekhi Mat. Nauk, 8,3 (1953), 3-20.

[6] Khinchin A. Ya., *On basic theorems of information theory*, Uspekhi Mat. Mauk, 11,1 (1956), 17-75.

[7] *The theory of transmission of electrical signals in the case of noise: a collection of translations*, Moscow, Foreign Language Publishers, 1953.

[8] A. A. Kharkevich, *Sketches on communications theory*, Moscow, Gostekhizdat, 1955.

[9] S. Goldman, *Information theory*, London, Constable and Co.,, 1953.

[10] C. E. Shannon, W. Weaver, *The mathematical theory of communications*, Urbana: Univ. III. Press (1949), 3-89.

[11] H. Cramèr, *Mathematical methods of statistics*, Princeton Univ. Press, 1963.

[12] V. A. Kotelnikov, *On the channel capacity of "ether" and of a wire in electricity*, Materials to the 1st All-Union Congress on questions of technical reconstructions of communications and development of weak current technology, Moscow, 1933.

[13] I. M. Gelfand, A. N. Kolmogorov, A. M. Yaglom, *The amount of information and entropy for continuous distributions*, Works of the 3rd All-Union Mathematical Congress, vol. 3, Moscow, Academy of Science Publishers, 1958, pp. 300-320.

[14] A. M. Obukhov, *Normal correlation of vectors*, Nauch. Zapiski Mosk. Univ. 45 (1940), 73-92.

[15] M. Rosenblat-Rot, *The notion of entropy in probability theory and its application to the theory of transmission channels*, Works of the 3rd All-Union Mathematical Congress, vol. .2,, Moscow, Academy of Science Publishers, 1956, pp. 132-133.

[16] M. S. Pinsker, *The amount of information about one stationary random process contained in another stationary random process*, Works of the 3rd All-Union Mathematical Congress, vol. 1, Moscow, Academy of Sciences Publishers, 1956, pp. 125.

[17] A. N. Kolmogorov, *Stationary sequences in the Hilbert space*, Bulletin of Moscow University. Mathematics. 2,6 (1941), 1-40.

[18] M. S. Pinsker, *The amount of information in a Gauss random stationary process contained in a second process stationarily related to the other one*, Dokl. Akad. Nauk SSSR 99,2 (1954,), 213-216.

INTERVENTION AT THE SESSION

The disagreement between A. A. Lyapunov and myself on the question of cybernetics is more of verbal character than of fundamental nature. Undoubtedly, the mathematical theory of finite algorithms with bounded memory involving the total inclusion of the time factor (number of steps until the result is obtained, distinction between rapid and slow memory, etc.) must stabilize in the form of an independent and systematically developing mathematical discipline. In our country certain comrades, in particular A. A. Lyapunov, feel that it is precisely such a theory, which unites mathematical questions related to the functioning of computers and controlling devices working discretely in time, that should be called cybernetics. However, the word cybernetics was introduced by N. Wiener, who gave it a much wider significance. Cybernetics according to Wiener must include the entire mathematical apparatus of the theory of controlling devices of discrete and continuous action. This means that cybernetics according to Wiener includes an important part of the theory of stability for systems of differential equations, classical control theory, the theory of random processes with their extrapolation and filtration, the theory of information, game theory with applications to operations research, the technical aspects of logic algebra, the theory of programming and many other topics. It is easy to understand that as a mathematical discipline cybernetics in Wiener's understanding lacks unity, and it is difficult to imagine productive work in training a specialist, say a postgraduate student, in cybernetics in this sense. If A. A. Lyapunov can clearly formulate the programme of development of a narrower and more unified discipline, then in my opinion it would be better not to call it cybernetics.

I agree with many points of V. S. Pugachov's intervention. I should only like to point out that the notion of ε-entropy was essentially introduced by Shannon himself, who used the term rate of message creation. The change of terminology we have proposed stresses the fundamental role of this notion and, it seems to me, does not lack importance in the sense of correctly directing further work, but there is nothing especially new here. Much more original work in my opinion was done by Soviet mathematicians concerning the development of spectral methods. I think that the work done in this direction by M. S. Pinsker and myself will find sufficiently interesting continuation. An basically complete mathematical foundation of the remarkable ideas of V. A. Kotelnikov on the equivalence of the width of the frequency strip and the number of degrees of freedom per unit time is of great importance.

A few words about the question of scientific personnel. At the present time we lack mathematicians capable of creatively using probabilistic methods. In applied technical fields, perhaps only specialists in shooting theory possess a sufficient amount of qualified personnel doing research with some creativity and an active knowledge of probabilistic methods. A series of other traditional applications of probability theory via mathematical statistics in the domains of economics, medicine, agriculture and biology have long ago come to nothing and are not always justified. Statistical methods in questions of mass production are being developed, but the group of researchers involved is much too small. Now, in connection with the increasing interest in information theory, in statistical methods in the study of control devices, in the Monte Carlo method in computer applications and in series of other new fields, we are trying, at Moscow University, to widen the curriculum of mathematics students intending to work in the technical applications of probability theory. Among the students who have completed their graduate work at Moscow University recently and began to work in questions related to information theory at scientific research institutes, I should mention M. S. Pinsker and R. L. Dobrushin. In 1957 we shall be able to train for research work approximately twenty alumni of the university who will be well versed in probability theory, in the theory of random processes, and in mathematical statistics. Although at present they are still students, some of them have begun their careers as independent young researchers.

In connection with L. V. Kantorovich's intervention about mathematical methods of planning, I should like to point out that the centre of gravity of the question here is not the application of fast computers. Of course in this situation computers are very useful, but at present not enough is being done in our country in this interesting direction along the traditional lines of computation.

Concerning the question of mathematical translation, I agree with I. S. Bruk in the respect that for practical contacts, the knowledge of languages or the use of live translators will be replaced in the near future by computers. But already today it is doubtless that work in computer translations yields interesting results, in particular for people designing computers and is especially interesting to linguists who, in the process of this work, are forced to specify their formulations of the laws of language. These researchers can now use the criterion of experience, so important for the development of the theory: if some formulation, say, of the rules of disposition of words in some language is sufficient to obtain a good literary translation by means of a computer, then they have the assurance that this rule is formulated in a conclusive way. Now it is impossible to use vague phrases and present them as being "laws", something that unfortunately people working in the humanities tend to do. Now, to conclude on the lighter side, allow me to express the hope that in the coming year of 1957, young people of different countries at the Youth Festival in Moscow will be able to talk without the help of computers and will find simpler and more humane forms of communication.

4
AMOUNT OF INFORMATION AND ENTROPY FOR CONTINUOUS DISTRIBUTIONS*[1]

(Jointly with I. M. Gelfand and A. M. Yaglom)

§1. Definition of the amount on information contained in one random object about another.

Suppose $\mathfrak{A}$ is a certain trial which may have n different outcomes $A_1, A_2, \ldots, A_n$ with probabilities $P(A_1), P(A_2), \ldots, P(A_n)$; let $\mathfrak{B}$ be another trial which has outcomes $B_1, B_2, \ldots, B_m$ with probabilities $P(B_1), P(B_2), \ldots, P(B_m)$. If by $P(A_iB_k)$ we denote the probability of the simultaneous outcome A_i of the trial $\mathfrak{A}$ and outcome B_k of the trial $\mathfrak{B}$, then according to Shannon [1], the mean amount of information contained in the outcomes of the trial $\mathfrak{U}$ with respect to the outcome of the trial $\mathfrak{B}$ is equal to

$$I(\mathfrak{U}, \mathfrak{B}) = \sum_{i,k} P(A_iB_k) \log \frac{P(A_iB_k)}{P(A_i)P(B_k)}. \tag{1.1}$$

The amount of information contained in the results of the trial $\mathfrak{A}$ with respect to itself (i.e., the total amount of information contained in the results of the trial $\mathfrak{A}$) is called the *entropy* of the trial $\mathfrak{A}$ and is denoted by $H(\mathfrak{A})$:

$$H(\mathfrak{A}) = I(\mathfrak{A}, \mathfrak{A}) = - \sum_i P(A_i) \log P(A_i). \tag{1.2}$$

---

*In the book Works of the Third All-Union Mathematical Congress. Moscow. Academy of Science Publishers, 1958, vol.3, pp. 300-320.(In Russian).

[1]The report is devoted basically to the definition, general properties and methods of computation of the expressions $I(\xi, \eta)$ and $H_W(\xi)$. The general definition of $I(\xi, \eta)$ is contained in Appendix 7 to Shannon's paper [1] (using the Russian translation of this paper, in which Appendix 7 was omitted, we did not notice this in time, in particular, when writing the note [22]). However, before our note [22], there was apparently no detailed exposition of the principles of development of the general theory based on this definition in the literature. If the additional remark above is taken into account, then the paper [22] presents the history of the appearance of results presented in §1 of the present report. In §2 we give indications relating to the importance of the expressions $I(\xi, \eta)$ and $H_W(\xi)$ for the theory of data transmissions along connecting channels (concerning this question also see the paper [21]; the greater part of the results and general considerations developd in [21] are reproduced in English in the paper [24]). Results given in §3 without reference to other authors were obtained by I. M. Gelfand and A. M. Yaglom (see [23]) and in §4,5 by A. N. Kolmogorov (except the constants in the formulas (4.13) and (5.12) calculated by A. M. Yaglom). The text of the report was written by A. N. Kolmogorov and A. M. Yaglom.

If we denote by $H(\mathfrak{A}/B_k)$ the "conditional entropy" of the trial $\mathfrak{A}$ obtained by replacing the probability $P(A_i)$ by the conditional probability $P_B(A_i) = P(A_i/B_k)$ in formula (1.2) and denote by $H(\mathfrak{A}/\mathfrak{B})$ the random variable assuming the value $H(\mathfrak{A}/B_k)$ with probability $P(B_k)$, then $I(\mathfrak{A}, \mathfrak{B})$ may also be represented in the form

$$I(\mathfrak{A}, \mathfrak{B}) = H(\mathfrak{A}) - \mathbf{E}H(\mathfrak{A}/\mathfrak{B}), \tag{1.3}$$

where $\mathbf{E}$ denotes expectation.

In order to generalize definition (1.1) to the case of arbitrary probabilistic objects, it is convenient at first to reformulate it in the language of normed Boolean algebras. Namely, we shall start from the normed Boolean algebra $\mathfrak{G}$ of all possible "random events" $A, B, C, \ldots$ supplied with the operations of taking the complement A' to any element A, the union ("sum") $A \cup B$, the product AB, the unit element E and the zero element N; the norm here is the probability distribution $P(A)$, which is a non-negative additive function on $\mathfrak{G}$, normed by the condition $P(E) = 1$. Here the notion of "trial" is naturally identified with the notion of subalgebra of the algebra $\mathfrak{G}$ (the subalgebra of all those events whose outcome becomes known after the trial has been carried out).

It is known that each finite Boolean algebra $\mathfrak{A}$ coincides with the system of all possible sums of its "atoms" $A_1, A_2, \ldots, A_n$. If both subalgebras $\mathfrak{A}$ and $\mathfrak{B}$ are finite, then we shall define the amount of information contained in the outcomes of the trial $\mathfrak{A}$ with respect to the outcomes of the trial $\mathfrak{B}$, by using Shannon's formula (1.1), where $A_1, \ldots, A_n$ are atoms of $\mathfrak{A}$ while $B_1, \ldots, B_m$ are atoms of $\mathfrak{B}$.

In the case of arbitrary subalgebras $\mathfrak{A}$ and $\mathfrak{B}$, it is natural to set

$$I(\mathfrak{A}, \mathfrak{B}) = \sup_{\mathfrak{A}_1 \subseteq \mathfrak{A}, \mathfrak{B}_1 \subseteq \mathfrak{B}} I(\mathfrak{A}_1, \mathfrak{B}_1), \tag{1.4}$$

where the least upper bound is taken over all possible subalgebras $\mathfrak{A}_1 \subseteq \mathfrak{A}$ and $\mathfrak{B}_1 \subseteq \mathfrak{B}$. Symbolically, the definition (1.4) may be written in the form

$$I(\mathfrak{A}, \mathfrak{B}) = \int_{\mathfrak{A}} \int_{\mathfrak{B}} P(d\mathfrak{A}\,d\mathfrak{B}) \log \frac{P(d\mathfrak{A}\,d\mathfrak{B})}{P(d\mathfrak{A})P(d\mathfrak{B})}. \tag{1.5}$$

In the case of finite algebras $\mathfrak{A}$ and $\mathfrak{B}$, the amount of information $I(\mathfrak{A}, \mathfrak{B})$ is always real and non-negative (see [1]); this clearly means, in the general case, that $I(\mathfrak{A}, \mathfrak{B})$ may either be real and non-negative or equal to $+\infty$. It is also clear that in the general case the following properties of the expression I, well known in the finite case, remain valid (here by $[\mathfrak{M}]$ we denote the smallest subalgebra of the algebra $\mathfrak{G}$ containing the set $\mathfrak{M}$):

a) $I(\mathfrak{A}, \mathfrak{B}) = I(\mathfrak{B}, \mathfrak{A})$;

b) $I(\mathfrak{A}, \mathfrak{B}) = 0$ if and only if the systems of events $\mathfrak{A}$ and $\mathfrak{B}$ are independent;

c) if $[\mathfrak{A}_1 \cup \mathfrak{B}_1]$, and $[\mathfrak{A}_2 \cup \mathfrak{B}_2]$ are independent systems of events, then

$$I([\mathfrak{A}_1 \cup \mathfrak{A}_2], [\mathfrak{B}_1 \cup \mathfrak{B}_2]) = I(\mathfrak{A}_1, \mathfrak{B}_1) + I(\mathfrak{A}_2, \mathfrak{B}_2);$$

d) if $\mathfrak{A}_1 \subseteq \mathfrak{A}$ then $I(\mathfrak{A}_1, \mathfrak{B}) \leqslant I(\mathfrak{A}, \mathfrak{B})$.

Further let us indicate three more properties of information which, in certain cases, allow passing to the limit in the computation or estimate of the expression I

for a concrete systems of events: e) if $\mathfrak{A}_1 \subseteq \mathfrak{A}$ is everywhere dense in $\mathfrak{A}$ in the sense of the metric $\rho(A, B) = P(AB' \cup A'B)$, then

$$I(\mathfrak{A}_1, \mathfrak{B}) = I(\mathfrak{A}, \mathfrak{B});$$

f) if $\mathfrak{A}_1 \subseteq \mathfrak{A}_2 \subseteq \cdots \subseteq \mathfrak{A}_n \subseteq \ldots$ and $\mathfrak{A} = \cup_n \mathfrak{A}_n$, then

$$I(\mathfrak{A}, \mathfrak{B}) = \lim_{n \to \infty} (\mathfrak{A}_n, \mathfrak{B});$$

g) if the sequence of probability distributions $P^{(n)}$ converges in $[\mathfrak{A} \cup \mathfrak{B}]$ to the distribution P, i.e. if $\lim_{n \to \infty} P^{(n)}(C) = P(C)$ for $C \in [\mathfrak{A} \cup \mathfrak{B}]$, then

$$I(\mathfrak{A}, \mathfrak{B}) \leqslant \varliminf_{n \to \infty} I^{(n)}(\mathfrak{A}, \mathfrak{B}),$$

where $I(\mathfrak{A}, \mathfrak{B})$ is the amount of information for the distribution P, while $I^{(n)}(\mathfrak{A}, \mathfrak{B})$ is this amount for the distribution $P^{(n)}$.

All these properties follow directly from definition (1.4) of the amount of information and are almost obvious.

So far we have only been talking about the notion of amount of information contained in the outcomes of a trial with respect to the outcomes of another; in the applications, however, one more often deals with the notion of amount of information contained in one random object (for example, a random variable, or a random vector, or a random function) about another object of the same type. In order to describe this new notion, we shall assume that the main Boolean algebra $\mathfrak{G}$ is a σ-algebra (see [2]) and consider it together with a certain "measurable space" X, i.e. a pair consisting of the set X and of a σ-algebra S_X (with unit X) whose elements are subsets of X, called "measurable" (thereby modifying the terminology of [2], where S_X may be a σ-ring). Here ξ will be called a *random element* of the space X (or a *random variable* or *random object* from X) if to each subset $A \in S_X$ corresponds an event $B \in \mathfrak{G}$ (the event consisting in that $\xi \in A$); in other words, a random element ξ is a homomorphism ξ^* of the σ-algebra S_X into the Boolean algebra $\mathfrak{G}$ such that

$$\xi^*(A) = B. \tag{1.6}$$

Here the entire algebra S_X is taken to a certain subalgebra of the algebra $\mathfrak{G}$ denoted by

$$\mathfrak{G}_\xi = \xi^*(S_X), \tag{1.7}$$

and the amount of information contained in the random object ξ about the random object η is naturally defined as

$$I(\xi, \eta) = I(\mathfrak{G}_\xi, \mathfrak{G}_\eta). \tag{1.8}$$

The formula

$$P_\xi(A) = P(\xi^*(A)) \tag{1.9}$$

defines a measure in the space X, called the distribution of the random object ξ; in accordance to the intuitive meaning of the event $\xi^*(A)$, we can write

$$P_\xi(A) = \mathbf{P}\{\xi \in A\} \tag{1.9'}$$

(where the right-hand side denotes the "probability of the event $\xi \in A$"). A similar random object η from Y is assigned the measure

$$P_\eta(B) = \mathbf{P}\{\eta \in B\} \tag{1.10}$$

in the space Y. If we now define the measurable space $X \times Y$ from the given spaces X, Y in the ordinary way (see [2]), then the condition

$$(\xi, \eta)^*(A \times B) = \xi^*(A)\eta^*(B) \tag{1.11}$$

uniquely determines a homomorphism $(\xi, \eta)^*$ of the Boolean σ-algebra $S_{X \times Y}$ into $\mathfrak{G}$, i.e., a new random object (ξ, η) from $X \times Y$, namely the pair (ξ, η) of random objects ξ and η. This new random object is assigned the distribution

$$P_{\xi,\eta}(C) = \mathbf{P}\{(\xi, \eta) \in C\}, \tag{1.12}$$

which is a measure on $X \times Y$; further, on $X \times Y$ we can define another measure $P_\xi \times P_\eta$ by putting

$$P_\xi \times P_\eta(A \times B) = P_\xi(A)P_\eta(B), \quad A \in S_X, \ B \in S_Y. \tag{1.13}$$

The information $I(\xi, \eta)$ may be expressed in terms of the measures $P_{\xi,\eta}$ and $P_\xi \times P_\eta$ by means of the formula

$$I(\xi, \eta) = \int_X \int_Y P_{\xi,\eta}(dx\,dy) \log \frac{P_{\xi,\eta}(dx\,dy)}{P_\xi(dx)P_\eta(dy)} \tag{1.14}$$

(compare (1.5)), where the integral is understood in the general sense indicated in A. N. Kolmogorov's paper [3] (also see [4]).

Instead of formula (1.14), it is sometimes convenient to use a different notation for $I(\xi, \eta)$ in the form of an integral which can be understood in the ordinary Lebesgue sense. Namely, applying the Radon-Nikodim theorem to the measures $P_{\xi,\eta}$ and $P_\xi \times P_\eta$ on $X \times Y$, we can write

$$P_{\xi,\eta}(C) = \int \int_C a(x, y)P_\xi(dx)P_\eta(dy) + S(C), \tag{1.15}$$

where the measure $S(C)$ is singular with respect to $P_\xi \times P_\eta$; in the particular case when $S(X \times Y) = 0$, the function $a(x, y)$ coincides with the Stiltjes-Radon derivative of the measure $P_{\xi,\eta}$ with respect to the measure $P_\xi \times P_\eta$,

$$a(x, y) = \frac{P_{\xi,\eta}(dx\,dy)}{P_\xi(dx)P_\eta(dy)}. \tag{1.16}$$

We then have the following

Theorem 1. *The amount of information* $I(\xi, \eta)$ *may be finite only in the case when* $S(X \times Y) = 0$. *In this case*

$$I(\xi, \eta) = \int_X \int_Y a(x, y) \log a(x, y) P_\xi(dx) P_\eta(dy) =$$
$$= \int_X \int_Y \log a(x, y) P_{\xi, \eta}(dx\, dy). \tag{1.17}$$

Note also that formuala (1.17) may be rewritten in the form

$$I(\xi, \eta) = \mathbf{E} \log a(\xi, \eta), \tag{1.18}$$

where the right-hand side involves the expectation of the new real random variable $\log a(\xi, \eta)$, a function of the pair (ξ, η).[2]

If the distributions P_ξ, P_η and $P_{\xi, \eta}$ are expressed in terms of densities

$$P_\xi(A) = \int_A p_\xi(x) dx, \quad P_\eta(B) = \int_B p_\eta(y) dy,$$
$$P_{\xi, \eta}(C) = \int \int_C p_{\xi, \eta}(x, y) dx\, dy, \tag{1.19}$$

where dx and dy denote integration over the measures given in X and Y, then formulas (1.17) and (1.14) become the expression well-known from the applications

$$I(\xi, \eta) = \int_X \int_Y p_{\xi, \eta}(x, y) \log \frac{p_{\xi, \eta}(x, y)}{p_\xi(x) p_\eta(y)} dx dy. \tag{1.20}$$

It is curious to note that if we introduce the "differential entropy"

$$h(\xi) = -\int p_\xi(x) \log p_\xi(x) dx \tag{1.21}$$

and the "conditional differential entropy" $h(\xi|\eta)$ determined by the conditional distribution $P_{\xi|\eta}$ of the random element[3] ξ for fixed η, then formula (1.20) may be written in a form similar to (1.3)

$$I(\xi, \eta) = h(\xi) - \mathbf{E} h(\xi|\eta). \tag{1.21'}$$

---

[2] A function $\eta = f(x)$ of the random element ξ, where $y = f(x)$, is a Borel measurable map of the space X into another space Y that can be naturally defined as the random element of Y determined by the map

$$\eta^*(A) = \xi^* f^{-1}(A).$$

[3] $P_{\xi|\eta}(A)$ is a real random function $f(\eta)$ of η satisfying the requirement

$$P_{\xi\eta}(A \times B) = \int_B f(y) P_\eta(dy), \ B \in S_Y.$$

However, unlike (1.3), $h(\xi)$ here does not coincide with $I(\xi, \xi)$ (the latter expression is infinite).

It is easy to deduce from properties a)-d) of the expression $I(\mathfrak{A}, \mathfrak{B})$ the following properties of the amount of information $I(\xi, \eta)$:

a) $I(\xi, \eta) = I(\eta, \xi)$;

b) $I(\xi, \eta) \geqslant 0$; $I(\xi\eta) = 0$ if and only if ξ and η are independent;

c) if the pairs (ξ_1, η_1) and (ξ_2, η_2) are independent, then

$$I((\xi_1, \xi_2), (\eta_1, \eta_2)) = I(\xi_1, \eta_1) + I(\xi_2, \eta_2);$$

d) if ξ_1 is a function of ξ (in the sense indicated), then

$$I(\xi_1, \eta) \leqslant I(\xi, \eta).$$

In particular, we always have

$$I((\xi, \eta), \zeta) \geqslant I(\eta, \zeta).$$

If, similarly to (1.2), we define the entropy of a random variable ξ as

$$I(\xi, \xi) = H(\xi),$$

then $H(\xi)$ turns out to be equal to infinity under very general conditions (for example, whenever the distribution P_ξ is continuous and determined by a density p_ξ). Therefore the notion of entropy will be practically useful only for a narrow class of random variables (first of all for discrete variables assuming a finite number of values). It follows from properties c)-d) of the amount of information $I(\xi, \eta)$ that

$$H(\xi, \eta) \geqslant H(\xi), \quad H(\xi, \eta) \geqslant H(\eta) \tag{1.22}$$

and that if ξ and η are independent, then

$$H(\xi, \eta) = H(\xi) + H(\eta). \tag{1.23}$$

It is well known (see, for example, [1]) that for the entropy $H(\xi, \eta)$, besides the estimate (1.22) from below, there is also a simple estimate from above: we always have

$$H(\xi, \eta) \leqslant H(\xi) + H(\eta). \tag{1.24}$$

It should be stressed that no similar estimate exists for the amount of information in the pair (ξ, η) with respect to the variable ζ: using elementary examples, it is easy to show that

$$I(\xi, \zeta) = 0, \; I(\eta, \zeta) = 0$$

does not necessarily imply

$$I((\xi, \eta), \zeta) = 0$$

(as a counterexample, we can cite any example of three random variables which are all pairwise independent, but none of which is independent of the other two).

Let us indicate two more properties of the amount of information $I(\xi, \eta)$ which follow from properties e)-g) of the expression $I(\mathfrak{A}, \mathfrak{B})$:

e) if $\xi = (\xi_1, \xi_2, \ldots, \xi_n, \ldots)$ (this is a random element of the space $X = \times_{n=1}^{\infty} X_n$), then

$$I(\xi, \eta) = \lim_{n \to \infty} I((\xi_1, \ldots, \xi_n), \eta);$$

f) if X and Y are complete metric spaces in which for S_X and S_Y we take algebras of Borel sets and if the distributions $P_{\xi, \eta}^{(n)}$ converge weakly to the distribution $P_{\xi, \eta}$, then for the corresponding amounts of information we have the inequality

$$I(\xi, \eta) \leqslant \lim_{n \to \infty} I^{(n)}(\xi, \eta).$$

In conclusion let us dwell especially on the case when X and Y are the spaces of all real functions on certain sets A and B. Random elements ξ and η will be random functions $\xi(a), \eta(b)$ on the sets A and B, i.e. certain families of real random variables depending on the parameters $a \in A, b \in B$. Here $I(\xi, \eta)$ will have the meaning of the amount of information contained in one family of random variables about the other family. The spaces X and Y here are powers R^A and R^B of the space $\mathbf{R}$ of real numbers (naturally with the system of its Borel sets in the role of S_R). It is known that the algebra S_X is generated by all possible cylindrical sets ("quasi-intervals" of the space X) defined by a finite number of systems of inequalities of the form

$$A_i \leqslant x(a_i) < B_i, \ a_i \in A, \ i = 1, 2, \ldots, n. \tag{1.25}$$

Therefore the definition of the amount of information $I\{\xi, \eta\}$ for arbitrary systems of random variables reduces to the definition of the amount of information $I\{\xi(a_1, \ldots, a_n), \eta(b_1, \ldots, b_m)\}$ for finite-dimensional random vectors

$$\xi(a_1, \ldots, a_n) = \{\xi(a_1), \ldots, \xi(a_n)\} \text{ and } \eta(b_1, \ldots, b_m) = \{\eta(b_1), \ldots, \eta(b_m)\},$$

namely

$$I\{\xi(a), \eta(b)\} = \sup I\{\xi(a_1, \ldots, a_n), \eta(b_1, \ldots, b_m)\} \tag{1.26}$$

(the least upper bound is taken over all integers $n, m, a_i \in A, i = 1, 2, \ldots, n$, and $b_j \in B, j = 1, 2, \ldots, m$). Thus in studying the amount of information $I\{\xi(a), \eta(b)\}$, we can limit ourselves to considering only finite-dimensional probability distributions, avoiding the use of distributions in function spaces.

A special case of a "family of random variables" are the so-called generalized random processes, introduced recently by K. Ito [5] and I. M. Gelfand [6]. In this case the set A is the space Φ of infinitely differentiable functions $\varphi(t)$ on the set T of real numbers[4] each of which vanishes outside a certain finite interval (the space Φ

---

[4]It goes without saying that instead of generalized random processes (random functions of a real argument t) one can also consider "generalized random processes on an arbitrary differential manifold T"; however, such a generalization does not lead to any serious changes and will not be considered here.

of principal functions in the terminology of L. Schwartz [7]). A generalized random process $\xi(\varphi)$ is given by a family of multidimensional probability distributions for all possible finite families of random variables $\xi(\varphi_i), \varphi_i \in \Phi, i = 1, \ldots, n$, i.e., by the family of probabilities of all possible "quasi-intervals"

$$A_i \leqslant x(\varphi_i) < B_i, \ \varphi_i \in \Phi, \ i = 1, \ldots, n. \tag{1.25'}$$

If $\xi(\varphi)$ and $\eta(\psi)$ are two such generalized processes corresponding to the spaces of principal functions Φ and Ψ (differing for example in the choice of the set T), then the amount of information $I\{\xi(\varphi), \eta(\psi)\}$ according to (1.8) and (1.4) is defined as

$$I\{\xi(\varphi), \eta(\psi)\} = \sup I\{\xi(\varphi_1, \ldots, \varphi_n), \eta(\psi_1, \ldots, \psi_m)\}, \tag{1.26'}$$

where

$$\xi(\varphi_1, \ldots, \varphi_n) = \{\xi(\varphi_1), \ldots, \xi(\varphi_n)\},$$
$$\eta(\psi_1, \ldots, \psi_m) = \{\eta(\psi_1), \ldots, \eta(\psi_m)\}$$

and the upper bound is taken over all possible integers n, m and $\varphi_i \in \Phi, i = 1, \ldots, n, \psi_j \in \Psi, j = 1, \ldots, m$. Definition (1.26') can also be applied to any ordinary continuous stochastic and stochastically integrable random process $\xi(t)$ by putting

$$\xi(\varphi) = \int_{-\infty}^{\infty} \xi(t)\varphi(t)dt; \tag{1.27}$$

it can be proved that in this case definition (1.26') leads to the same result as the direct definition based on formula (1.26). Note that often even for ordinary (not generalized) processes the definition (1.26') is more convenient than (1.26).

§2. Abstract exposition of the foundations of Shannon theory.

Shannon considers the transmission of messages according to the scheme

$$\xi \to \eta \to \eta' \to \xi', \tag{2.1}$$

where the transmitting device

$$\eta \to \eta'$$

is characterized by the conditional distribution

$$P_{\eta'|\eta}(B'|y) = \mathbf{P}\{\eta' \in B'|\eta = y\} \tag{2.2}$$

of the "output signal" η' for a given "input signal" η and a certain restriction

$$P_\eta \in V \tag{2.3}$$

of the distribution P_η^1 of the input signal. The "coding operation"

$$\xi \to \eta$$

and the "decoding operation"

$$\eta' \to \xi'$$

are characterized by the conditional distributions

$$P_{\eta|\xi}(B|x) = \mathbf{P}\{\eta \in B | \xi = x\}, \tag{2.4}$$

$$P_{\xi'|\eta'}(A'|y') = \mathbf{P}\{\xi' \in A' | \eta' = y'\}. \tag{2.5}$$

The main problem considered by Shannon is the following. The spaces X, X', Y, Y' of possible values of input messages ξ, output messages ξ', input signals η and output signals η' are given, as well as the characteristics of the transmitting device, i.e. the conditional distribution $P_{\eta|\eta'}$ and the class V of transmissible distributions P_η of the input signal; finally, the distributions

$$P_\xi(A) = \mathbf{P}\{\xi \in A\} \tag{2.6}$$

of messages on the input and the requirements to the precision of transmission

$$P_{\xi\xi'} \in W \tag{2.7}$$

are given, where W is a certain class of joint distributions

$$P_{\xi\xi'}(C) = \mathbf{P}\{(\xi, \xi) \in C\} \tag{2.8}$$

of input and output messages. It is asked if is it possible, and if so, how is it possible, to determine the coding and decoding rules, i.e. conditional distributions $P_{\eta|\xi}$ and $P_{\xi'|\eta'}$, so that in the calculation of the distribution $P_{\xi\xi'}$ from the distributions $P_\xi, P_{\eta|\xi}, P_{\eta'|\eta}, P_{\xi'|\eta'}$, under the assumption that the sequence

$$\xi, \eta, \eta', \xi'$$

is Markov, we obtain

$$P_{\xi\xi'} \in W?$$

Following Shannon, let us define the capacity of the transmitting device by the formula

$$C = \sup_{P_\eta \in V} I(\eta, \eta') \tag{2.9}$$

and introduce the expression

$$H_W(\xi) = \inf_{P_{\xi\xi'} \in W} I(\xi, \xi') \tag{2.10}$$

which (when calculated per unit time) Shannon calls the "rate of message creation". Then the important property, $I((\xi, \eta), \zeta) = I(\eta, \xi)$ iff the sequence ξ, η, ζ is Markov, of the amount of information I, immediately implies the following necessary condition for the transmission to be possible

$$H_W(\xi) \leqslant C. \tag{2.11}$$

As we already noted, an uncomparably deeper idea of Shannon is that condition (2.11), when applied to a sufficiently long work interval of the "transmission channels" is in a certain sense and under certain very general conditions also "almost sufficient". From the mathematical point of view, here we are talking about the proof of limiting theorems of the following type. It is assumed that the spaces X, X', Y, Y', the distributions P_ξ and $P_{\eta'|\eta}$, the classes V and W and therefore the expressions C and $H_W(\xi)$ depend on the parameter T (which in the applications plays the role of the duration of work of the transmitting devices). It is required to establish, under certain very general assumptions, that the condition

$$\lim_{T \to \infty} \frac{C^T}{H_W^T(\xi)} > 1 \qquad (2.12)$$

is sufficient to make possible a transmition that would satisfy the conditions stated above for sufficiently large T. Naturally in such a setting the problem is somewhat vague (as is, for example, the general problem of studying possible limiting distributions for sums of a large number of "small" summands); a more rigorous formulation may be obtained by assuming, for example, that ξ and ξ' are stationary random processes. However, we purposely omit the use of the terminology of stationary random processes, since the young Rumanian mathematician Rosenblat-Rot Mill in his report at the present congress [8] (see also [9]) showed that it is possible to obtain interesing results in this direction without assuming stationarity.

The deduction of limiting theorems of the type indicated above has been carried out in many remarkable papers; let us cite recent work by A. Ya. Khinchin [16] and M. S. Pinsker [17], which contains references to earlier research. We feel that much must still be done in this direction. We stress that it is precisely this kind of result that can be used to justify the general belief that the expression $I(\xi, \eta)$ is not simply one of the possible ways of measuring the "amount of information", but is the measure of the amount of information that has indeed an advantage in principle over the others. Since by its original nature "information" is not a scalar magnitude, we feel that the axiomatic study allowing to characterize $I(\xi, \eta)$ uniquely (or uniquely characterize the entropy $H(\xi)$) by means of simple formal propeties have in this respect a lesser importance. Here we believe that the situation is similar to the one related to the fact that of all the methods for obtaining a foundation of the normal distribution law for errors proposed by Gauss, today we favour the method based on limiting theorems for sums of large numbers of small summands. Other methods (for example, the method based on the arithmetical mean) only explains why no other distribution law for errors can be as pleasant and convenient as the normal one, but does not answer the question why the normal distribution law so often appears in real problems. Exactly in the same way, the beautiful formal properties of the expressions $H(\xi)$ and $I(\xi, \eta)$ cannot explain why they suffice in many problems for finding a complete solution (at least from the asymptotical point of view).

§3. **Calculation of the amount of information in the case of Gaussian distributions.**

In the present section we shall consider the question of computing the amount of information $I(\xi, \eta)$, limiting ourselves only to the special (but extremely important) cases when the distributions P_ξ, P_η and $P_{\xi\eta}$ are Gaussian.

Let us begin with the case when ξ and η are random vectors

$$
\begin{aligned}
\xi &= (\xi_1, \ldots, \xi_m), \\
\eta &= (\eta_1, \ldots, \eta_n) \equiv (\xi_{m+1}, \ldots, \xi_{m+n}),
\end{aligned}
\tag{3.1}
$$

normally distributed with central second moments

$$
s_{ij} = \mathbf{E}[(\xi_i - \mathbf{E}\xi_i)(\xi_j - \mathbf{E}\xi_j)].
\tag{3.2}
$$

If the determinant

$$
\det C = \det \|s_{ij}\|_{1 \leqslant i, j \leqslant m+n}
\tag{3.3}
$$

is non-zero, then the calculation of the amount of information $I(\xi, \eta)$ by means of formula (1.20) reduces to computing an elementary integral and yields the result

$$
I(\xi, \eta) = \frac{1}{2} \log \frac{\det A \det B}{\det C},
\tag{3.4}
$$

where

$$
A = \|s_{ij}\|_{1 \leqslant i, j \leqslant m}, \quad B = \|s_{ij}\|_{m+1 \leqslant i, j \leqslant m+n}.
\tag{3.5}
$$

For specific computations it is sometimes more convenient to use, instead of formula (3.4), the formula

$$
\begin{aligned}
I(\xi, \eta) &= -1/2 \log \det(E - DB^{-1}D'A^{-1}) = \\
&= -1/2 \log \det(E - D'A^{-1}DB^{-1}),
\end{aligned}
\tag{3.6}
$$

where E is the unit matrix (this notation will be used in the sequel as well) and

$$
D = \|s_{ij}\|_{\substack{1 \leqslant i \leqslant m \\ m < j \leqslant m+n}}, \quad D' = \|s_{ij}\|_{\substack{m < i \leqslant m+n \\ 1 \leqslant j \leqslant m}}.
\tag{3.7}
$$

In the simplest case of one-dimensional random variables ξ and η, formula (3.4) (or (3.7)) yields

$$
I(\xi, \eta) = -1/2 \log[1 - r^2(\xi, \eta)],
\tag{3.8}
$$

where $r^2(\xi, \eta)$ is the correlation coefficient between ξ and η; the last formula is obviously also true without assuming that the two-dimensional probability distribution for ξ and η is non-degenerate (that the matrix C is non-degenerate). Formula (3.8) may be generalized to the multi-dimensional case (also without restricting ourselves to non-degenerate matrices C). Indeed, as is well known [11, 12], it is possible, by using an appropriate linear transformation of coordinates in the spaces X and Y, to assume that all the second moments s_{ij}, vanish except those for which $j = i$ and $j = m + i$. For such a choice of coordinates, using formula (3.8) and property c), we obtain[5] the amount of information in the form

$$
I(\xi, \eta) = -\frac{1}{2} \sum_k \log[1 - r^2(\xi_k, \eta_k)],
\tag{3.9}
$$

---

[5]Since $\eta_k = \xi_{m+k}$.

where the sum is taken over all $k \leqslant \min(m, n)$ for which the correlation coefficient between ξ_k and $\eta_k = \xi_{m+k}$ is non-zero.

Also note that the expression obtained for $I(\xi, \eta)$ has a simple geometric significance. Let us consider real random variables (to be more precise, classes of equivalent random variables equal to each other with probability 1) with finite dispersion as vectors of the Hilbert space $\mathfrak{H}$ with scalar product $(\xi_1, \xi_2) = \mathbf{E}[(\xi_1 - \mathbf{E}\xi_1)(\xi_2 - \mathbf{E}\xi_2)]$. Then to each random vector $\xi = (\xi_1, \ldots, \xi_m)$ corresponds a finite-dimensional linear subspace H_1 of the space $\mathfrak{H}$ spanning the elements $(\xi_1, \ldots, \xi_m)$ of this space; similarly, to the vector $\eta = (\eta_1, \eta_n)$ will correspond another finite-dimensional linear subspace H_2. Now consider the linear transformation B_1 in the space H_1 under which each vector from H_1 is first projected onto H_2 and then the vector obtained is projected back onto H_1; in the space H_2 a similar role is played by the linear transformation B_2 consisting of the projection of each vector from H_2 onto H_1 and then back onto H_2. The transformations B_1 and B_2 may be expressed in terms of the projection operators P_1 and P_2 on the subspaces H_1 and H_2 by means of the formulas

$$B_1 = P_1 P_2 \quad \text{in } H_1; \quad B_2 = P_2 P_1 \quad \text{in } H_2; \tag{3.10}$$

we can also extend the definitions of B_1 and B_2 to obtain operators in the entire space $\mathfrak{H}$ assuming that they are zero operators on the orthogonal complements to the subspace H_1 and H_2 respectively, and write

$$B_1 = P_1 P_2 P_1, \quad B_2 = P_2 P_1 P_2. \tag{3.11}$$

Clearly B_1 and B_2 will be non-negative self-adjoint operators whose norm is less than or equal to 1, so that all their eigenvalues are real and contained in the interval between 0 and 1. It then turns out that the squares of the correlation coefficients $r^2(\xi_k, \eta_k)$ in formula (3.9) coincide with the eigenvalues of the operators B_1 and B_2 (which have the same eigenvalues), so that this formula may be rewritten in the form

$$I(\xi, \eta) = -1/2 \log \det(E - B_1) = -1/2 \log \det(E - B_2). \tag{3.12}$$

The expression (3.12) for the information $I(\xi, \eta)$ shows that this magnitude depends only on the subspaces H_1 and H_2 and not on the choice of basis in these subspaces and that it is a geometrical invariant of the pair of subspaces indicated. But a complete system of geometrical invariants for a pair of finite-dimensional linear subspaces is the system of angles between the subspaces (speaking more prcisely, of the "stationary angles" between them, see for example [13]). This implies that $I(\xi, \eta)$ must be expressed in terms of the angles between H_1 and H_2. And indeed it can be shown that two linear subspaces H_1 and H_2 (one no more than m-dimensional, the second no more than n-dimensional) form a total of $k \leqslant \min(m, n)$ different stationary angles, and the eigenvalues of the operators $\alpha_1, \ldots, \alpha_k$ are equal to $\cos^2 \alpha_1, \ldots, \cos^2 \alpha_k$, so that

$$I(\xi, \eta) = -\log \sin |\sin \alpha_1 \ldots \sin \alpha_k|. \tag{3.13}$$

It is easy to generalize formula (3.12) to the case when $\xi = \xi(a)$ and $\eta = \eta(b)$ are two arbitrary Gaussian families of real random variables, i.e. two families of random variables such that finite-dimensional probability distributions for any finite number of variables belonging to one of the families or the other are all Gaussian distributions. The family $\xi(a)$ can also be assigned a linear subspace H_1 of the space $\mathfrak{H}$ (as a rule, the subspace will be infinite-dimensional), spanning elements $\xi(a)$ of this space; similarly the family $\eta(b)$ is assigned a linear subspace H_2 spanning elements $\eta(b) \in \mathfrak{H}$. Here we have the following

Theorem 2. *Suppose $\xi = \xi(a)$ and $\eta = \eta(b)$ are two Gaussian families of real random variables, H_1 and H_2 are the corresponding linear subspaces of the space $\mathfrak{H}$, and P_1 and P_2 are the projection operators in $\mathfrak{H}$ on H_1 and H_2 respectively, while B_1, B_2 are the operators (3.11). In order that the amount of information $I\{\xi, \eta\}$ be a finite magnitude in this case, it is necessary and sufficient that at least one of the operators B_1, B_2 be completely continuous and have finite trace; in this case the second of the operators B_1, B_2 will also possess the same properties and the amount of information $I\{\xi, \eta\}$ will be given by formula (3.12), where E is the unit operator.*

The application of Theorem 2 is especially simple in the case when $\xi = \xi(\varphi)$ is a random process (generally speaking, a generalized one) on a certain interval (possibly infinite) T, while η is a scalar random variable. The projection $\bar{\eta} = P_1\eta$ of η on the subspace H_1 here will coincide with the best approximation to η linearly depending on $\xi(\varphi)$, i.e., will be the solution of the problem of linear filtration for the process $\xi(\varphi)$; a large amount of special literature is devoted to this problem (see for example, the review [14]). If we denote by $\sigma - [\mathbf{E}|\eta - \bar{\eta}|^2]^{1/2}$ the mean quadratic filtration error and by $\sigma_1 = \sigma/[\mathbf{E}\eta^2]^{1/2}$ the relative mean quadratic error, then, as can easily be seen, we have the formula

$$I\{\xi(\varphi), \eta\} = -\log \sigma_1. \tag{3.14}$$

The case when $\xi = \xi(\varphi)$ and $\eta = \eta(\psi)$ are both random processes is more complicated; here we shall deal only with the problem of computing $I\{\xi(\varphi), \eta\}$ for stationary (and stationarily related) processes $\xi(\varphi)$ and $\eta(\varphi)$ given on the same finite interval[6] $-T/2 \leqslant t \leqslant T/2$. As is well known (see [5, 6]), a Gaussian stationary generalized process $\xi(\varphi)$ is uniquely characterized by the generalized correlation function $b_{\xi\xi}(\varphi)$, a linear functional on the space Φ of basic functions $\varphi(t)$ (in the case of an ordinary process $\xi = \xi(t)$,

$$b_{\xi\xi}(\varphi) = \int_{-\infty}^{\infty} b_{\xi\xi}(\tau)\varphi(\tau)d\tau,$$

where $b_{\xi\xi}(\tau) = \mathbf{E}\xi(t+\tau)\xi(t)$ is the correlation function of the process $\xi(t)$). To determine the pair of processes $\xi(\varphi)$ and $\eta(\varphi)$, we must choose generalized correlation functions (distributions) $b_{\xi\xi}(\varphi)$ and $b_{\eta\eta}(\varphi)$ and also a generalized mutual correlation function $b_{\xi\eta}(\varphi) \equiv b_{\eta\xi}(\varphi^*)$, where $\varphi^*(t) = \varphi(-t)$.

---

[6]Speaking of a process $\xi(\varphi)$ given on a finite interval $-T/2 \leqslant t \leqslant T/2$, we assume that of all the random variables $\xi(\varphi, \varphi) \in \Phi$ constituting the generalized process, we are given only those magnitudes which correspond to functions $\varphi(t)$ identically vanishing outside the interval under consideration.

Now in the space Φ_T of functions $\varphi(t)$ on the segment $-T/2 \leqslant t \leqslant T/2$ consider the following operators

$$A\varphi = b_{\xi\xi}(\tau_t\varphi), \quad B\varphi \qquad\qquad = b_{\eta\eta}(\tau_t\varphi),$$
$$D\varphi = b_{\xi\eta}(\tau_t\varphi), \quad D'\varphi = b_{\eta\xi}(\tau_t\varphi) = b_{\xi\eta}(\tau_{-t}\varphi), \qquad (3.15)$$

where τ_t is the shift operator: $\tau_t\varphi(s) = \varphi(s+t)$ (the operators (3.15) may be viewed as operators in Hilbert space $L^2(-T/2, T/2)$ determined on an everywhere dense set of functions $\varphi(t)$). Thus, in the case considered here, Theorem 2 assumes the following form:

Theorem 2'. *For the amount of information $I\{\xi_T, \eta_T\}$ contained in a stationary Gauss random process $\xi(\varphi)$ defined on a segment of length T about another similar process $\eta(\varphi)$ on the same segment (also stationary and Gauss related to the first) to be finite, it is necessary and sufficient that at least one of the operators*

$$B_1 = DB^{-1}D'A^{-1}, \qquad B_2 = D'A^{-1}DB^{-1}, \qquad (3.16)$$

where A, B, D and D' are defined as in (3.15), be completely continuous and have finite trace. Under this condition the second of the two operators will possess the same property and the information $I\{\xi_T, \eta_T\}$ will be given by relation (3.12).

Theorem 2' may be used, for example, to obtain explicit expressions for the amount of information $I\{\xi_T, \eta_T\}$ in the case when the Fourier transforms of the distributions $b_{\xi\xi}, b_{\eta\eta}$ and $b_{\xi\eta}$ ("the spectral densities") exist and are rational functions. In the important particular case when $\eta = \eta(t)$ is an ordinary stationary random process with correlation function $B(\tau)$, while $\xi = \eta + \zeta$, where $\zeta = \zeta(\varphi)$ is a "white noise", non-correlated with η (a generalized stationary process with "independent values" for which

$$b_{\zeta\zeta} = 2\pi f\delta, \quad \delta(\varphi) = \varphi(0), \qquad (3.17)$$

where δ is the Dirac δ-function , while f is the constant spectral density of the "white noise") , formula (3.12) gives

$$I\{\xi_T, \eta_T\} = \frac{1}{2}\log\det(E + \frac{1}{2\pi f}B), \qquad (3.18)$$

where B is the Fredholm operator with kernel $B(t - t')$

$$B\varphi = \int_{-T/2}^{T/2} B(t - t')\varphi(t')dt'. \qquad (3.19)$$

Thus in this case the calculation of the amount of information $I\{\xi_T, \eta_T\}$ reduces to finding the determinant of the integral operator

$$E + B/2\pi f.$$

If the operator $(1/2\pi f)B$ in formula (3.18) is only a small addition to the unit operator E, i.e. if the spectral density f of the noise is much greater than the characteristic value f_0 of the spectral density of the process $\eta(t)$ or if T is much smaller than the characteristic time T_0 when the correlation function $B(\tau)$ decays, then the determinant in the right-hand side of (3.18) may be approximately computed by means of the perturbation theory series

$$\log \det(E + \frac{1}{2\pi f}B) = Sp \log(E + \frac{1}{2\pi f}B) =$$
$$\frac{1}{2\pi f}SpB - \frac{1}{2(2\pi f)^2}SpB^2 + \frac{1}{3(2\pi f)^3}SpB^3 - \dots, \tag{3.20}$$

where Sp denotes trace, so that

$$I\{\xi_T, \eta_T\} = \frac{B(0)}{4\pi f}T - \frac{T}{16\pi^2 f^2}\int_{-T}^{T}(1 - \frac{|\tau|}{T})B^2(\tau)d\tau + \dots \tag{3.21}$$

In the case when the correlation function $B(\tau)$ is the Fourier transform of the rational spectral density $f(\lambda)$:

$$B(\tau) = \int_{-\infty}^{\infty} e^{i\tau\lambda}f(\lambda)d\lambda, \tag{3.22}$$

where

$$f(\lambda) = \frac{|b_0(i\lambda)^m + \dots + b_m|^2}{|a_0(i\lambda)^n + \dots + a_n|^2} = \frac{|Q(i\lambda)|^2}{|P(i\lambda)|^2}, \quad m < n, \tag{3.23}$$

then we can obtain an explicit formula for $I\{\xi_T, \eta_T\}$; in this case the eigenvalues of the operator 3.19 coincide with the eigenvalues of a certain Sturm-Liouville problem for equations with constant coefficients, i.e. are zeros of a certain entire transcendental function, and the determinant in the right-hand side in (3.18) may be expressed in terms of the value of this entire function at a fixed point of the complex plane. In particular, for a Markov stationary process $\eta(t)$ with

$$B(\tau) = Ce^{-\alpha|\tau|}, \quad f(\lambda) = \frac{C\alpha}{\pi(\lambda^2 + \alpha^2)}, \tag{3.24}$$

we obtain

$$I\{\xi_T, \eta_T\} = \frac{1}{2}\log\{\frac{\sqrt{1+q} + (1+k/2)}{2\sqrt{1+q}}e^{\tau(\sqrt{1+q}-1)} +$$
$$\frac{\sqrt{1+q} + (1+k/2)}{2\sqrt{1+q}}e^{-\tau(\sqrt{1+q}+1)}\}, \tag{3.25}$$

where

$$k = f(0)/f = C/\pi\alpha f, \quad \tau = \alpha T. \tag{3.26}$$

When $k \ll 1$ ("strong noise") or $\tau \ll 1$ ("small observation time") formula (3.25) becomes the result provided by the perturbation theory series (3.21), as expected. In the other extreme case, for $\tau \gg 1$ ("large observation time") we shall have

$$I\{\xi_T, \eta_T\} \approx \frac{\sqrt{1+k} - 1}{2}\tau. \tag{3.27}$$

Finally, when $k \to 0$ ("weak noise"), formula (3.25) gives

$$I\{\xi_T, \eta_T\} \approx 1/2\sqrt{k\tau}. \tag{3.28}$$

For rational spectral densities $f(\lambda)$ more complicated than those in (3.24), the explicit formula for $I\{\xi_T, \eta_T\}$, as a rule, turns out to be fairly cumbersome; hence we shall limit ourselves here to writing out certain asymptotic results relating to the case of general rational spectral density (3.23). As $T \to 0$ or $f \to \infty$, the asymptotic behaviour of $I\{\xi_T, \eta_T\}$ in all cases is given by the perturbation theory series (3.21), so that we need not consider this case. As $T \to \infty$, the principal term of $I\{\xi_T, \eta_T\}$ will be

$$I\{\xi_T, \eta_T\} \approx \frac{1}{2} \sum_{k=1}^{n} (\Lambda_k - \Lambda_k^0)T, \tag{3.29}$$

where Λ_k, $k = 1, \dots, n$ are the roots of the equation

$$P(\Lambda)P(-\Lambda) + f^{-1} + f^{-1}Q(\Lambda)Q(-\Lambda) = 0, \tag{3.30}$$

situated in the right half-plane, while the Λ_k^0 are the corresponding roots for $f = \infty$. Finally, as $f \to 0$, we have asymptotically

$$I\{\xi_T, \eta_T\} \approx \frac{T}{2} \sum_{k=1}^{n} \Lambda_k, \tag{3.31}$$

where Λ_k have the same meaning as in (3.29); this easily implies that, as $f \to 0$,

$$I\{\xi_T, \eta_T\} \approx cT f^{-1/2(n-m)}, \tag{3.32}$$

where c is a constant depending on the coefficients of the polynomials $P(\Lambda)$ and $Q(\Lambda)$.

Note that the exponent $-[2(n-m)]^{-1}$ in formula (3.32) characterizes the "degree of smoothness" of the correlation functions $B(\tau)$ (and therefore of the random process $\eta(t)$ itself); in the case of spectral density (3.23), the correlation function $B(\tau)$ has $2(n-m)-2$ continuous derivatives, while the $(2n-2m-1)$th derivative of this function has a jump at the point $\tau = 0$. It can be shown that the fact that the asymptotic behaviour of $I\{\xi_T, \eta_T\}$, where $\xi = \eta + \zeta$, ζ is "white noise" with spectral density f, is determined as $f \to 0$ by the degree of smoothness of the function $B(\tau)$, is not related with the special form (3.23) of the spectral density, but holds under fairly general conditions, so that for a relatively wide class of correlation functions $B(\tau)$ possessing $(2n - 2)$ continuous derivatives and a $(2n - 1)$th derivative with a jump at $\tau = 0$, we have

$$I\{\xi_T, \eta_T\} \approx cT f^{-1/2n} \text{ as } f \to 0. \tag{3.33}$$

§4. Computation and estimates of ε-entropy (the rate of creation of messages) in certain particular cases.

We now pass to the consideration of the "rate of creation of messages"

$$H_W(\xi) = \inf_{P_{\xi\xi'} \in W} I(\xi, \xi').$$

If the condition $P_{\xi\xi'} \in W$ is chosen to mean that there is a strictly exact coincidence between ξ and ξ'

$$\mathbf{P}\{\xi = \xi'\} = 1, \tag{4.1}$$

then

$$H_W(\xi) = H(\xi), \tag{4.2}$$

i.e. the rate of creation of messages becomes the ordinary entropy of the random element ξ. As we have already indicated, for continuous P_ξ this last expression is always infinite: a continuously distributed variable ξ contains an infinite amount of information about itself. However, since continuous messages are always observed with limited precision, it follows that in the transmission of such messages only an amount of information that can be made as close as we wish to the generally finite expression $H_W(\xi)$ is transmitted. Hence, in the theory of transmission of continuous messages, the main role should be played precisely by the expression $H_W(\xi)$ (and not $H(\xi)$) which, in view of its analogy to entropy, may be called "the entropy of the random object ξ for exactly reproduced W".

The condition $P_{\xi\xi'} \in W$ must guarantee that the element ξ' is close to ξ in a certain sense. If we assume that X is a metric space, while X' coincides with X (i.e., if we study the approximate transmission of messages concerning the position of points in the space X by indicating points in the same space) then it seems natural to require that

$$\mathbf{P}\{\rho(\xi, \xi') \leqslant \varepsilon\} = 1 \tag{W_ε^0}$$

or that

$$\mathbf{E}\rho^2(\xi, \xi') \leqslant \varepsilon^2. \tag{W_ε}$$

These two types of ε-entropy H_ε with distribution P_ξ will be denoted by the symbol

$$H_{W_\varepsilon^0}(\xi) = H_\varepsilon^0(\xi), \tag{4.3}$$

$$H_{W_\varepsilon}(\xi) = H_\varepsilon(\xi). \tag{4.4}$$

First let us indicate some estimates for the expressions

$$H_\varepsilon^0(X) = \sup_{P_\xi} H_\varepsilon^0(\xi), \tag{4.5}$$

where the upper bound is taken over all probability distributions P_ξ on the space X. When $\varepsilon = 0$, as is well known, we have

$$H_0^0(X) = \sup_{P_\xi} H(\xi) = \log N_X, \tag{4.6}$$

where N_X is the number of elements of the set X (this upper bound is achieved for the distribution under which all the elements of X are equally probable). When $\varepsilon > 0$, we have

$$\log N_X^c(2\varepsilon) \leqslant H_\varepsilon^0(X) \leqslant \log N_X^a(\varepsilon), \tag{4.7}$$

where $N_X^a(\varepsilon)$ and $N_X^c(\varepsilon)$ are characteristics of the space X introduced in the recent note by A.N.Kolmogorov [15]. The asymptotic properties of the function $N_X(\varepsilon)$ as $\varepsilon \to 0$, studied for a series of concrete spaces X in [15], are interesting analogues of results on asymptotic behaviour of the function $H_\varepsilon(X)$ developed further on.

If X is a line segment and the distribution P_ξ, is given by a continuous density $p(x)$ positive everywhere and satisfying a Lipschitz condition, then as $\varepsilon \to 0$ we have

$$H_\varepsilon^0(\xi) = \log \frac{1}{2\varepsilon} + h(\xi) + O(\varepsilon \log \frac{1}{\varepsilon}), \tag{4.8}$$

where $h(\xi) = -\int p(x) \log p(x) dx$ is the "differential entropy" of the distribution P_ξ which we have already met in §1 (see 1.21). In the case of an n-dimensional space X, under similar conditions, we obtain

$$H_\varepsilon^0(\xi) = n \log \frac{1}{d_n \varepsilon} + h(\xi) + O(\varepsilon \log \frac{1}{\varepsilon}), \tag{4.9}$$

where d_n is a certain constant which varies relatively little with n.

Nothing will change here also if instead of distributions concentrated on a segment (or in a cube), we take sufficiently smooth more general distributions, decaying rapidly enough at infinity.

Similar results can also be obtained for the entropy $H_\varepsilon(\xi)$: it turns out that for sufficiently smooth distributions P_ξ

$$H_\varepsilon(\xi) = \inf_{P_{\xi\xi'} \in W_\varepsilon} I(\xi, \xi') = \log \frac{1}{\varepsilon} - \log \sqrt{2\pi e} + h(\xi) + O(\varepsilon). \tag{4.10}$$

For continuous distributions in n-dimensional space X (with sufficiently smooth density $p(x)$) under general assumptions we obtain in a similar way

$$H_\varepsilon(\xi) = n \log \frac{1}{\varepsilon} + \left[h(\xi) - n \log \sqrt{\frac{2\pi e}{n}} \right] + O(\varepsilon). \tag{4.11}$$

Thus we see that the asymptotic behaviour of $H_\varepsilon(\xi)$ in the case of sufficiently smooth continuous distributions in n-dimensional space is first of all determined by the dimension of the space and only after that does the differential entropy $h(\xi)$ enter in the estimate in the form of a second term. However, since the first (principal) term $H_\varepsilon(\xi)$ does not depend on the probability distribution, the asymptotic behaviour of $H_\varepsilon(\xi)$ for the given specific distribution is naturally characterized by the expression $h(\xi)$; this circumstance (together with formula (1.21)) determines the role of differential entropy for information theory.

Until now we have only considered probability distributions in finite-dimensional spaces. It is natural to expect that for typical distributions in infinite-dimensional spaces the growth of $H_\varepsilon(\xi)$ as $\varepsilon \to 0$ will be much faster. As a simplest example consider the Wiener random function $\xi(t)$ on the interval $0 \leqslant t \leqslant 1$ with normally distributed independent increments

$$\Delta\xi = \xi(t + \Delta t) - \xi(t),$$

for which

$$\xi(0) = 0, \quad \mathbf{E}\Delta\xi = 0, \quad \mathbf{E}(\Delta\xi)^2 = \Delta t. \tag{4.12}$$

In this case it can be shown that, in the metric of L^2, i.e., under the condition

$$\mathbf{E}\{\int_0^1 [\xi(t) - \xi'(t)]^2 \, dt\} \leqslant \varepsilon^2, \tag{W_ε}$$

the asymptotic behaviour of $H_\varepsilon(\xi)$ is determined by the relation

$$H_\varepsilon(\xi) = \frac{4}{\pi}\frac{1}{\varepsilon^2} + o(\frac{1}{\varepsilon^2}). \tag{4.13}$$

Hence for a more general Markov process of diffusion type in the segment of time $t_0 \leqslant t \leqslant t_1$ satisfying

$$\begin{aligned} \mathbf{E}\Delta\xi &= A(t, \xi(t))\Delta t + o(\Delta t), \\ \mathbf{E}(\Delta\xi)^2 &= B(t, \xi(t))\Delta t + o(\Delta t) \end{aligned} \tag{4.14}$$

one can obtain, under certain natural assumptions, the formula

$$H_\varepsilon(\xi) = \frac{4}{\pi}\chi\frac{1}{\varepsilon^2} + o(\frac{1}{\varepsilon^2}), \tag{4.15}$$

where

$$\chi(\xi) = \int_{t_0}^{t_1} \mathbf{E}B(t, \xi(t))dt. \tag{4.16}$$

The fact that in formulas (4.13), (4.15) the expression $H_\varepsilon(\xi)$ is of order ε^{-2} may be generalized in the following way: for a process $\xi(t)$ controlled by the differential equation

$$d[A_0\xi^{(n)}(t) + A_1\xi^{(n-1)}(t) + \cdots + A_n\xi(t)] = d\psi(t),$$

where $\psi(t)$ is the Wiener function and $A_0 \neq 0$, the entropy $H_\varepsilon(\xi)$ is of order

$$H_\varepsilon(\xi) \asymp \varepsilon^{-1/(n+1/2)}.$$

In the case of a normal distribution in n-dimensional Euclidean or Hilbert space, the ε-entropy H_ε may be computed exactly. The n-dimensional vector ξ, after a suitable orthogonal transformation of coordinates, acquires the form

$$\xi = (\xi_1, \xi_2, \ldots, \xi_n),$$

where the coordinates ξ_k are mutually independent and are normally distributed. For a fixed ε, in order to find $H_\varepsilon(\xi)$, we must only determine the parameter θ from the equation

$$\varepsilon^2 = \sum_k \min(\theta^2, \mathbf{D}\xi_k); \tag{4.17}$$

after that we shall have

$$H_\varepsilon(\xi) = \frac{1}{2} \sum_{\mathbf{D}\xi_k > \theta^2} \log \frac{\mathbf{D}\xi_k}{\theta^2}. \tag{4.18}$$

The approximating vector

$$\xi' = (\xi_1', \xi_2', \ldots, \xi_n')$$

must be chosen normally distributed and such that $\mathbf{D}\xi_k \leqslant \theta^2$ implies

$$\xi_k' = 0,$$

$\mathbf{D}\xi_k > \theta^2$ implies

$$\xi_k = \xi_k' + \Delta_k, \quad \mathbf{D}\Delta_k = \theta^2, \quad \mathbf{D}\xi_k' = \mathbf{D}\xi_k - \theta^2$$

while the vectors ξ_k' and Δ_k are mutually independent. The infinite-dimensional case does not differ from the finite-dimensional one.

Finally, it is important to note that the maximal value of $H_\varepsilon(\xi)$ for the vector ξ (n-dimensional or infinite-dimensional) for given second central moments is achieved in the case of a normal distribution. This result may be obtained directly or from the following proposition due to M. S. Pinsker (compare [19]):

Theorem. *Suppose we are given a positive definite symmetric matrix with entries $s_{ij}, 0 \leqslant i, j \leqslant m + n$ and the distribution P_ξ of the vector*

$$\xi = (\xi_1, \xi_2, \ldots, \xi_m),$$

whose central moments equal s_{ij} (for $0 \leqslant i, j \leqslant m$). Suppose the condition W on the mutual distribution $P_{\xi\xi'}$ of the vector ξ and the vector

$$\xi' = (\xi_{m+1}, \xi_{m+2}, \ldots, \xi_{m+n})$$

consists in the central second moments of the expression

$$\xi_1, \xi_2, \ldots, \xi_{m+n}$$

being equal to s_{ij} (for $0 \leqslant i, j \leqslant m + n$). Then

$$H_W(\xi) \leqslant 1/2 \log(AB/C). \tag{4.19}$$

The notations of formula (4.19) correspond to our exposition in §3. Comparing with the text in §3, we see that inequality (4.19) becomes an equality in the case of a normal distribution P_ξ.

§5. Amount of information per unit time for stationary processes.

Consider two stationary and stationarily related processes $\xi(t), \eta(t)$, $-\infty < t < \infty$, or, in a more general way, two stationary and stationarily related generalized processes $\xi(\varphi), \eta(\varphi)$, $\varphi \in \Phi$. We shall denote by ξ_T and η_T the intervals of the processes ξ and η during time $0 < t \leqslant T$ (see the footnote following formula (3.14)) and by ξ_- and η_- the values of the same processes on the negative semi-interval $-\infty < t \leqslant 0$. To choose a pair (ξ, η) of stationarily related processes means to choose a probability distribution, invariant with respect to shifts along the t-axis, in the space of pairs of functions $\{x(t), y(t)\}$, or in a more general way, pairs of generalized functions (distributions) $\{x(\varphi), y(\varphi)\}$.

It is natural to call the expression

$$\bar{I}(\xi, \eta) = \lim_{T \to \infty} \frac{1}{T} I(\xi_T, \eta_T) \tag{5.1}$$

the amount of information in the process ξ about the process η per unit time. Since $I(\xi_T, \eta_T) = I(\eta_T, \xi_T)$, we always have

$$I(\xi, \eta) = I(\eta, \xi). \tag{5.2}$$

We shall assume that the distribution $P_{\xi\eta}$ is normal. In this case, using a definition of the limit $\bar{I}(\xi, \eta)$ somewhat different from (5.1) (but apparently giving the same magnitude), M. S. Pinsker [19] established that at least in the case of ordinary (not generalized) processes one of which is regular in the sense of [18] , $\bar{I}(\xi, \eta)$ is given by the formula

$$\bar{I}(\xi, \eta) = -\frac{1}{4\pi} \int_{-\infty}^{\infty} \log[1 - r^2(\lambda)] \, d\lambda, \tag{5.3}$$

where

$$r^2(\lambda) = \frac{|f_{\xi\eta}(\lambda)|^2}{f_{\xi\xi}(\lambda) f_{\eta\eta}(\lambda)}, \tag{5.4}$$

while $f_{\xi\xi}$, $f_{\eta\eta}$ and $f_{\xi\eta}$ are the corresponding spectral densities of the Fourier transform of the correlation functions $b_{\xi\xi}$, $b_{\eta\eta}$ and $b_{\xi\eta}$. It can be checked that formula (5.3) is valid as well in many cases when the processes ξ and η (or at least one of them) are generalized (see in particular formula (3.29) above). It is possible even that in all cases (for ordinary as well as generalized processes) the expression (5.1) will be given by formula (5.3); however at this time this conjecture has not been proved or disproved by anyone.

In the case of stationary processes $\xi(t)$, the precision of reproduction of ξ by means of a stationary process ξ' or a process ξ' stationary related to ξ, is naturally characterized by the expression

$$\sigma^2 = \mathbf{E}[\xi(a) - \xi'(a)]^2. \tag{5.5}$$

If for the condition W we take

$$\sigma^2 \leqslant \epsilon^2, \tag{5.6}$$

then the expression

$$\bar{H}_W(\xi) = \bar{H}_\varepsilon(\xi) \tag{5.7}$$

should naturally be called ε-*entropy* per unit time of the process ξ. From the corresponding proposition about finite-dimensional distributions (see §4) it can be deduced that for a given spectral density $f_{\xi\xi}(\lambda)$ the expression $\bar{H}_\varepsilon(\xi)$ achieves its maximum in the case of a normal process ξ. In this same normal case the expression $\bar{H}_\varepsilon(\xi)$ may be easily computed from the spectral density $f_{\xi\xi}(\lambda)$ just as the expression $H_\varepsilon(\xi)$ for finite-dimensional distributions was computed in §4. To do this, we need only determine the parameter θ from the equation

$$\varepsilon^2 = \int_{-\infty}^{\infty} \min(\theta^2, f_{\xi\xi}(\lambda)) \, d\lambda; \tag{5.8}$$

after that the expression $\bar{H}_\varepsilon(\xi)$ can be found by the formula

$$\bar{H}_\varepsilon(\xi) = \frac{1}{2} \int_{f_{\xi\xi}(\lambda) > \theta^2} \log \frac{f_{\xi\xi}(\lambda)}{\theta^2} \, d\lambda, \tag{5.9}$$

and $\bar{I}(\xi, \xi') = \bar{H}_\varepsilon(\xi)$ is achieved when

$$\xi = \xi' + \Delta,$$

where ξ' and Δ are normally distributed, independent and (see Figure 1) we have

$$f_{\Delta\Delta}(\lambda) = \min(\theta^2, f_{\xi\xi}(\lambda)), \tag{5.10}$$

$$f_{\xi'\xi'}(\lambda) = \begin{cases} f_{\xi\xi}(\lambda) - \theta^2 & \text{for } f_{\xi\xi}(\lambda) > \theta^2, \\ 0 & \text{for } f_{\xi\xi}(\lambda) \leqslant \theta^2. \end{cases} \tag{5.11}$$

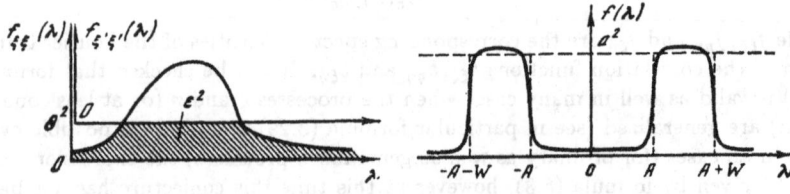

FIGURE 1   FIGURE 2

In many cases (for example, in the case of spectral density (3.24)), formulas (5.8) and (5.9) allow the explicit computation of $\bar{H}_\varepsilon(\xi)$ as a function of ε. The most interesting, however, is the asymptotic behaviour of $\bar{H}_\varepsilon(\xi)$ as $\varepsilon \to 0$. In particular, for the density (3.24), this asymptotic behaviour is described by the formula

$$\bar{H}_\varepsilon(\xi) = 16 C \alpha / \pi \varepsilon^2 + O(1). \tag{5.12}$$

In the more general case of spectral density decreasing as $|\lambda|^{-k}$ when $|\lambda| \to \infty$ (for example, the spectral density (3.23)), we shall have

$$\bar{H}_\varepsilon(\xi) \approx c_1 \varepsilon^{-2/(k-1)}. \tag{5.13}$$

Of great practical interest are spectral densities of the form shown on Figure 2, which are very well approximated by the formula

$$f_{\xi\xi}(\lambda) = \begin{cases} a^2 & \text{for } A \leqslant |\lambda| \leqslant A + W, \\ 0 & \text{in other cases.} \end{cases} \tag{5.14}$$

In this case, for values of ε which are not too small (much larger than the part of the energy of the process ξ contained outside the main spectral strip $A \leqslant |\lambda| \leqslant A + W$) in a normal process, it is easy to calculate that we shall have (with a high degree of precision)

$$\theta^2 \approx \varepsilon^2/2W, \ \bar{H}_\varepsilon(\xi) \approx W \log(2Wa^2/\varepsilon^2) \tag{5.15}$$

(note that $\bar{H}_\varepsilon(\xi)$ here increases much slower as $\varepsilon \to 0$ than in the case of a relatively slow polynomial decrease of the spectral density at infinity). Formula (5.15) of course is none other than the well-known Shannon formula

$$R = W \log(Q/N) \tag{5.16}$$

(see [1], Theorem 22). The new development in principle here, however, is that we now see why and within what limits (for ε not too small) this formula may be applied to processes with an unbounded spectrum such as all practically interesting message transmission processes in the theory of data transmission. Rewriting (5.15) in the form

$$\bar{H}_\varepsilon(\xi) \sim 2W[\log(a\sqrt{2W}) + \log(1/\varepsilon)] \tag{5.17}$$

and comparing it with (4.11), we see that the double width $2W$ of the strip of frequencies appearing here plays the role of dimension. This idea of the equivalence of the double width of the strip of frequencies and the dimension (arising in a certain sense per unit time) was apparently first stated by V. A. Kotelnikov [20]. To justify this idea, V. A. Kotelnikov indicated the fact that a function whose spectrum is contained in a strip of width $2W$ must be uniquely determined by its values at the points

$$\ldots, -\frac{2}{2W}, -\frac{1}{2W}, 0, \frac{1}{2W}, \frac{2}{2W}, \ldots, \frac{k}{2W}, \ldots$$

This same argument can be found in the work of Shannon, who uses the representations obtained in this way to deduce formula (5.16). Since, however, a function with bounded spectrum is always singular in the sense of [18] and the observation of such a function in general is not related to the stationary influx of new information, the meaning of such an argument remains rather unclear, so that the new deduction of the approximate formula (5.15) presented here is possibly not without some interest.

REFERENCES

[1] C. E. Shannon, *A mathematical theory of communication.*, Parts I and II.-Bell, Syst. Techn.J. **27,4** (1948), 623-656.

[2] P. R. Halmos, *Measure theory*, Van Nostrand, 1966.

[3] A. Kolmogoroff, *Untersuchungen uber den Integralbegriff.*, Math. Ann. **103** (1930), 654-696.

[4] V. I. Smirnov, *A course in higher mathematics*, vol. . 5, 90, 91, Moscow, Leningrad, Gostekhizdat, 1947.

[5] K. Ito, *Stationary random distributions.*, Mem. Coll. Sci. Univ. Kyoto, A, **28,3** (1954,), 209-223.

[6] I. M. Gelfand, *Generalized random proceses*, Doklady Akad. Nauk SSSR, **100,5** (1955,), 853-856.

[7] L. Schwartz, *Theorie des distributions*, vol. 1-2, Paris, 1950-1951.

[8] M. Rosenblat-Rom, *Notion of entropy in the probability theory and its application to data transmission channels*, Works of the Third All-Union Mathematical Congress, vol. 2, Moscow, Academy of Science Publication,, 1956, pp. 132-133.

[9] M. Rosenblat-Rom, *Entropy of stochastic processes*, Doklady Akad. Nauk SSSR, **112,** 1 (1957,), 16-19.

[10] A. G. Vitushkin, *Concerning Hilbert's thirteenth problem*, Doklady Akad. Nauk SSSR, **95,4,** (1954,), 701-704.

[11] H. Hotelling., *Relation between two sets of variates*, Biometrica **28**, (1956), 321-377.

[12] A. M. Obukhov, *Normal correlation of vectors*, Izvestia Akad. Nauk SSSR, **2** (1938), 339-370.

[13] P. A. Shirokov, *Tensor calculus*, Moscow, Leningrad, GONTI, 1934.

[14] A. M. Yaglom, *The theory of extrapolation and filtration of random processes*, Ukrainian Math Journal **6**, 1, (1954), 43-57.

[15] A. N. Kolmogorov, *On certain asymptotic characteristics of completely bounded metric spaces*, Doklady Akad. Nauk SSSR, **108,3**, (1956), 385-388.

[16] A. Ya. Khinchin, *On the basic theorems of information theory*, Uspekhi Akad. Nauk SSSR, **11,1** (1956,), 17-75.

[17] M. S. Pinsker, *The computation of the speed of creation of messages for stationary random processes and the throughput of a stationary channel*, Doklady Akad. Nauk SSSR, **111,4** (1956), 753-756.

[18] A. N. Kolmogorov, *Stationary sequences in Hilbert space*, Bulletin Moscow University. Mathematics **2,6** (1941,), 1-40.

[19] M. S. Pinsker, *The amount of information about a Gaussian random stationary process contained in a second process, stationarily related to the first one*, Doklady Akad. Nauk SSSR, **99,2** (1954,), 213-216.

[20] V. A. Kotelnikov, *On the capacity of "ether" and of wires in electrical transmission*, Materials to the 1st All-Union Congress on questions of technical reconstructions of communication systems and development of weak electricity technology, Moscow, 1933.

[21] A. N. Kolmogorov, *The theory of transmission of information*, Publications of the Academy of Sciences of the USSR, (1957,), 66-99, The Session of the Academy of Sciences of the USSR on scientific problems of industrial automation, October 15-20, 1956. Plenary Sessions, Moscow.

[22] I. M. Gelfand, A. N. Kolmogorov, A. M. Yaglom, *Concerning the general definition of the amount of information*, Doklady Akad. Nauk SSSR, .111,4, (1956,), 745-748.

[23] I. M. Gelfand, A. M. Yaglom., *On the computation of the amount of information on a random function contained in another such function*, Uspekhi Akad. Nauk SSSR **12,1**, (1957), 3-52.

[24] A. Kolmogorov, *To the Shannon theory of information transmission in the case of continuous signals*, IRE Trans. Inform. Theory. **IT-2,4** (1956), 102-108.

NEW METRIC INVARIANT OF
TRANSITIVE DYNAMICAL SYSTEMS
AND AUTOMORPHISMS OF LEBESGUE SPACES*

It is well known that a considerable part of the metric theory of dynamical systems may be developed as an abstract theory of "flows" $\{S_t\}$ on "Lebesgue spaces" M with measure μ in terms invariant with respect to "isomorphisms modulo zero" (see V.A. Rokhlin's survey [1], whose definitions and notations are used extensively here). The measure on M will be assumed normed by the condition

$$\mu(M) = 1 \qquad (1)$$

and nontrivial (i.e. we shall assume the existence of sets $A \subset M$ such that $0 < \mu(A) < 1$). There are many examples of transitive automorphisms and transitive flows with so-called countable Lebesgue spectra (for the automorphisms see [1, §4] , for flows see [2-5]). From the spectral point of view we have here one type of automorphism $\mathcal{L}_0^\omega$ and one type of flow $\mathcal{L}^\omega$. The quesiton as to whether all automorphisms of type $\mathcal{L}_0^\omega$ (respectively flows of type $\mathcal{L}^\omega$) are isomorphic to each other mod 0 is still an open question. We shall show in §§3,4 that the answer to this question is negative, both in the case of automorphisms and in the case of flows. The new invariant which allows to split the class of automorphisms $\mathcal{L}_0^\omega$ (and the class of flows $\mathcal{L}^\omega$) into an uncountable amount of invariant subclasses is entropy per unit time. In §1 we develop the necessary facts from information theory (the notions of conditional entropy and conditional information introduced here and their properties apparently have a wider interest, although the entire exposition is directly related to the definition of amount of information from [7] and numerous papers developing this definition). In §2 we give the definition of the characteristic h and prove its invariance. In §§3,4 we indicate examples of automorphisms and flows with arbitrary values of h in the limits $0 < h \leqslant \infty$. In the case of automorphisms, we use examples constructed long ago, while in the case of flows the constructions of examples with finite h is a more delicate problem related to certain curious questions of the theory of Markov processes.

§1. Properties of conditional entropy and the conditional amount of information.

Following [1], denote by γ the Boolean algebra of measurable sets of the space M considered mod 0. Suppose $\mathfrak{D}$ is a subalgebra of the algebra γ closed in the

---

*Dokl. Akad. Nauk SSSR, 1958, vol.119, No.5, p. 861-864.

metric $\rho(A, B) = \mu((A - B) \cup (B - A))$. It determines a certain partition $\xi_{\mathfrak{D}}$ mod 0 of the space M, defined by the condition that $A \in \mathfrak{D}$ if and only if all of A may be constituted by entire elements of the partition $\xi_{\mathfrak{D}}$. On elements C of the partition $\xi_{\mathfrak{D}}$ one can define the "canonical system of measures μ_C"[1]. For any $x \in C$ we shall assume that

$$\mu_x(A|\mathfrak{D}) = \mu_C(A|C). \tag{2}$$

From the point of view of probability theory (where any measurable function of the element $x \in M$ is called a "random variable") the random variable $\mu_x(A|\mathfrak{D})$ is the "conditional probability" of the event A when the outcome of the test is known $\mathfrak{D}$ [6, Chapter 1, §7].

For three subalgebras $\mathfrak{A}$, $\mathfrak{B}$ and $\mathfrak{D}$ of the algebra γ and $C \in \xi_{\mathfrak{D}}$, put

$$I_C(\mathfrak{A}, \mathfrak{B}|\mathfrak{D}) = \sup \sum_{i,j} \mu_x(A_i \cap B_j) \log \frac{\mu_x(A_i \cap B_j)}{\mu_x(A_i)\mu_x(B_j)}, \tag{3}$$

where the upper bound is taken over all finite decompositions

$$M = A_1 \cup A_2 \cup \cdots \cup A_n, \; M = B_1 \cup B_2 \cup \cdots \cup B_n$$

for which

$$A_i \cap A_j = N, \; B_i \cap B_j = N, \; i \neq j, \; A_i \in \mathfrak{A}, \; B_j \in \mathfrak{B}$$

(N is the empty set). If $\mathfrak{D}$ is a trivial algebra $\mathfrak{R} = \{N, M\}$, then (3) becomes the definition of the (unconditional) information $I(\mathfrak{A}, \mathfrak{B})$ from Appendix 7 to [7][1] The expression (3) itself can be interpreted as the "amount of information in the results of the test $\mathfrak{A}$ about the test $\mathfrak{B}$ for a known outcome C of the test $\mathfrak{D}$". If we do not fix $C \in \xi_{\mathfrak{D}}$, then it is natural to consider the random variable $I(\mathfrak{A}, \mathfrak{B}|\mathfrak{D})$ which is equal to $I_x(\mathfrak{A}, \mathfrak{B}|\mathfrak{D}) = I_C(\mathfrak{A}, \mathfrak{B}|\mathfrak{D})$ for $x \in C$. Further we shall consider the expectation

$$\mathbf{E}I(\mathfrak{A}, \mathfrak{B}|\mathfrak{D}) = \int_M I_x(\mathfrak{A}, \mathfrak{B}|\mathfrak{D})\mu(dx). \tag{4}$$

No special explanations of the definition of conditional entropy and mean conditional entropy

$$H(\mathfrak{A}|\mathfrak{D}) = I(\mathfrak{A}, \mathfrak{A}|\mathfrak{D}), \; \mathbf{E}H(\mathfrak{A}|\mathfrak{D}) = \int_M H_x(\mathfrak{A}|\mathfrak{D})\mu(dx).$$

are needed.

Let us note those properties of the conditional amount of information and conditional entropy which we shall need in the sequel. Property (α) and (δ) for the case of unconditional amount of information and entropy are well-known, property (ε) for an unconditional amount of information is the content of Theorem 2 from the note [8]. Properties (β) and (γ) can be proved without difficulty. Concerning property (β), we should only remark that the statement that a similar proposition

---

[1]At the time when the note [8] was written, its authors were not aware of the contents of Appendix 7 in [7],which was not included in the Russian translation [9]. The note [8] should have begun with a reference to this Appendix from [7].

for the amount of information ($\mathfrak{D} \supseteq \mathfrak{D}'$ implies $I(\mathfrak{A}, \mathfrak{B}|\mathfrak{D}) \leqslant I(\mathfrak{A}, \mathfrak{B}|\mathfrak{D}')$) would already be erroneous. This is related to the fact that property (ζ) involves a lower limit and the sign $\geqslant$: the corresponding limit may not exist, while the lower limit in certain cases may turn out to be larger than $EI(\mathfrak{A}, \mathfrak{B}|\mathfrak{D})$.

(α) $I(\mathfrak{A}, \mathfrak{B}|\mathfrak{D}) \leqslant H(\mathfrak{A}|\mathfrak{D})$, the equality certainly holding when $\mathfrak{B} \supseteq \mathfrak{A}$.

(β) If $\mathfrak{D} \supseteq \mathfrak{D}'$, then $H(\mathfrak{A}|\mathfrak{D}) \leqslant H(\mathfrak{A}|\mathfrak{D}')$, mod 0.

(γ) If $\mathfrak{B} \supseteq \mathfrak{B}'$, then $EI(\mathfrak{A}, \mathfrak{B}|\mathfrak{D}) = EI(\mathfrak{A}, \mathfrak{B}'|\mathfrak{D}) + EI(\mathfrak{A}, \mathfrak{B}|\mathfrak{D} \vee \mathfrak{B}')$, where $\mathfrak{D} \vee \mathfrak{B}'$ is the minimal closed σ-algebra containing $\mathfrak{D}$ and $\mathfrak{B}'$.

(δ) If $\mathfrak{B} \supseteq \mathfrak{B}'$, then $EI(\mathfrak{A}, \mathfrak{B}|\mathfrak{D}) \geqslant EI((\mathfrak{A}, \mathfrak{B}'|\mathfrak{D})$.

(ε) If $\mathfrak{A}_1 \subseteq \mathfrak{A}_2 \subseteq \cdots \subseteq \mathfrak{A}_n \ldots, \cup_n \mathfrak{A}_n = \mathfrak{A}$, then

$$\lim_{n \to \infty} EI(\mathfrak{A}, \mathfrak{B}|\mathfrak{D}) = EI(\mathfrak{A}, \mathfrak{B}|\mathfrak{D}).$$

(ζ) If $\mathfrak{D}_1 \supseteq \mathfrak{D}_2 \supseteq \cdots \supseteq \mathfrak{D}_n \supseteq \ldots, \cap_n \mathfrak{D}_n = \mathfrak{D}$, then

$$\lim_{n \to \infty} \inf EI(\mathfrak{A}, \mathfrak{B}|\mathfrak{D}_n) \geqslant EI(\mathfrak{A}, \mathfrak{B}|\mathfrak{D}).$$

§2. Definition of the invariant h.

We shall say that a flow $\{S_t\}$ is quasiregular (has type $\mathcal{R}$), if[2] there exists a closed subalgebra γ_0 of the algebra γ whose shifts $\gamma_t = S_t \gamma_0$ possess the following properties

$$(I) \gamma_t \supseteq \gamma_{t'} \text{ if } t \leqslant t'. (II) \cup_t \gamma_t = \gamma. \cap_t \gamma_t = \mathfrak{R}.$$

In the interpretation of a flow as a stationary random process γ_t, we can consider the algebra of events "depending only on the process until the moment of time t". It is easy to prove that flows of type $\mathcal{R}$ are transitive and from the results of Plesner [10,11], we can deduce that they have a homogeneous Lebesgue spectrum. If the multiplicity of the spectrum is equal to $\nu(\nu = 1, 2, \ldots, \omega)$, we shall say that the flow is of type $\mathcal{R}^\nu$. Obviously $\mathcal{R}^\nu \subseteq \mathcal{L}^\nu$, where $\mathcal{L}^\nu$ is the class of flows with Lebesgue spectrum of homogeneous multiplicity ν. It is possible, however, that all the $\mathcal{L}^\nu$ (and therefore $\mathcal{R}^\nu$), except $\mathcal{L}^\omega(\mathcal{R}^\omega)$, are empty and $\mathcal{L}^\omega = \mathcal{R}^\omega$.

Theorem 1. *If for any flow $\{S_t\}$ there exists a γ_0 satisfying conditions (I)-(III), then when $\Delta > 0$ we have $EH(\gamma_{t+\Delta}|\gamma_t) = h\Delta$, where h is a constant lying in the interval $0 < h \leqslant \infty$.*

Theorem 2. *The constant h for a given flow $\{S_t\}$ does not depend on the choice of γ_0 satisfying condition (I)-(III).*

Let us sketch the proof of Theorem 2 here. Suppose that to the values of γ_0 and γ_0' correspond the invariants $h < \infty$ and h'. By Theorem 1 and Lemmas (α) and (ε), for any $\varepsilon > 0$ we can find a k such that

$$h = EI(\gamma_{t+1}|\gamma_t) = EI(\gamma_{t+1}, \gamma|\gamma_t) \leqslant EI(\gamma_{t+1}, \gamma'_{t+k}|\gamma_t) + \varepsilon. \tag{5}$$

It follows from (5) by Lemma (ζ) that there exists an m such that

$$h \leqslant EI(\gamma_{t+1}, \gamma'_{t+k}|\gamma_t \vee \gamma'_s) + 2\varepsilon \text{ as } t - s \geqslant m. \tag{6}$$

---

[2]This condition is considerably weaker than the "regularity" condition usually used in the theory of random processes. This is discussed at the end of §4.

From (6) and $(\delta),(\gamma),(\alpha),(\beta)$ (which should be applied in the order indicated) we get

$$nh \leqslant \sum_{t=0}^{n-1} \mathbf{E}I(\gamma_{t+1}, \gamma'_{t+k}|\gamma_t \vee \gamma'_{-m}) + 2n\varepsilon \leqslant \qquad (\delta)$$

$$leqslant \sum_{t=0}^{n-1} \mathbf{E}I(\gamma_{t+1}, \gamma'_{n+k}|\gamma_t \vee \gamma'_{-m}) + 2n\varepsilon = \qquad (\gamma)$$

$$= \mathbf{E}I(\gamma_n, \gamma'_{n+k}|\gamma_0 \vee \gamma'_{-m}) + 2n\varepsilon \leqslant \mathbf{E}H(\gamma'_{n+k}|\gamma_0 \vee \gamma'_{-m}) + 2n\varepsilon \leqslant \qquad (\alpha)$$

$$\leqslant \mathbf{E}H(\gamma'_{n+k}|\gamma'_{-m}) + 2n\varepsilon \leqslant (n+k+m)h' + 2n\varepsilon, \qquad (\beta)$$

$$h \leqslant \frac{n+k+m}{n}h' + 2\varepsilon. \qquad (7)$$

Since $\varepsilon > 0$ and n are arbitrary (n is chosen after k and m are fixed), it follows from (7) that we have the inequality $h \leqslant h'$. This inequality can be proved in a similar way in the case $h = \infty$. The converse inequality $h' \leqslant h$ is proved by arguing as in the end of the proof of Theorem 2.

§3. Invariants of automorphisms.

If we assume in §2 that t takes on only integer values, then $\{S_t\}$ is uniquely determined by the automorphism $T = S_1$. By Theorems 1 and 2, there exists an invariant $0 < h(T) \leqslant \infty$.

It is easy to prove that any automorphism of type $\mathcal{R}_0$ (the subscript is written to distinguish this case from the case of flows with continuous time) has a countable Lebesgue spectrum, i.e. among the classes $\mathcal{R}_0^\nu$ the only non-empty one is $\mathcal{R}_0^\omega \subseteq \mathcal{L}_0^\omega$. This class is split according to the values of $h(T)$ into classes $\mathcal{R}_0^\omega(h)$.

Theorem 3. *For any h, $0 < h \leqslant \infty$, there exists an automorphism belonging to $\mathcal{R}_0^\omega(h)$.*

The corresponding examples are well-known and are obtained, for example, from the scheme of independent random tests $\mathcal{L}_{-1}, \mathcal{L}_0, \mathcal{L}_1, \ldots, \mathcal{L}_t, \ldots$ with probability distribution outcomes ξ_t for the test $\mathcal{L}_t$

$$\mathbf{P}\{\xi_t = a_i\} = p_i, \quad -\sum_{i=1}^{\infty} p_i \log p_i = h. \qquad (8)$$

The space M is constituted by sequences $x = (\ldots, x_{-1}, x_0, x_1, \ldots, x_t, \ldots), x_t = a_1, a_2, \ldots$, while the shift $T_x = x'$ is determined by the formula $x'_t = x_{t-1}$. The measure μ on M is defined as the direct product of probability measures (8).

§4. Invariants of flows.

Theorem 4. *For any h, $0 < h < \infty$, there exists a flow of class $\mathcal{R}^\omega(h)$, i.e. a flow with a countable Lebesgue spectrum and the given value of the constant h.*

As in §3, one naturally comes to the idea of using for the proof of Theorem 4, instead of the scheme of discrete independent trials, the scheme of "processes with independent increments" or generalized processes with "independent values"[12,

13]. However, this approach only yields flows of class $\mathcal{R}^\omega(\infty)$ [5]. To obtain finite values of h, one must use more artificial constructions. In the present note it is possible to give only one of these constructions.

Define mutually independent random valuables ξ_n, corresponding to all integers n, by choosing the distribution of their values $P\{\xi_0 = k\} = 3 \cdot 4^{-k}$, $k = 1, 2, \ldots$, and, for $n \neq 0$, $P\{\xi_n = k\} = 2^{-k}$, $k = 1, 2, \ldots$. In the case $\xi_0 = k$ the point τ_0 on the t-axis must be chosen with uniform probability distribution on the segment $-u2^{-k} \leqslant \tau_0 \leqslant 0$, while the points τ_n for $n \neq 0$ are defined from the relation $\tau_{n+1} = \tau_n + u2^{-\xi_n}$.

Put $\varphi(t) = \xi_n$ for $\tau_n \leqslant t < \tau_{n+1}$. It is easy to check that the distribution of the random function $\varphi(t)$ is invariant with respect to shifts $S_t\varphi(t_0) = \varphi(t_0 - t)$. An easy computation gives $h\{S_t\} = 6/u$ (per unit of time there are in the mean $3/u$ points τ_n, while for each ξ_n the entropy is $\sum_{k=1}^{\infty} k2^{-k} = 2$).

One can obtain a more intuitively clear idea of our random process if we include in the description of its state $\omega(t)$ at the moment of time t, besides the function $\varphi(t)$, also the value $\delta(t) = t - \tau^*(t)$ of the difference between t and the nearest point τ_n to t to its left. For this description of our process, the latter turns out to be a stationary Markov process. It only deserves the name of "quasiregular" since, although the corresponding dynamical system is transitive, the value of the difference $f(\omega(t), t) = \tau^*(t) = t - \delta(t)$ is determined up to a binary rational summand describing the behaviour of the realization of the process arbitrarily far back in time.

January 21, 1958

REFERENCES

[1] V. A. Rokhlin, *Selected topics in the metric theory of dynamical systems*, Uspekhi Mat. Nauk **4,2(30)** (1949), 57-128.

[2] I. M. Gelfand, S. V. Fomin, *Geodesic flows on manifolds of constant negative curvature*, Uspekhi Mat. Nauk **7,1(47)** (1952), 118-137.

[3] S. V. Fomin, *On dynamical systems in the space of functions.*, Ukrainian Math Journal **2,2** (1950), 25-47.

[4] K. Itô, *Complex multiple Wiener integral*, Jap. J. Math. **22** (1952), 63-86.

[5] K. Itô, *Spectral type of the shift transformation of differential processes with stationary increments*, Trans. Amer. Math. Soc. **81,2** (1956), 253-263.

[6] J. L. Doob, *Stochastic processes*, Wiley, 1953.

[7] C. E. Shannon, W. Weaver, *The mathematical theory of communication*, Urbana, Univ. Ill. Press, 1949.

[8] I. M. Gelfand, A. N. Kolmogorov, A. M. Yaglom, *To the general definition of the amount of information*, Dokl. Akad. Nauk SSSR **111,4** (1956), 745-748.

[9] C. Shannon., *The statistical theory of the transmission of electric signals.*, The theory of transmission of electrical signals in the presence of noise., Moscow, Foreign Language Editions, 1953, pp. 7-87.

[10] A. I. Plesner, *Functions of the maximal operator*, Dokl. Akad. Nauk SSSR **23,4** (1939), 327-330.

[11] A. I. Plesner, *On semiunitary operators*, Dokl. Akad. Nauk SSSR **25,9** (1939), 708-710.

[12] K. Itô, *Stationary random distributions*, Mem. Coll. Sci. Univ. Kyoto. A **28,3** (1954), 209-223.

[13] I. M. Gelfand, *Generalized random processes*, Dokl. Akad. Nauk SSSR **100,5** (1955), 853-856.

6
TO THE DEFINITION OF ALGORITHMS*[1]

Jointly with V. A. Uspenski

By an algorithm one usually means a system of computations that, for a certain class of mathematical problems, given a text

$$A$$

expressing the "condition" of the problem, obtains, by means of a well defined sequence of operations, carried out "mechanically" (without the intervention of the creative possibilities of a human), the text

$$B$$

expressing the "solution" of the problem. How this traditional very approximate description of an algorithm can be made more precise is the topic of a very complete discussion in the introductory sections of A. A. Markov's article [1]. Giving a clear and developed idea of the essence of the question, A. A. Markov's article (and the monograph which followed [2]) limits itself to algorithms which process "words" and among such algorithms considers only those called "normal" in the article. Our goal is to indicate the possibility of a wider approach to the topic and at the same time to show as clearly as possible that the most general notion of algorithm

---

*Uspekhi Mat. Nauk 1958, vol. 13, No.4, p. 3-28

[1]This article, except for unimportant style editing, was written when the authors, trying to understand the meaning of the notion of "computable function" and "algorithms", tried to find a new approach to various versions of definitions found in the literature and finally to convince themselves that there is no possibility to widen the content of the notion of "computable function" itself. Thus the result turned out to be negative. It is not without hesitation that the authors publish this description of their search, since from the point of view of the subject matter it gives nothing new as compared to simpler definitions. However, it appears that the approach to the subject that we developed here corresponds to the line of thought of many mathematicians whom the publication of our article will possibly save from going through the same steps of hesitation and thought. Let us also note that we also discussed a version of the definitions in which, instead of a passive unbounded volume of memory in the computer, it is possible to work with an unbounded amount of "active regions" working in parallel, each of which has a bounded complexity of structure. Naturally nothing new appeared in this direction, except partially recursive functions.

in the present state of our science is quite naturally related to the notion of partially recursive function.

In all cases that interest mathematicians, the texts of the conditions A which are to be processed by the given algorithm can easily be included in a sequence numbered by of non-negative integers

$$A_0, A_1, A_2, \ldots, A_n, \ldots,$$

while the text of the possible solution B is included in a sequence

$$B_0, B_1, B_2, \ldots, B_n, \ldots,$$

also numbered by non-negative integers.[2] If we denote by G the set of integers n of those conditions A_n which the algorithm is capable of transforming into solutions, then the result of the work of the algorithm carrying out the processing

$$A_n \to B_m,$$

is uniquely determined by a numerical function given on G

$$m = \varphi(n).$$

Thus an arbitrary algorithm is reduced to an algorithm computing the values of a certain numerical function (by number here and in the sequel we always mean non-negative integer).

Conversely, if for the function φ there exists an algorithm which, when applied to the standard notation[3] of the value of the argument n from the domain of definition of the function φ, yields the standard notation of the value $m = \varphi(n)$ of the function, then the function φ should naturally be called *algorithmically computable*, or briefly a *computable function*. Hence the question of defining an algorithm is essentially equivalent to that of defining a computable function. In §1 we give a review of existing definitions of computable functions and algorithms and discuss their generality and logical completeness. In §2 we develop a new definition of algorithm[4]. In §3 we show that any algorithm which satisfies this new definition, which is very general in form, nevertheless reduces to an algorithm computing the values of a partially recursive function. In Addendum I we give a list of definitions and facts related to recursive functions and in Addendum II we consider an example of an algorithm.

---

[2]The method for finding the number of a text from the text itself, as well as to recover the text from its number, is usually very simple (so that the existence of an algorithm taking the text to its number and an algorithm taking the number to the corresponding text does not raise any doubts).

[3]For numbers greater than zero, the standard notation may be defined as written in the form $1 + 1 + \cdots + 1$, or in the form of decimal notation, etc. Without fixing a standard method for denoting numbers, it is impossible to speak of the algorithm computing $m = \varphi(n)$ from n.

[4]The main ideas of this definition, which was found in 1952, were first published in [3].

§1

Let us consider the following versions of the mathematical definition of a computable function or algorithm.

A) The definition of a computable function as a function whose values can be deduced in a certain logical calculus (Gödel [4], Church [5])[5]

B) The definition of a computable function as a function whose values can be obtained by means of Church's λ-conversion calculus [5,7].

C) The definition of a computable function as a partially recursive function (see Kleene's work [8])[6] or, in the case of everywhere defined functions, as a recursive function (Kleene [10]). (The terms "partially recursive" and "recursive" are understood here in the sense of Appendix I.)

D) Turing's computational machine [11][7].

E) The finite combinatorial process of Post [13].

F) A. A. Markov's normal algorithm [1,2].

Let us first look at the definitions listed above from the point of view of the natural requirement in the definition of algorithm saying that there must exist a well-determined sequence of operations "transforming" the condition A into the solution B, or the value n into the value $m = \varphi(n)$. It is immediately clear that definitions A), B) and C) are not satisfactory from this point of view, since such a well-defined sequence of operations is not indicated in them. It is true, however, that the definition of partially recursive function essentially contains all the prerequisites required for the construction of a well-defined algorithm computing its values from the given value of its argument, but in explicit form such an algorithm is not indicated.

A Turing machine carries out a well defined sequence of operations that leads to the successive development of expression for the values

$$\varphi(0), \ \varphi(1), \ \varphi(2), \ldots$$

It is sufficient to add to it a simple device that would stop its work after $\varphi(n)$ is obtained for the given n, and then it can be viewed as a method for implementing a true algorithm for finding $\varphi(n)$ from a given n.[8]

Definitions E) and F) entirely satisfy the requirement of well-definedness of the algorithmic computation process from a given condition A.

Now let us discuss the definitions under consideration from the point of view of subdividing the algorithmic process into elementary steps which can be carried out "mechanically". In the requirement that the computable process be subdivided into elementary steps of *bounded* (within limits fixed in the given algorithm) *complexity*, we see the second no less essential trait of the notion of algorithm. In order to

---

[5]Concerning definition A) also see [6, §§62, 63].

[6]Concerning definition C) also see [6, §§62, 63].

[7]Here we have in mind the original definition of a computable function by means of a Turing machine proposed by Turing himself [11]. A very closely related exposition of this can be found in R. Peter's book [12].

[8]The computation of functions on Turing machines is described in the spirit in Kleene's book [6]. Starting from the standard notation for the number n, the machine obtains the standard notation for the number $\varphi(n)$ and stops.

understand the main difficulties involved in specifying the notion of algorithm so as to satisfy the requirement of bounded complexity of its separate steps, let us note that those algorithms which are interesting to mathematicians are usually applied to an infinite class of problems. Such are for instance the simple algorithms of addition and multiplication of natural numbers written in decimal form. Hence the comparison of the computation process carried out by an algorithm to that done by a certain "machine" should not be understood too literally. A real machine possesses only a finite number of clearly delimited "states" and in this connection the number of different "conditions" A which a real machine is capable of processing into "solutions" B is also necessarily finite. Hence the construction of the theory of algorithms along the lines of the theory of computing machines requires at least a certain idealization of the notion of "machine".

Nevertheless, the intuitive notions and ideas that come from the development of the theory of discretely working computers may be useful in the general approach to the rational definition of the notion of algorithm. All the idealization necessary to pass from real computers to mathematical algorithms are contained in the assumption that the machine has an unbounded volume of "memory". We can also illustrate the exposition of the theory of algorithms by more traditional images related to computation written on paper. As compared to the real process of such computations, the idealization consists in assuming that it is possible to do an unbounded amount of writing, successively saving and again using the results in further computations according to a specified system.

If we assume an unbounded volume of "memory", the notion of mathematical algorithm must include, among the properties of real discretely working computers and real computations carried out on paper, the following two properties.

1. The computational operations themselves must be carried out in discrete steps and at each step use only a bounded amount of data stored during the previous steps. If the amount of data available at the beginning of the step is greater than the maximum volume of the "active domain", then the data is "pulled in" into this active domain step by step, when free place appears in it (and this must happen during subsequent steps). It is possible, for example, to imagine the situation in the following way: the sheets of papers used to do the calculations carried out at each step must fit on a table and can only contain a strictly limited number of symbols, whereas the folders in which these sheets of paper are stored and from which they can be pulled out when necessary are of unbounded volume.

2. The unboundedness of the "volume of memory" or that of the sheets of paper that are stored, is understood from a purely quantitative point of view: it is possible to store an unbounded number of elements (symbols, elementary parts of the machine) only of a strictly bounded number of types, these elements being related to each other by relations of bounded complexity that also belong to a bounded number of different types. The process of getting to the data saved in the "memory" of the machine or putting aside the filled-in sheets of paper is carried out in sequence: from an element that has entered the active domain, we pass on to other elements related to it and pull them into the active domain.

The properties stated above are completely satisfied only by the Turing machine and Post's finite combinatorial process. In the definitions A), B), C), the elementary and "mechanical" character of certain operations, such as the substitution of variables into certain expressions whose complexity (in the sense of the number of elementary symbols constituting them) is not limited by the rules of the computation, remains unclear. Similar objections may be made to the definition of normal algorithm in the sense of A. A. Markov, although there the possibility of overcoming these objections by means of a simple additional construction is especially obvious. In the definition of A. A. Markov's normal algorithm, the only unsufficiently subdivided operation, in the sense of the requirement that we have described above, is the operation of finding the "first occurence" of the word P in the word X being processed. Since the words P being processed may be as long as we wish, the process of finding the first occurence in P of the given word X may be considered as a process which requires, in general, a sequence of operations. We can say that in the definition of normal algorithm in the sense of A. A. Markov, one assumes an already constructed special "algorithm for finding the first occurence of words". The reader will have no difficulty in constructing such an algorithm in the terms specified in §2 below.

One can suppose, however, that A. A. Markov, as well as the authors of the different definitions of the notions of computable function we have discussed here, leave certain operations of unbounded volume without subdividing them into elementary steps, simply because the possibility of doing that is quite obvious.

The last and most difficult question is whether the definitions of algorithm and computable function proposed are *general* enough. At first glance it would seem that definition A) is quite general. However, this generality is only apparent. Actually in order to formulate the definition of a function computable in the sense of A), we necessarily limit ourselves to a certain fixed calculus with fixed rules of inference. If the definition appeals to the deduction in an arbitrary calculus with arbitrary rules of inference, one must keep in mind that the notion of deduction in a logical calculus in its most general form can only be specified within the framework of the notion of algorithm being defined.

Concerning the more rigorous (from the formal point of view) definitions B),C), D),E),F), we can say that in a certain sense they are all equivalent to each other, i.e. essentially define classes of computational processes (algorithms) of equal power. The equivalence of definitions B), C), E) are established in [5, 14-16][9]. Definitions E) and F) are particular cases of algorithms in the sense of our §2, while the equivalence of our definition to definition C) is established in §3 of our paper. Note that in fact the very exact formulation of the equivalence of the definitions which, from a formal point of view, are constructed in a totally different way and determine formally different objects (computable functions; in the Turing case, computable functions which assume only two values 0 and 1; algorithms processing words, etc.), involves overcoming certain difficulties.

Whatever the case may be, it is precisely a series of established or implicitly understood statements on the equivalence of all the definitions of algorithm or computable function proposed since 1936 which constitute the empirical basis of the presently existing general consensus on the conclusive character of these definitions. In the application to computable functions (to be definite, let us say to functions assuming positive integer values with a positive integer argument), the generally

---

[9]Also see the translator's note on page 341 of the book [6].

accepted views existing at present may be described as follows.

The class of algorithmically computable numerical functions defined for all natural n coincides with the class of *recursive* functions (the definition of the latter is given in Appendix I). Concerning functions which are defined only on a certain subset G of the entire set of natural numbers, their algorithmic presentation may be understood in two different senses. In the first more restrictive understanding, the procedure prescribed by the algorithm must, for any natural n, after a finite number of steps, either lead to finding $m = \varphi(n)$ or to establishing the fact that n does not belong to G. In the second, more general understanding of the situation, one only requires that the procedure prescribed by the algorithm will yield the correct result $m = \varphi(n)$ for any n from G, while if it is applied to a natural number n not belonging to G, will not yield the erroneous equality $\varphi(n) = m$, where m is some definite natural number (since φ is not defined outside of G). Otherwise in the second (wider) understanding of the situation, the character of the algorithm's work when applied to a number n not belonging to G is not restrictive: in this case the algorithmic process is allowed to stop without result after a finite number of steps, or can continue endlessly without yielding any result.

It is expedient to assume (and we do this in our further exposition) that the second (wider) understanding of algorithmic computability of numerical functions defined on the set of natural numbers is the fundamental one. At the present time this wider understanding of algorithmically computable functions is identified with the notion of *partially recursive function* (see Appendix I). In that case, the possible domains of definition G of an algorithmically computable function $\varphi(n)$ turn out to be *recursively enumerable sets* of natural numbers (for the definition of the latter see Appendix I)[10]. The general notion of algorithm processing "condition" A of some general form or other into the "solution" B is defined in the corresponding way. For any algorithm Γ, the class $\mathfrak{A}(\Gamma)$ of formally admissible expressions for the "conditions" A and the class $\mathfrak{B}(\Gamma)$ of possible expressions of the "solutions" B are defined so that it is possible to distinguish whether an expression belongs or does not belong to these sets. Here we do not require that the algorithms Γ applied to any expression A of the class $\mathfrak{A}(\Gamma)$ necessarily leads, after a finite number of steps, to an expression B from the class $\mathfrak{B}(\Gamma)$. Instead of this, we allow both the algorithmic procedure to stop without obtaining a solution, or to continue indefinitely. The set $\mathfrak{S}(\Gamma)$ of those expressions A from $\mathfrak{A}(\Gamma)$ that the algorithm Γ processes to a "solution", i.e. to an expression B from $\mathfrak{B}(\Gamma)$, will be called the *domain of applicability* of the algorithm.

The most interesting results of the theory of algorithms are precisely those about algorithms possessing a nontrivial domain of applicability; the main results obtained up to the present time concerning the impossibility of algorithms may be reduced to the following form: for certain algorithm Γ one proves the impossibility of a "deciding" algorithm Φ which would "distinguish" whether a given expression A from $\mathfrak{A}(\Gamma)$ belongs to $\mathfrak{S}(\Gamma)$ or does not[11].

The classes $\mathfrak{A}(\Gamma)$ and $\mathfrak{B}(\Gamma)$ for all the existing definitions of algorithm are easily

---

[10]An exposition of certain basic facts of the theory of computable functions based on an intuitive understanding of algorithms is given in the article [17].

[11]I.e., the impossibility of an algorithm Φ that would process any notation A from $\mathfrak{A}(\Gamma)$ into 1 when A belongs to $\mathfrak{S}(\Gamma)$ and into 0 when A belongs to $\mathfrak{S}(\Gamma)$.

enumerated, as indicated at the beginning of this article. This allows to assign to each algorithm Γ a numerical function φ_Γ defined for all numbers n such that condition A belongs to $\mathfrak{S}(\Gamma)$ and assume as their value $m = \varphi(n)$ the number of the solution obtained when the condition with number n has been processed. The general conviction about the degree of generality of the notion of algorithm developed at the present time may be stated briefly by saying that for any algorithm Γ the function φ_Γ must be partially recursive. This is the situation for all the definitions which have been proposed up to now. Our goal is to give an explanation of this state of affairs that would be as complete as possible. To do this, we have attempted to construct a definition of algorithm that would be as general as possible in form and would satisfy the requirement of bounded complexity at each step described above. The result, as should be expected, is that despite all our attempts to make the construction as general as possible, we do not go beyond the limits of algorithms described in the manner indicated by partially recursive functions.

§2

An algorithm Γ is given by means of instructions indicating the method for passing from the "state" S of the process of calculations to the "following" state S^*, i.e. by means of the operator $\Omega_\Gamma(S) = S^*$, which will be called the operator of "direct processing". In the class $\mathfrak{S}(\Gamma) = \{S\}$ of possible states, we single out the class $\mathfrak{A}(\Gamma)$ of "initial" states which are the expressions for the "conditions" of the problem, and the class $\mathfrak{U}(\Gamma)$ of "final states". When the algorithmic process yields a state from the class $\mathfrak{U}(\Gamma)$ the process ends and the "final" state yields the "solution". The domains $\mathfrak{D}(\Gamma) \subseteq \mathfrak{S}(\Gamma)$ on which the operator Ω is defined is naturally assumed to have an empty intersection with $\mathfrak{U}(\Gamma)$. The class of states not entering into $\mathfrak{U}(\Gamma)$ nor into $\mathfrak{D}(\Gamma)$ is denoted by $\mathfrak{R}(\Gamma)$.

The algorithmic process

$$S^0 = A \in \mathfrak{A}(\Gamma),$$
$$S^1 = \Omega_\Gamma(S^0),$$
$$S^2 = \Omega_\Gamma(S^1),$$
$$\cdots$$
$$S^{t+1} = \Omega_v(S^t)$$

may evolve in one of the following three ways.

1. For some $t = \bar{t}$ the process reaches the final state

$$S^{\bar{t}} \in \mathfrak{T}(\Gamma),$$

which yields the solution B.

2. For some $t = \bar{t}$ the process ends without giving a result

$$S^{\bar{t}} \in \mathfrak{R}(\Gamma).$$

3. The process continues indefinitely without yielding any result.

The class $\mathfrak{S}(\Gamma)$ of $A \in \mathfrak{A}(\Gamma)$ that result in the first type of evolution of the process is called the "domain of applicability" of the algorithm.

Now let us specify the notions of "state" and "active part" of the state in more detail. The aim of the considerations that we shall now develop is simply to give an intuitively clear justification of the sufficient generality of the formal definition of these notions presented a few pages further.

It is natural to assume that the state S is defined by the presence of a certain finite number of elements each of which belongs to one of the types

$$\bigcirc_1, \bigcirc_2, \ldots, \bigcirc_v$$
$$T_0, T_1, T_2, \ldots, T_n,$$

and the presence of certain relationships between certain groups of elements belonging to one of the types

$$R_1, R_2, \ldots, R_m.$$

For each type of relationship R_i, we shall assume fixed the number k_i of elements participating in the relation. The set of types of elements and the set of types of relationships (and therefore the numbers $n, m, k_1, \ldots, k_m$) are chosen once and for all for the given algorithm.

Elements will be denoted by little circles with the type number placed *inside* the circle. As to the indices placed *near* the circles, they only concern our consideration about the algorithm and have nothing to do with the state of the corresponding element. Since the number of elements that, together with the relationships they are involved in, constitute the state itself, is unbounded, it follows that the variety of these external indices is not limited apriori in any way: it is possible, for example, to use natural numbers as large as we wish for these indices.

Finally, according to our previous remarks, we can assume that within a fixed algorithm the number of relationships which can simultaneously (i.e. in the construction of one definite state) enter into one element is bounded once and for all by a certain chosen number.

Generally speaking, the order in which certain elements appear in the relationship may be essential or unessential. Further we shall consider only symmetrical relationships between pairs of elements. However, at the beginning, in order not to lose the possibility of constructing powerful algorithms, we shall assume that the notion of state essentially uses the order in which its elements paticipate in relationships; i.e., for example, the presence of the pairwise relationship

$$R(\bigcirc_1, \bigcirc_2)$$

will be assumed different from the presence of the pairwise relationship

$$R(\bigcirc_2, \bigcirc_1).$$

We shall also assume as essential in the notion of state the order of the relationships in which the given element participates. Thus the participation of the elements $\bigcirc_1, \bigcirc_2, \ldots, \bigcirc_k$ in relationship of type R will be understood as taking place in accordance to the diagram shown on Figure 1.

In this diagram the indices p_j indicate the place occupied by the relationship of type R under consideration in the ordered sequence of relations into which the element $\bigcirc_j$ enters. The indices p_j here may only assume the values $1, 2, \ldots, s$. As to the indices $1, 2, \ldots, k$, they assume, in general, values from the sequences $1, 2, \ldots, K$, where K is equal to the largest of the numbers k_i ($i = 1, 2, \ldots, m$).

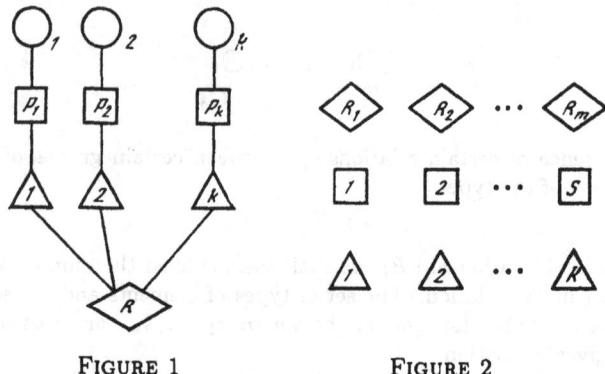

FIGURE 1   FIGURE 2

Instead of considering states consisting of elements involved in various relationship $R_1, R_2, \ldots, R_m$ (in accordance to the general diagram indicated above), it is expedient to introduce additional types of elements (Figure 2).

Obviously this allows to replace the algorithm under consideration by a new one, which differs only in the notation for its states: in this notation only one type of symmetric relationship between pairs of elements is used. Such a relationship between the elements $\bigcirc_1$ and $\bigcirc_2$ can simply be indicated by joining them by a line

For algorithms modified in this way, the following condition holds automatically.

Condition α). All elements related to some fixed element are of different types.

Naturally, when condition α) holds, a special numeration of the order in which the elements enter into different relationship is redundant: each relationship in which the given element participates is automatically supplied with the number of the element with which it is related.

Without loss of generality, we can standardize the method of singling out the "active part" of the state in a similar way; the structure of the active part is uniquely determined by the transformations which process S into $S^* = \Omega(S)$. Namely, we shall assume that the active part of the state has a distinguished "initial" element. When we require that the complexity of the active part be bounded, we mean that all the elements which enter into it must be joined to the initial element $\bigcirc$ by chains

$$\underset{}{\bigcirc}\!\!-\!\!\overset{(1)}{\bigcirc}\!\!-\!\!\overset{(2)}{\bigcirc}\!\!-\cdots-\overset{(\lambda)}{\bigcirc}$$

of bounded length $\lambda \leqslant N$, where N is a characteristic number, constant for the given algorithm.

The operator

$$S^* = \Omega_\Gamma(S),$$

which changes the state S to the "next" state S^* must be such that S^* is constructed exclusively on the basis of the information about the active part of S. Exactly in the same way, on the basis of the information concerning the form of the active part of S, it must be possible to distinguish whether the final state has been reached or not. It is simplest (and obviously does not lead to loss of generality) to use the initial element to this end: we shall assume that the initial element and only this element is of the types T_0 and T_1 and the passage of the initial element from type T_0 to type T_1 means that the final state has been achieved.

Not all the final states can be viewed as solutions. Thus, when computations on paper end, not everything that has been written down is the solution: most of the writing usually consists of intermediate calculations. In the same way, when the computer stops, only a small part of its full "state" is a solution of the problem being solved by the computer. It is therefore natural to agree that the solution is only a part of the "final state" related to the initial element. We shall take for the "solution" the set of all elements related with the initial element by chains of arbitrary length.

Now let us pass to the formulation of the definition of algorithm.

1. An algorithm Γ presupposes the presence of an ordered family $\mathfrak{T} = \mathfrak{T}(\Gamma)$ of classes

$$T_0, T_1, \ldots, T_\Gamma,$$

unbounded in volume, consisting of "elements" from which the "states" will be constructed. It is assumed that these classes are pairwise nonintersecting. Their union is denoted by T. Elements of T are denoted by circles $\bigcirc$ with various indices near the circles. In order to denote the fact that the element $\bigcirc$ belongs to the class T_i, we put the number i inside the circle. An element belonging to the class T_i is also called an element of type T_i.

2. A complex over the set $\mathfrak{T}$ is defined as an ordinary one-dimensional complex with vertices from T, i.e. the union $K = K_0 \cup K_1$ of a finite set

$$K_0 = \Big\{ \bigcirc \Big\}$$

of certain elements of T, called the "vertices" of the complex, and the set K_1 (obviously also finite)

$$K_1 = \Big\{ \underset{\alpha'}{\bigcirc}\!\!-\!\!\underset{\alpha''}{\bigcirc} \Big\}$$

of certain pairs of elements of K_0. These pairs are called "edges" of the complex K.

3. A certain set $\mathfrak{G}_{\mathfrak{T}}$ consisting of elements K over $\mathfrak{T}$ possessing the following properties is distinguished:

a) the vertices joined by edges to a fixed vertex belong to various types T_i;

b) in K there exists one and only one vertex of type T_0 or T_1, called "initial"; the other vertices belong to type T_i with $i \geqslant 2$.

4. In the set $\mathfrak{G}_{\mathfrak{T}}$, a subset $\mathfrak{A}_{\mathfrak{T}}$ of complexes whose initial vertex is of type T_0 is distinguished, as well as the subset $\mathfrak{U}_{\mathfrak{T}}$ of complexes whose initial vertex is of type T_1.

5. The complexes from $\mathfrak{G}_{\mathfrak{T}}$ are considered to be the "states" of the algorithm Γ, where the complexes from $\mathfrak{A}\mathfrak{T}$ are the "initial" states, or "conditions", while the complexes from $\mathfrak{U}_{\mathfrak{T}}$ are the "final" states:

$$\mathfrak{G}_\Gamma = \mathfrak{G}_{\mathfrak{T}}, \ \mathfrak{A}(\Gamma) = \mathfrak{A}_{\mathfrak{T}}, \ \mathfrak{S}(\Gamma) = \mathfrak{G}_{\mathfrak{T}}.$$

6. By the active part $U(S)$ of the state S we mean the subcomplex of the complex S consisting of vertices and edges belonging to chains of length $\lambda \leqslant N$ that contain the initial vertex. Here N is an arbitrary number which is, however, fixed for the algorithm Γ.

The term "subcomplex" is understood in the usual set-theoretic sense: a complex K' is a subcomplex of the complex K if $K_0' \subseteq K_0$, $K_1' \subseteq K_1$. A chain is a complex of the form

The length of the chain is the number λ of the edges which it contains. A complex is called connected, if any two of its vertices can be joined by a chain.

7. The external part $V(S)$ of the state S is by definition the subcomplex consisting of vertices not joined to the initial vertex by chains of length $\lambda < N$, and of edges not belonging to chains that contain the initial vertex and have length $\lambda \leqslant N$. The intersection

$$L(S) = U(S) \bigcap V(S)$$

is said to be the boundary of the active part of the state S. It is easy to see that $L(S)$ is a zero-dimensional complex (i.e., it does not contain any edges) and consists of those vertices that can be joined to the initial one by chains of length $\lambda = N$, but cannot be joined to it by shorter chains (of length $\lambda < N$).

Before we conclude the statement of the definition of an algorithm, let us make a few remarks.

The requirement 1 stated in §1 may be understood as follows: the operator

$$\Omega_\Gamma(S) = S^*$$

must be such that the processing of S into S^* only requires work on a certain new complex from the active part of $U(S)$ and does not change the structure of the external part $V(S)$; here the character of the processing of $U(S)$ must be determined exclusively by the structure of the complex $U(S)$ itself.

Two complexes K' and K'' will be considered *isomorphic*, if, as sets $K' = K_0' \cup K_1'$ and $K'' = K_0'' \cup K_1''$ they can be put into one-to-one correspondence so that:

1) vertices correspond to vertices, edges to edges,

2) to an edge joining the vertices $\bigcirc_\alpha'$ and $\bigcirc_\beta'$ corresponds the edge joining the corresponding vertices $\bigcirc_\alpha''$ and $\bigcirc_\beta''$,

3) corresponding vertices have the same type.

Obviously the number of non-isomorphic active parts $U(S)$ possible for a given algorithm is bounded. Therefore, the rules of their processing can easily be formulated in finite terms. We now return to our description of the definition of algorithm.

8. The rules of direct processing of the algorithm are given by indicating pairs of states

$$U_1 \to W_1,$$
$$U_2 \to W_2,$$
$$\cdots$$
$$U_r \to W_r$$

together with an isomorphic map φ_i of the subcomplex $L(U_i)$ for each i onto a certain subcomplex $\widetilde{L}(W_i)$ of the state W_i. If the vertices $\bigcirc_\mu \in L(U_i)$ and $\bigcirc_\nu \in \widetilde{L}(W_i)$ correspond to each other under the isomorphism φ_i, then an arbitrary non-initial vertex of the complex W_i related to $\bigcirc_\nu$ has the type of one of the vertices of the complex U_i related to $\bigcirc_\mu$[12]. Each complex U_i is a state of the class $\mathfrak{A}(\Gamma)$ (a condition) satisfying the requirement $U(U_i) = U_i$. All the $U_1, U_2, \ldots, U_r$ are assumed non-isomorphic to each other. As to W_i, this is an arbitrary state which may be both initial and final.

9. The domain of definition $\mathfrak{D}(\Gamma)$ of the operator $\Omega_\Gamma(S) = S^*$ consists of those states S for which the active part $U(S)$ is isomorphic to one of the complexes $U_1, U_2, \ldots, U_r$.

10. Suppose the subcomplex $U(S)$ is isomorphic to U_i. Since $U(S)$ and U_i are connected complexes belonging to $\mathfrak{B}_\mathfrak{T}$, it follows that if they are isomorphic there exists only one isomorphism between them. This isomorphism uniquely determines the isomorphism between $L(S)$ and $L(U_i)$, and since we are given an isomorphism φ_i of the complex $L(U_i)$ onto $\widetilde{L}(W_i)$, we thereby induce a well-determined isomorphism θ_i^S between $L(S)$ and $L(W_i)$.

11. Suppose $S \in \mathfrak{D}(\Gamma)$, where the subcomplex $U(S)$ is isomorphic to the complex U_i. Obviously we can construct a complex $\widetilde{W}$ isomorphic to the complex W_i and satisfying the following conditions:

1) $\widetilde{W} \cap V(S) = L(S)$,

---

[12] Because of this requirement, property a) from part 3 of the definition will not be contradicted in the process of the algorithm's work.

2) the isomorphism θ_i^S between the subcomplexes $L(S) \subseteq \widetilde{W}$ and $\widetilde{L}(W_i) \subseteq W_i$ can be extended to an isomorphism between the complexes $\widetilde{W}$ and W_i.

The result of applying the operator Ω_Γ to the complex S is determined (uniquely up to isomorphism) as a complex of the form

$$S^* = \widetilde{W} \cup V(S).$$

12. If S is the final state, then the connected component of the initial vertex is taken to be the "solution" (by connected component of any vertex we mean the maximal connected subcomplex which contains this vertex)[13]; the family of connected components of the initial vertices of the final states may be denoted, following our previous explanations, by $\mathfrak{B}(\Gamma)$.

Now the definition of algorithm can be concluded exactly as indicated at the beginning of this section (with relation to the three types of development of the algorithmic process and the definition of their "domain of applicability" $\mathfrak{G}(\Gamma)$).

An algorithm given in this way, characterized by the ordered set $\mathfrak{T}$ and the number N, will be said to belong to the class $\mathbf{A}_N(\mathfrak{T})$. Every algorithm of the class $\mathbf{A}_N(\mathfrak{T})$ will be said to belong to the class $\mathbf{A}_\mathfrak{T}$.

In order to make the description of the algorithm more intelligible, we have introduced certain conventions which are not related directly to the general underlying approach, but we feel that the sufficient generality of the definition proposed above remains convincing despite these conventions. We think it is convincing that an arbitrary algorithmic process satisfies our definition of algorithm. We would like to stress here that we are not arguing about the reducibility of any algorithm to an algorithm in the sense of our definition (thus in the following section we shall prove the reducibility of any algorithm to an algorithm for computing a partially recursive function), but about the fact that any algorithm essentially fits the proposed definition.

The difficulties which the reader may meet if he attempts to represent, say, the computations of the values of a partially recursive function in the form of an algorithm as defined above and involving complexes are only due to the fact that, as we have already pointed out in §1, the outline for computing the values of a partially recursive function is not specified directly in the form of an algorithm. If we do specify all these computations in the form of an algorithmic process (which of course is easy to do), then we shall automatically obtain a certain algorithm in the sense of the proposed definition. Let us formulate this statement in more detail.

Consider the set of types $\mathfrak{T}_6 = \{T_0, T_1, T_2, T_3, T_4, T_5, T_6\}$. Let us call "$A$-representation" of a system of q numbers $(j_1, j_2, \ldots, j_q)$ the complex $A_{j_1, j_2, \ldots, j_q}$ (Figure 3), and "B-representation" of the same system, the complex $B_{j_1, j_2, \ldots, j_q}$ obtained from $A_{j_1, j_2, \ldots, j_q}$ by replacing the vertex ⓪ by the vertex ①

---

[13]Obviously we shall obtain the same "solution", if we pass to the connected component of the initial vertex at each step of the algorithm's work, i.e. we shall define the result S^* of a direct processing of the complex S as the connected component of the initial vertex of the complex $\widetilde{W} \cup V(S)$. For this method of defining the operator Ω_Γ all the states of the algorithmic process, except perhaps the initial state, will turn out to be connected complexes and therefore the "solution" will coincide with the final state. (*Author's note added in the preparation of this book*).

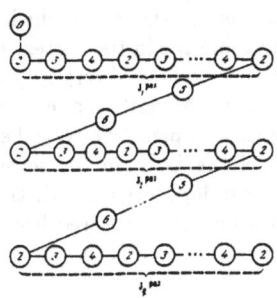

FIGURE 3

For every partially recursive function $f(x_1, x_2, \ldots, x_q)$ there exists a set of types $\mathfrak{T} \supseteq \mathfrak{T}_6$ and an algorithm Γ of class $\mathbf{A}(\mathfrak{T})$ applicable to the complex $A_{j_1, j_2, \ldots, j_q}$ if and only if f is defined for the system of numbers $j_1, j_2, \ldots, j_q$ and in this last case the algorithm transforms this complex into the complex B_t, where

$$t = f(j_1, j_2, \ldots, j_q).$$

All that we have said about recursive functions remains entirely true if we consider all the definitions A)-F). As soon as one of these definitions can be used to yield an algorithmic process, then this process turns out to be an algorithm in the sense indicated above.

§3

Any algorithm can be reduced to the computation of the values of a certain partially recursive function. This is proved by the so-called method of arithmetization.

First let us note that every complex S from $\mathfrak{G}_{\mathfrak{T}}$ may be given in the form of a table. To do this let us order the set of vertices of the complex S in the following way[14].

To each chain

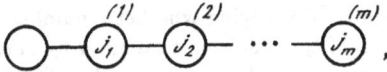

joining the initial vertex $\bigcirc$ to an arbitrary vertex $\bigcirc^{(m)}$ let us assign the row

$$j_1, j_2, \ldots, j_m.$$

By condition α) the correspondence between the chains originating at the initial vertex $\bigcirc$ of the complex S and a certain set of rows of natural numbers will

---

[14] A. V. Gladky in his review of the present article, published in Ref.J.Matematika, 1959, No.7, p.9 (Ref. 6527), correctly remarks: "The method of arithmetization presented in §3 needs certain corrections since the method of ordering the set of vertices used there works only for connected complexes. The necessary corrections may be carried out without difficulty." *Editor's note.*

be bijective. If we order the rows of numbers in lexicographic order, then we automatically obtain an order among the chains originating at the initial vertex. To each vertex of S let us now assign the chain which is minimal in the sense of the order established above, and join the initial vertex to the one chosen by us; this correspondence induces an order in the set of vertices (the initial vertex will be the first one in this order). Thus the set of vertices of each complex from S is ordered.

It is natural to assign to each complex from $\mathfrak{G}_{\mathfrak{T}}$ the symmetric table in which i_k is the number of the type of the k-th (in the sense of the order established) vertex from S, while α_{kl} is 1 or 0 depending on whether the k-th and l-th vertex are joined or not. The complex may be recovered from the table in a unique way up to isomorphism.

	i_1	i_2	$\cdots$	i_v
i_1	α_{11}	α_{12}	$\cdots$	α_{1v}
i_2	α_{21}	α_{22}	$\cdots$	α_{2v}
$\vdots$	$\vdots$	$\vdots$	$\ddots$	$\vdots$
i_v	α_{v1}	α_{v2}	$\cdots$	α_{vv}

This construction transforms each algorithm of class $\mathbf{A}(\mathfrak{T})$ into the language of tables.

To each table let us assign its "Gödel number"; to do this we shall write the table as a row

$$\alpha_{11}\alpha_{12}\cdots\alpha_{1v}\alpha_{21}\alpha_{22}\cdots\alpha_{2v}\cdots\alpha_{v1},\alpha_{v2},\cdots\alpha_{vv},$$

and to the row we assign the number

$$p_1^{i_1}p_2^{i_2}\cdots p_v^{i_v}p_{v+1}^{\alpha_{11}}p_{v+2}^{\alpha_{12}}\cdots p_{v^2+v}^{\alpha_{vv}},$$

where p_r is the rth prime number.

In the series of natural numbers, we thus obtain an infinite subset Q of all numbers which are "Gödel numbers" of tables corresponding to complexes from $\mathfrak{G}_{\mathfrak{T}}$. There is no difficulty in verifying whether a number belongs to Q or not. Let us number the elements of Q in their natural order by natural numbers. Let us call *number of the table* the number of its corresponding "Gödel number". The *number of a complex* from $\mathfrak{G}_{\mathfrak{T}}$ will be by definition the number of the table representing this complex.

From a complex S belonging to $\mathfrak{G}_{\mathfrak{T}}$, one can effectively and uniquely find its number s, while the number s effectively and uniquely (up to isomorphism) allows us to recover the complex S itself.

We now have the situation that was described in §1. The set $\mathfrak{G}_{\mathfrak{T}}$ is numbered by natural numbers

$$S_0, S_1, S_2, \ldots, S_p, \ldots;$$

and in this sequence of complexes we can include both the set $\mathfrak{A}_{\mathfrak{T}}$ of "conditions" and the set $\mathfrak{B}_{\mathfrak{T}}$ of "solutions". Each algorithm Γ of class $\mathbf{A}(\mathfrak{T})$ uniquely determines its own domain of applicability $\mathfrak{S}(\Gamma)$ in the set $\mathfrak{B}_{\mathfrak{T}}$; this map induces a numerical function

$$m = \varphi_\Gamma(n),$$

given on the set G of all numbers n for which $S_n \in \mathfrak{S}(\Gamma)$. Let us show that the function φ_Γ is partially recursive.

Indeed it is easy to prove that the function $s^* = \sigma(s)$ induced on the set of numbers by the operator of direct processing $S^* = \Omega_\Gamma(S)$ is primitive recursive (for the definition of primitive recursive functions see Appendix I). This immediately implies the primitive recursivity of the function $\rho(n, t)$ that gives the number of the complex S^t obtained on the tth step of work of the complex having number n. The process is continued until the complex $S^{\bar{t}}$ arising at step $\bar{t}$ turns out to be "final", i.e., a complex from $\mathfrak{U}(\Gamma)$. The numbers of the complexes from $\mathfrak{U}(\Gamma) = \mathfrak{U}_{\mathfrak{T}}$, i.e. the numbers of the complexes whose initial vertex belongs to the class T_1, satisfy the equation

$$\gamma(x) = 0,$$

where the function γ, as can easily be shown, may always be chosen among primitive recursive functions. The process can be continued in this way to the first t satisfying the condition

$$\gamma(\rho(n, \bar{t})) = 0,$$

i.e. to the number

$$\bar{t} = \mu t[\gamma(\rho(n, t)) = 0]$$

(about the operator μ see Appendix I). The complex $S^{\bar{t}}$ belongs to $\mathfrak{T}(\Gamma)$, its number $\rho(n, \bar{t})$ can be found from the relation

$$\rho(n, \bar{t}) = \rho(n, \mu t[\gamma(\rho(n, t)) = 0]).$$

The number $m = \varphi_\Gamma(n)$ of the solution is obtained from the number of the complex $S^{\bar{t}}$ by means of the function β (also primitive recursive) which finds, given the number of a complex in $\mathfrak{G}_{\mathfrak{T}}$, the number of the connected component of the initial vertex of this complex. Namely

$$\varphi_\Gamma(n) = \beta(\rho(n, \mu t[\gamma(\rho(n, t)) = 0])), \tag{1}$$

which implies (see Appendix I) that the function $\varphi_\Gamma(n)$ is partially recursive.

This means in particular that the function $f(x_1, \ldots, x_q)$ given by the algorithm transforming A-representations of q-tuples of numbers into B-representations of numbers is partially recursive.

Indeed, we can construct primitive recursive functions ξ and η satisfying the following conditions:

1) for every string of numbers $j_1, \ldots, j_q$, the number $\xi(j_1, \ldots, j_q)$ is the number of the A-representation of the string $j_1, \ldots, j_q$,

2) for every number n which is the number of the B-representation of the number t the value $\eta(n)$ is equal to t.

If we now denote by Γ the algorithm which carries out the computation of the function f, then obviously

$$f(x_1, \ldots, x_q) = \eta(\varphi_\Gamma(\xi(x_1, \ldots, x_q))), \tag{2}$$

which means that f is a partially recursive function.

Since for every partially recursive function f there exists an algorithm Γ computing it (in the sense indicated at the end of §2), it follows that each partially recursive function will possess the representation (2). Using formula (1), we obtain

$$f(x_1,\ldots,x_q) == \eta(\beta(\rho(\xi(x_1,\ldots,x_q),\mu t[\gamma(\rho(\xi(x_1,\ldots,x_q),t)) = 0]))).$$

For the sake of brevity, we put

$$\eta(\beta(\rho(\xi(x_1,\ldots,x_q),v)))) = \pi(x_1,\ldots,x_q,v),$$
$$\gamma(\rho(\xi(x_1,\ldots,x_q),t)) = \tau(x_1,\ldots,x_q,t),$$

and finally obtain that each partially recursive function can be represented in the form

$$f(x_1,\ldots,x_q) = \pi(x_1,\ldots,x_q,\mu t[\tau(x_1,\ldots,x_q,t) = 0]), \qquad (3)$$

where π and τ are primitive recursive functions.

APPENDIX I

SOME FACTS FROM THE THEORY OF RECURSIVE FUNCTIONS

Here we consider functions of one, two or more variables, and each variable and the function itself assume values in the set of natural numbers. We agree to consider each natural number as a function of zero variables. We do not require that a function of n variables be defined for all systems of n natural numbers. The equality of two functions

$$f(x_1,x_2,\ldots,x_n) = g(x_1,x_2,\ldots,x_n)$$

means that for every system of n natural numbers $(j_1,j_2,\ldots,j_n)$ such that one of these functions is defined the other one is also defined and

$$f(j_1,j_2,\ldots,j_n) = g(j_1,j_2,\ldots,j_n).$$

Let us introduce some operators on the set of functions. Arguments of each operator are functions or strings of functions.

The superposition operator. The domain of definition of the superposition operator is a system of functions of the form

$$\psi(x_1,x_2,\ldots,x_n); \ \ \varphi_1(x_1,\ldots,x_m), \ \ \varphi_2(x_1,\ldots,x_m),\ldots,\varphi_n(x_1,\ldots,x_m)$$
$$(n = 1,2,\ldots; \ m = 1,2,\ldots).$$

Being applied to such a system, the superposition operator yields the function $\theta(x_1,\ldots,x_m)$ which is the result of substituting the functions $\varphi_1,\varphi_2,\ldots,\varphi_n$ into the function ψ in place of its variables

$$\theta(x_1,\ldots,x_m) = \psi(\varphi_1(x_1,\ldots,x_m),\varphi_2(x_1,\ldots,x_m),\ldots,\varphi_n(x_1,\ldots,x_m)).$$

The function $\theta(x_1, \ldots, x_m)$ is defined for all strings of natural numbers $(j_1, \ldots, j_m)$ possessing the following properties:

1) for $(j_1, \ldots, j_m)$ all the functions $\varphi_i (i = 1, 2, \ldots, n)$ are defined,

2) if $\varphi_i(j_1, \ldots, j_m) = t_i (i = 1, 2, \ldots, n)$, then the function ψ is defined for the system $(t_1, t_2, \ldots, t_n)$.

The primitive recursion operator. The domain of definition consists of pairs of functions of the form

$$h(x_1, \ldots, x_{n_1}), \; g(x_1, \ldots, x_{n-1}, x_n, x_{n+1}) \quad (n = 1, 2, \ldots).$$

Applied to such a pair, the recursion operator gives the function $f(x_1, \ldots, x_{n-1}, x_n)$ related to h and g by the equalities

$$f(x_1, \ldots, x_{n-1}, 0) = h(x_1, \ldots, x_{n-1}),$$
$$f(x_1, \ldots, x_{n-1}, x_{n+1}) = g(x_1, \ldots, x_{n-1}, x_n, f(x_1, \ldots, x_{n-1}, x_n)).$$

The least number operator. The domain of definition is the set of all functions of one or several variables. Being applied to the function $\theta(x_1, \ldots, x_{n-1}, x_n)$, the least number operator yields the function $\psi(x_1, \ldots, x_{n-1})$ determined by the following rule: for every string of natural numbers $j_1, \ldots, j_{n-1}$, the value of the function

$$\psi(j_1, \ldots, j_{n-1})$$

is a number t such that:

1) $\theta(j_1, \ldots, j_{n-1}, t) = 0$;

2) for every number j less than t, the value $\theta(j_1, \ldots, j_{n-1}, j)$ is defined and is non-zero.

The result of applying the least number operator to the function $\theta(x_1, \ldots, x_{n-1}, x_n)$ is written ordinarily in the form

$$\mu x_n [\theta(x_1, \ldots, x_{n-1}, x_n) = 0],$$

where we remember that the notation

$$\mu x_n [\theta(x_1, \ldots, x_{n-1}, x_n) = 0]$$

means that we are dealing with a function depending only on the variables $x_1, \ldots$
$\ldots, x_{n-1}$.

The least number operator, unlike the superposition and recursion operator, may be applied to an everywhere defined function and give a function which is not everywhere defined (or even nowhere defined).

Let us introduce three groups of functions:

1) The function $O_n(x_1, \ldots, x_n)$ for $n = 0, 1, 2, \ldots$; for each n, the function $O_n(x_1, \ldots, x_n)$ is identically equal to zero.

2) The function $I_{nk}(x_1, \ldots, x_n)$ for $n = 1, 2, \ldots$ and $k = 1, 2, \ldots, n$; for every string $(j_1, \ldots, j_k, \ldots, j_n)$, we have $I_{nk}(j_1, \ldots, j_k, \ldots, j_n) = j_k$.

3) The function $\lambda(x)$, equal to $x + 1$.

All these functions will be called *basis* functions.

The class of *primitive recursive functions* is defined as the minimal class containing the basis functions and closed with respect to the application of the superposition operator and the primitive recursion operator [18,6]. Each primitive recursive function, as can be easily seen, is everywhere defined. All the simple arithmetical functions such as

$$x + y, \ xy \ x^y, \ |x - y|, \ \left[\frac{x}{y}\right], \ [\sqrt{y}], \ [\log_x y], \ x!$$

and many others (for example the exponent that a given prime number has in the decomposition of the number y into prime factors) is primitive recursive.

The class of *partially recursive functions* is defined as the minimal class containing the basis functions and closed with respect to the applications of the superposition operator, the primitive recursion operator and the least number operator [19,6].

A partially recursive function in n variables defined for all systems of n natural numbers is said to be *recursive* [19,6]. In particular, every primitive recursive function is a recursive function. There exists recursive functions that are not primitive recursive.

Each partially recursive function is constructed from primitive recursive functions by means of the application of the superposition operator, the primitive recursion and least number operator. It is of interest to consider the least number of operators required to construct a partially recursive function from primitive recursive ones. Apriori this number may be as large as we wish. However, there exists an important theorem, due to Kleene [19], which claims that every partially recursive function $f(x_1, \ldots, x_n)$ may be obtained by applying the least number operator and the superposition operator to two primitive recursive functions (of which one is fixed once and for all, while the other depends on the function $f(x_1, \ldots, x_n))$[1]. Namely, Kleene's theorem states that there exists a primitive recursive function $U(x)$ such that for every partially recursive function $f(x_1, \ldots; x_n)$ there exists a primitive recursive function $\theta(x_1, \ldots, x_n, y)$ with the following property

$$f(x_1, \ldots, x_n) = U(\mu y[\theta(x_1, \ldots, x_n, y) = 0]).$$

The Kleene formula is also important in that it reduces to a minimum the use of the least number operator; the application of this operator is always unpleasant, since, as we have pointed out, it may transform an everywhere defined function into one which is not everywhere defined.

We shall say that the set R is generated by the function $f(x)$, if the set of values of $f(x)$ is R. The set is called *recursively enumerable* if it is either empty or is the set of values of a certain recursive function [20]. The class of recursively enumerable sets coincides with the class of sets generated by partially recursive functions. The class of non-empty recursively enumerable sets coincides with the

---

[1]The formula (3) from §3 already shows that any partially recursive function may be obtained from two primitive recursive ones by applying the operators of least number and superposition.

class of sets generated by primitive recursive functions [21]. The following sets are recursively enumerable: the sum and intersection of two recursively enumerable sets; the image and the inverse image of the recursively enumerable set under a map given by a partially recursive function; the domain of definition of a partially recursive function of one variable.

A set is called *recursive* if the set itself and its complement are recursively enumerable [20]. There exist recursively enumerable but not recursive sets [20].

By a characteristic function of a set, we mean a function which assumes the value 1 on the set and the value 0 outside the set. A characteristic function of a recursive set is a recursive function [20]. Conversely, if a characteristic function of a set is recursive, then the set itself is recursive [20].

If we identify the notion of everywhere defined algorithmically computable function with a notion of recursive function, then recursively enumerable sets are exactly those sets which can be presented as an algorithmically computable sequence (perhaps with repetitions), while recursive sets are precisely those sets for which there exists an algorithm which allows to decide whether a given natural number belongs to the set or does not.

APPENDIX II

EXAMPLE OF AN ALGORITHM

We shall present as an example the "copying" algorithm, which transforms any complex A_k

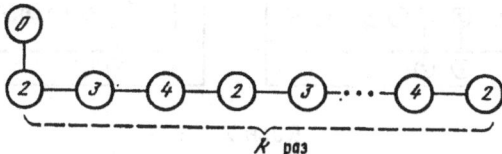

into the complex B_{2k}

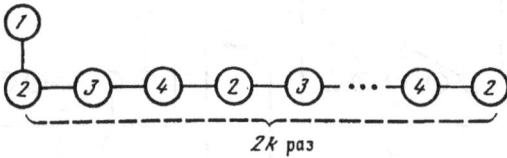

Note that the simplest algorithm (in the intuitive sense) which copies a linearly ordered sequence of points is the following procedure: we must look through all the points in turn and replace each point by two. If this procedure is formalized, then we obtain precisely the algorithm described below.

The algorithm that will be constructed belongs to the type $\mathbf{A}(\mathfrak{T}_5)$, more precisely $\mathbf{A}_4(\mathfrak{T}_5)$, where $\mathfrak{T}_5 = \{T_0, T_1, T_2, T_3, T_4, T_5\}$.

The algorithm is given by the rules of direct processing consisting of a finite number of pairs $U_i \to W_i$.

We shall agree to represent each pair graphically in the following way[1]. The complexes U_i and W_i will be drawn as non-intersecting. The complex U_i will be placed to the left, while the complex W_i to the right. In the complex U_i, we distinguish the boundary $L(U_i)$ by placing it in the little frame consisting of two concentric rectangles. Two other (right) concentric rectangles, congruent to the first (left), are used to distinguish the subcomplex $\tilde{L}(W_i)$ in W_i. In order to obtain an isomorphism φ_i between $L(U_i)$ and $\tilde{L}(W_i)$, we must bring into coincidence the right-hand-side system of rectangles with the let-hand side one and identify coinciding vertices placed between the rectangles. After these agreements, let us write ou the following five rules which define the copying algorithm.

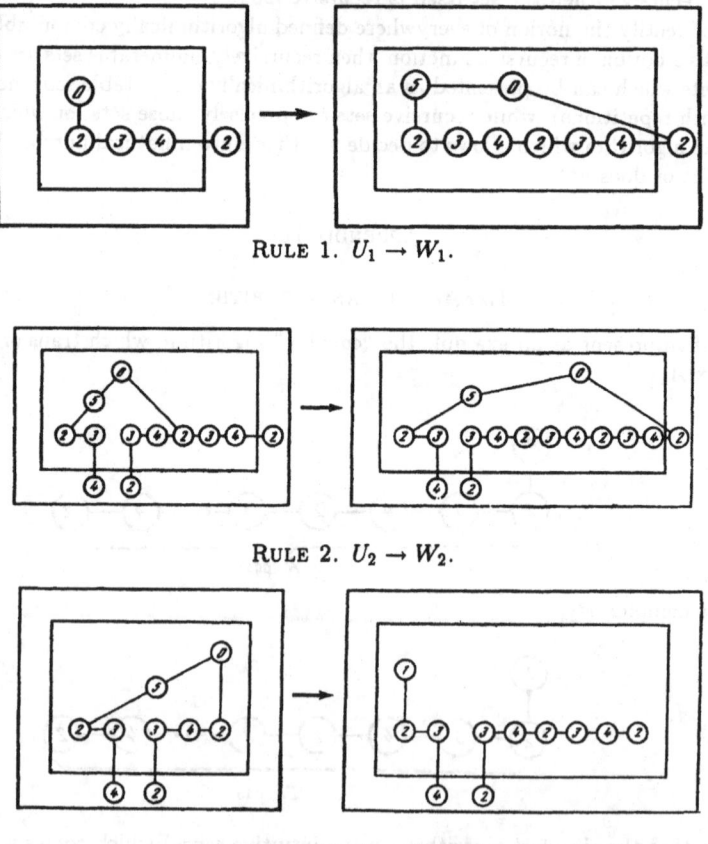

RULE 1. $U_1 \to W_1$.

RULE 2. $U_2 \to W_2$.

RULE 3. $U_3 \to W_3$.

The necessity of using Rule 4 arises when $k = 0$, while Rule 5 is applied when $k = 1$.

As an example, consider the application of our algorithm to the following complex $S^0 = A_3$:

---

[1]Each complex is shown on the picture up to isomorphism.

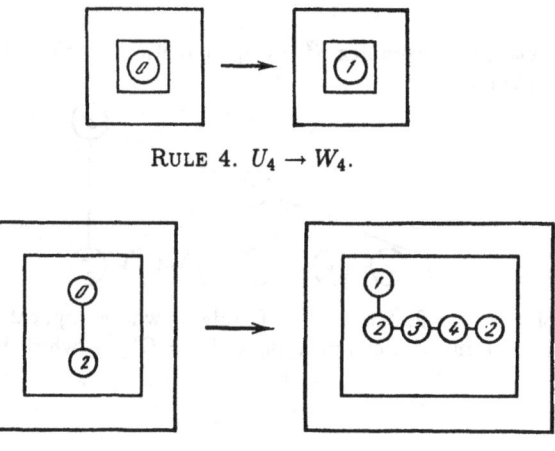

RULE 4. $U_4 \to W_4$.

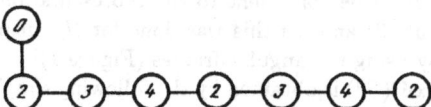

RULE 5. $U_5 \to W_5$.

By definition, we must apply the operator of direct transformation Ω given by the rules 1-5 to S^0. To do this, we distinguish the complex $U(S^0)$ in S^0. Since $N = 4$, it follows that the complex $U(S^0)$ is of the form

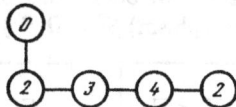

Now among the rules 1-5 we find the pair for which the first term is isomorphic to $U(S^0)$. Such a pair is given by Rule 1. Carrying out the direct transformation in accordance to this rule, we obtain the complex $S^1 = \Omega(S^0)$ of the form

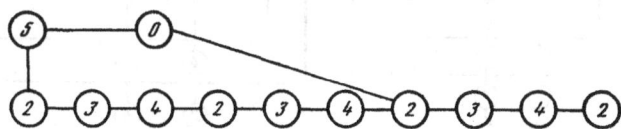

The complex $U(S^1)$ will have the following structure

It is isomorphic to the first term U_2 of Rule 2, therefore the complex $S^2 = \Omega(S^1)$ is the following one In the complex S^2, we again distinguish the subcomplex $U(S^2)$; the latter has the form

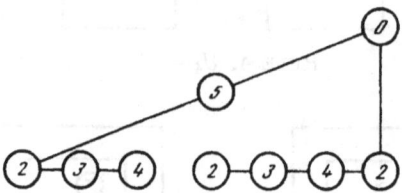

It is isomorphic to the first term U_3 of Rule 3, whose application yields us the "solution" in the form of a complex $B_6 = S^3 = \Omega(S^2)$, which has the following structure

Note that under our conventions concerning the graphic representation of the rules of direct transformation, the direct transformation itself has a graphically clear geometric interpretation. Let us obseve this in our example. We have seen that $U(S^2)$ is isomorphic to U_3. Let us draw the complex S^2 so that the subcomplex $U(S^2)$ turns out to be congruent to the representation of U_3(on the picture corresponding to Rule 3) and, as this was done for U_3, display the subcomplexes $U(S^2)$ and $L(S^2)$ by using rectangular frames (Figure 1).

Then the complex $\Omega(S^2)$ is obtained by the following simple method: the system of rectangles framing W_3 are placed on a congruent system of rectangles constructed for S^2; the coinciding vertices that lie between the rectangles are identified. Everything that lies inside the innermost rectangle constructed for S^2 is replaced by whatever is contained inside the innermost rectangle framing W_3. Indeed, if we carry out all these operations, we obtain the complex shown on Figure 2.

And this is (up to isomorphism) $S^3 = \Omega(S^2)$.

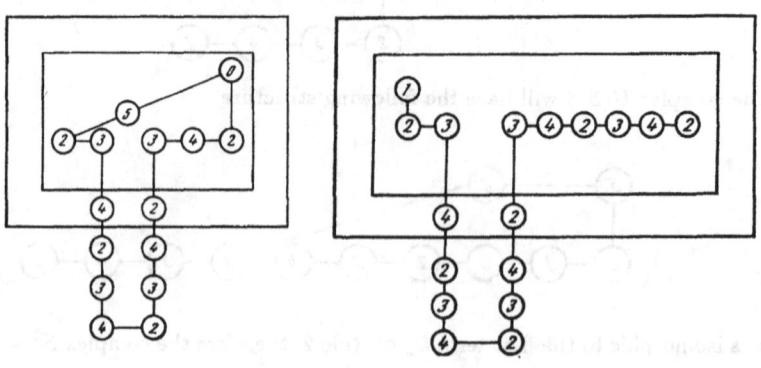

FIGURE 1   FIGURE 2

REFERENCES

[1] A. A. Markov, *Theory of algorithms*, Trudy MIAN USSR **38** (1951), 176-189.
[2] A. A. Markov, *Theory of algorithms*, Trudy MIAN SSSR **32** (1954), 3-375.
[3] A. N. Kolmogorov, *On the notion of algorithm*, Uspekhi Mat. Nauk **8,4** (1953), 175-176.
[4] K. Gödel, *On undecidable propositions of formal mathematical systems*, Princeton, 1934.
[5] A. Church, *An unsolvable problem of elementary number theory*, Amer. J. Math. **58, 2** (1936), 345-363.
[6] S. C. Kleene, *Introduction to metamatheematics*, Van Nostrand, N.Y., 1952.
[7] A. Church, *The calculi of lambda-conversion*, Princeton, 1941.
[8] S. C. Kleene, *On notation for ordinal numbers*, J. Symbol. Log. **3,4** (1938), 150-155.
[9] K. Gödel, *Uber die Lange von Beweisen*, Ergeb. math. Kolloq. (1934-1935), **7** (1936), 23-24.
[10] S. C. Kleene, *General recursive functions of natural numbers*, Math. Ann. **5** (1936,112), 727-742.
[11] A. M. Turing, *On computable numbers, with an application to the Entscheidungsproblem*, Proc. London Math. Soc. Ser. 2 **42,3-4** (1936), 230-265.
[12] R. Peter, *Rekursive Funktionen*, Akademischer Verlag, Budapest, 1949.
[13] E. L. Post, *Finite combinatory processes - formulation 1.*, J. Symbol. Log. **1,3** (1936), 103-105.
[14] S. C. Kleene, *λ-definability and recursiveness*, Duke Math. J. **2,2** (1936), 340-353.
[15] A. M. Turing., *Computability and λ-definability*, J. Symbol. Log. **2,4** (1937), 153-163.
[16] V. K. Detlos, *Normal algorithms and recursive functions*, Dokl. Mat. Nauk SSSR **90,5** (1953), 723-725.
[17] V. A. Uspensky, *On the theorem of uniform conversions*, Uspekhi Mat. Nauk SSSR **12,1** (1957), 99-142.
[18] R. M. Robinson, *Primitive recursive functions*, Bull. Amer. Math. Soc. **53,10** (1947), 925-942.
[19] S. C. Kleene, *Recursive predicates and quantifiers*, Trans. Amer. Math. Soc. **53,1** (1943), 41-73.
[20] E. L. Post, *Recursively enumerable sets of positive integers and their decision problem*, Bull. Amer. Math. Soc. **50,5**, (1944), 284-316.
[21] J. B. Rosser, *Extensions of some theorems of Gödel and Church*, J. Symbol. Log. **1,3** (1936), 87-91.
[22] E. L. Post, *Formal reduction of the general combinatorial decision problem*, Amer. J. Math. **65,2** (1943), 197-215.

7.
ε-ENTROPY AND ε-CAPACITY OF SETS IN FUNCTIONAL SPACES*)**)

(Jointly with V. M. Tikhomirov)

Introduction.

The article is mainly devoted to the systematic exposition of results that were published in the years 1954-1958 by K. I. Babenko [1], A. G. Vitushkin [2,3], V. D. Yerokhin [4], A. N. Kolmogorov [5,6] and V. M. Tikhomirov [7]. It is natural that when these materials were systematically rewritten, several new theorems were proved and certain examples were computed in more detail. This review also incorporates results not published previously which go beyond the framework of such a systematization, and belong to V. I. Arnold (§6) and V. M. Tikhomirov (§§4,7 and §8).

To obtain a general idea of the research under consideration, one can read §1, look through the examples in §2 and read §3, which is an introduction to the parts of this article that follow. The idea that it is possible to characterize the "mass" of sets in metric spaces by means of the order of increase of the number of elements of most economical coverings as $\varepsilon \to 0$ was developed in the paper by L. S. Pontryagin and L. G. Shnirelman (see [8], Appendix to the translation). Among earlier closely related constructions (which, however, are not convenient for our goals) we should indicate the definition of measures of fractional orders in the well-known paper by F. Hausdorff [9]. Interest in this group of ideas developed in Moscow again when A. G. Vitushkin [2] obtained an estimate from below of the functions $\mathcal{M}_\varepsilon(A)$ (in the notation of our §1) for classes of functions of n variables with bounded partial derivatives up to a given order p and applied this estimate to the proof of a theorem on the necessary decrease of smoothness in a representation of an arbitrary function of n variables by a superposition of functions of $m < n$ variables. A. N. Kolmogorov [5] showed that the second part of the proof of A. G. Vitushkin's theorem, carried out by the author by means of the theory of multidimensional variations, may be made very simple by using estimates of $\mathcal{M}_\varepsilon(A)$ from above (see Appendix I to our article). Using this paper and applying the general ideas of information theory that had become popular by that time,

---

*)Uspekhi Mat. Nauk, 1959, vol. 14, No.2, p. 3-86.

**)In this edition, subsections 8.4–8.7 of this paper (some eleven pages), which, for one reason or another, had been left out of the Russian edition of the selected papers of Kolmogorov, have been reinserted. *Series editors note.*

A. N. Kolmogorov [6] stated a general programme for studying the notions of ε-entropy and ε-capacity which are interesting from the point of view of the theory of functions on compact sets in functional spaces. The relationship of the entire subject-matter under consideration to probabilistic information theory is at the present time only on the level of analogies and parallelism (see Appendix II). For example, the results of V. M. Tikhomirov, developed in §8, are inspired by the "Kotelnikov theorem" from information theory.

It seems doubtless that the results obtained may be of interest to non-probabilistic information theory in studying the necessary volumes of memory and numbers of operations in computational algorithms (see the work of N. S. Bakhvalov [10,11] and A. G. Vitushkin [12]). However, to obtain practically applicable results, the estimates of the functions $\mathcal{E}_\varepsilon$ and $\mathcal{H}_\varepsilon$ known at the present time require considerable improvement. Our §2 contains some still very imperfect attempts in this direction. In the paper by L. S. Pontryagin and L. G. Shnirelman mentioned above, it was established that the asymptotics of $\mathcal{M}_\varepsilon$ and $\mathcal{N}_\varepsilon$ may lead to topological invariants (to the definition of topological dimension). In A. N. Kolmogorov's note [13], it is shown how in the same direction one may obtain certain definitions of linear dimension of topological vector spaces.

We feel that the relationships between the direction of research considered here and various topics in mathematics are interesting and varied. Let us note, in particular, that V. D. Yerokhin [14], solving the problem of improving estimates for the functions $\mathcal{E}_\varepsilon$ and $\mathcal{H}_\varepsilon$ for certain classes of analytic functions, discovered a new method for approximating analytic functions by linear forms $c_1\varphi_1(z)+\cdots+c_n\varphi_n(z)$ possessing interesting properties which may be stated without using the notions of ε-entropy and ε-capacity.

§1. Definition and main properties of the functions $\mathcal{H}_\varepsilon(A)$ and $\mathcal{E}_\varepsilon(A)$.

Suppose A is a non-empty set in a metric space R. Let us introduce the following definitions:

Definition 1. The system γ of set $U \subset R$ is said to be a ε-*covering* of the set A, if the diameter $d(U)$ of any $U \in \gamma$ is no greater than 2ε and

$$A \subseteq \bigcup_{U \in \gamma} U.$$

Definition 2. The set $U \subset R$ is said to be ε-*net* for the set A if any point of the set A is located at a distance no greater than ε from some point of the set U.

Definition 3. The set $U \subset R$ is said to be ε-*distinguishable*, if any two of its distinct points are located at distance greater than ε.

Further we shall see why in Definition 1 it is more convenient for us to write 2ε instead of the ordinary ε. Further the number ε will be assumed positive without special mention.

We shall always be dealing with completely bounded sets. It is useful to have in mind three equivalent definitions contained in the following theorem.

Theorem 1. *The following three properties of the set A are equivalent and depend on the metric of the set A itself (i.e. remain valid or remain non-valid if A is included with the given metric into any larger space R):*

α) *For any ε, there exists a finite ε-covering of the set A.*

β) *For any ε, there exists a finite ε-net for the set A.*

γ) *For any ε, any ε-distinguishable set is finite.*

The proof of Theorem 1 may be left to the reader (see [15, pages 312-320]), where property β is taken to be the definition. It is well known that all compact sets are completely bounded and, under the assumption that the space is complete, for the complete boundedness of A it is necessary and sufficient that the closure of A and R be a compact set (see [15, p. 315]).

For completely bounded A it is natural to introduce the following three functions which characterize in a certain sense how massive the set A is[1]:

$\mathcal{N}_\varepsilon(A)$ is the minimal number of sets in an ε-covering of A;

$\mathcal{N}_\varepsilon^R(A)$ is the minimal number of points in an ε-net of the set A;

$\mathcal{M}_\varepsilon(A)$ is the maximal number of points in an ε-distinguishable subset of the set A.

The functions $\mathcal{N}_\varepsilon(A)$ and $\mathcal{M}_\varepsilon(A)$ for a given metric on A do not depend on the choice of the underlying space $\mathbf{R}$ (for $\mathcal{M}_\varepsilon$ this is obvious by the definition itself, while for $\mathcal{N}_\varepsilon$ this may be proved without difficulty). On the contrary, the function $\mathcal{N}_\varepsilon^R(A)$ in general depends on R, as reflected in the notation.

Special names will be given to the binary logarithms of the functions defined above[2]:

$\mathcal{H}_\varepsilon(A) = \log_2 \mathcal{N}_\varepsilon(A)$ is the *minimal ε-entropy of the set A* or simply the *ε-entropy* of the set A;

$\mathcal{H}_\varepsilon^R(A) = \log_2 \mathcal{N}_\varepsilon^R(A)$ is the *ε-entropy of A with respect to R*;

$\mathcal{E}_\varepsilon(A) = \log \mathcal{M}_\varepsilon(A)$ is the *ε-capacity of A*[3].

These names are related to considerations from information theory. To clarify them, it suffices to give the following approximate indications: a) in information theory, the unit of "amount of information" is the amount of information in one binary digit (i.e. in indicating whether it is equal to 0 or to 1); b) the "entropy" of the set of possible "messages" which must be memorized or transmitted with given precision is by definition the number of binary digits necessary to transmit any of these messages with given precision (i.e. an h such that to each message x one can assign a sequence $s(x)$ of h binary digits by means of which x may be recovered with the required precision); c) the "capacity" of the transmitting or memorizing device is defined as the number of binary digits that it is capable to transmit or to memorize in suitable way.

If we look at A as the set of possible messages and assume that the choice of x' at a distance $\rho(x, x') \leqslant \varepsilon$ allows us to recover the message x with sufficient precision,

---

[1]In order to generalize the definition to the case when A is not completely bounded, it is necessary to change the definitions slightly (see [6]). Naturally, for a set A that is not completely bounded for a sufficiently small ε, all three functions $\mathcal{N}_\varepsilon, \mathcal{N}_\varepsilon^R$ and $\mathcal{M}_\varepsilon$ may assume infinite values.

[2]Since A is non-empty by assumption, we always have $\mathcal{N}_\varepsilon(A) \geqslant 1$, $\mathcal{N}_\varepsilon^R(A) \geqslant 1$, $\mathcal{M}_\varepsilon(A) \geqslant 1$ and therefore $\mathcal{H}_\varepsilon(A) \geqslant 1$, $\mathcal{H}_\varepsilon^R(A) \geqslant 0$, $\mathcal{E}_\varepsilon(A) \geqslant 0$.

[3]Here and everywhere further $\log N$ denotes the logarithm of the number N to the base 2.

then it is obvious that it suffices to use $\mathcal{N}_\varepsilon^R(A)$ different "signals" to transmit any of the messages $x \in A$; these signals may be taken as sequences of binary digits of length no greater than $\mathcal{H}_\varepsilon^R(A) + 1$. It is also obvious that it is no longer possible to manage this using sequences of length less than $\mathcal{H}_\varepsilon^R(A)$.

On the other hand, if we consider A as the set of possible signals and assume that two signals $x_1 \in A$ and $x_2 \in A$ differ in a reliable way whenever $\rho(x_1, x_2) > \varepsilon$, then it is obvious that within A one can choose $\mathcal{M}_\varepsilon(A)$ reliably different signals and use them to fix any binary sequence of length $\mathcal{E}_\varepsilon(A) - 1$ for memorizing or for transmission. On the other hand, it is impossible to choose a system of reliably differing signals in A for the transmission of any binary sequences of length $\mathcal{E}_\varepsilon(A) + 1$; this is also obvious.

Concerning the function $\mathcal{H}_\varepsilon(A)$, its role is the following: 1) the inequality $\mathcal{H}_\varepsilon(A) \leqslant \mathcal{H}_\varepsilon^R(A)$ (see Theorem IV further) is used to estimate $\mathcal{H}_\varepsilon^R(A)$ from below; 2) in the case when the space is *centrable* (see Definition 4 below), we have $\mathcal{H}_\varepsilon^R(A) = \mathcal{H}_\varepsilon(A)$ and $\mathcal{H}_\varepsilon(A)$ acquires the direct meaning of entropy. Any completely bounded metric space A may be embedded in a centralizable space R. Hence, following A. G. Vitushkin's idea, in a certain sense the words of $\mathcal{H}_\varepsilon(A)$ may be viewed as the entropy of A by itself ("absolute entropy"). The practical significance of this statement is not actually very clear, except in those cases when the precision in reproducing $x \in A$ actually required involves getting an x' so that $\rho(x, x') \leqslant \varepsilon$ in a certain centralizable space R, as it happens in the case of problems on the approximate determination of real functions $f(t)$ of an arbitrary argument t with a specified uniform precision ε. In this connection, see the sequel, in particular Theorems VI, VII.

Now let us state a few simple theorems which express the main properties of the functions introduced above.

Theorem II. *All six functions* $\mathcal{N}_\varepsilon(A)$, $\mathcal{N}_\varepsilon^R(A)$, $\mathcal{M}_\varepsilon(A)$, $\mathcal{H}_\varepsilon(A)$, $\mathcal{H}_\varepsilon^R(A)$, $\mathcal{E}_\varepsilon(A)$ *as functions of the set* A *are semi-additive, i.e. for them*

$$A \subset \bigcup_k A_k$$

implies

$$F(A) \leqslant \sum_k F(A_k).$$

The proof may be left to the reader. Semi-additivity and non-negativity of our function implies that, in the case $A' \subset A$, for each of them we have the inequality

$$F(A') \leqslant F(A).$$

Theorem III. *All six functions* $\mathcal{N}_\varepsilon(A)$, $\mathcal{N}_\varepsilon^R(A)$, $\mathcal{M}_\varepsilon(A)$, $\mathcal{H}_\varepsilon(A)$, $\mathcal{H}_\varepsilon^R(A)$, $\mathcal{E}_\varepsilon(A)$ *as functions of* ε *are non-increasing (when* ε *increases, i.e. are non-decreasing when* ε *decreases). The functions* $\mathcal{N}_\varepsilon(A)$, $\mathcal{M}_\varepsilon(A)$, $\mathcal{H}_\varepsilon(A)$, $\mathcal{E}_\varepsilon(A)$ *are continuous from the right.*

The first half of the theorem directly follows from definitions, so that the proof may be left to the reader.

It suffices to prove the second part of the theorem for the functions $\mathcal{M}_\varepsilon(A)$ or $\mathcal{N}_\varepsilon(A)$. Suppose $\mathcal{M}_{\varepsilon_0}(A) = n$ and $x_1, \ldots, x_n$ is the corresponding ε_0-distinguishable set. Then

$$\varepsilon_0 = 1 = \min_{i \neq j} \rho(x_i, x_j) > \varepsilon_0$$

and for all ε in the interval $\varepsilon_0 \leqslant \varepsilon < \varepsilon_1$, we must have $\mathcal{M}_\varepsilon(A) \geqslant n$; since the function is monotone, we have $\mathcal{M}_\varepsilon(A) = n$.

For $\mathcal{N}_\varepsilon(A)$ the proof is somewhat more difficult and proceeds by *reductio at absurdum*. If we assume the existence of a sequence

$$\varepsilon_1 > \varepsilon_2 > \cdots > \varepsilon_k > \cdots \to \varepsilon_0,$$

for which $\mathcal{N}_{\varepsilon_k}(A) = m < \mathcal{N}_{\varepsilon_0}(A) = n$, then for any $k \geqslant 1$ there must exist ε_k-coverings of A by sets $A_{k_1}, Ak_2, \ldots, A_{km}$. The closure $\bar{A}$ of the set A in the completion R^* of the space R is compact.

In the "discrepancy" metric $\alpha(F, F')$, the closed subsets of the compact set $\bar{A}$ constitute a compact set (see [16], p. 549-550]).

Consider the set

$$F_{ki} = \bar{A}_{ki} \cap \bar{A}.$$

From the above it follows that we can choose a sequence $k_s \to \infty$ such that

$$F_{k_s i} \to F_i \ (i = 1, 2, \ldots, m).$$

It is easy to check that the sets $F_1, \ldots, F_m$ constitute an ε_0-covering of the set A, which contradicts the assumption $n \geqslant m$.

Theorem IV. *For any completely bounded set A in a metric space $\mathbf{R}$, we have the following inequality*

$$\mathcal{M}_{2\varepsilon}(A) \leqslant \mathcal{N}_\varepsilon(A) \leqslant \mathcal{N}_\varepsilon^R(A) \leqslant \mathcal{N}_\varepsilon^A(A) \leqslant \mathcal{M}_\varepsilon(A), \tag{1}$$

and therefore

$$\mathcal{E}_{2\varepsilon}(A) \leqslant \mathcal{H}_\varepsilon(A) \leqslant \mathcal{H}_\varepsilon^R(A) \leqslant \mathcal{H}_\varepsilon^A(A) \leqslant \mathcal{E}_\varepsilon(A). \tag{2}$$

We shall prove inequality (1) from right to left. Suppose $x_1, \ldots, x_{\mathcal{M}_\varepsilon}$ is a maximal ε-distinguishable set in A. Then this set is obviously a ε-net, since in the converse case there would exist a point $x' \in A$ such that $\rho(x', x) > \varepsilon$; the latter contradicts the maximality of $x_1, \ldots, x_{\mathcal{M}_\varepsilon}$. Since we have $x_i \in A$ by definition, we obtain

$$\mathcal{N}_\varepsilon^A(A) \leqslant \mathcal{M}_\varepsilon(A).$$

Obviously, every ε-net consisting of points $x_i \in A$ is an ε-net consisting of points $x_i \in R \supset A$, i.e.

$$\mathcal{N}_\varepsilon^R(A) \leqslant \mathcal{N}_\varepsilon^A(A).$$

Suppose $y_1, \ldots, y_n$ is an ε-net with respect to A in R. Consider the sets $S_\varepsilon(y_i) \cap A = U_i$, where by $S_\varepsilon(Y_i)$ we denote the ball of radius ε with centre at the point y_i. The sets $\{U_i\}$ constitute an ε-covering of A, i.e. each ε-net generates an ε-covering, hence

$$\mathcal{N}_\varepsilon(A) \leqslant \mathcal{N}_\varepsilon^R(A).$$

And finally, for any ε-covering and 2ε-distinguishable set, the number of points in the latter is no greater than the number of points in the former, since in the converse case two points whose distance is $> 2\varepsilon$ would fit into one set of diameter $\leqslant \varepsilon$; hence

$$\mathcal{N}_{2\varepsilon}(A) \leqslant \mathcal{N}_{\varepsilon}(A).$$

As we already mentioned, the relationship between the functions introduced above is simplified in the case of a centralizable space.

Definition 4. The space R is called *centralizable* if in it, for any open set U of diameter $d = 2r$, there exists a point x_0 from which any point x is located at a distance no greater than r.

Theorem V. *For any completely bounded set A in a centralizable space R we have*

$$\mathcal{N}_{\varepsilon}^{R}(A) = \mathcal{N}_{\varepsilon}(A), \ \mathcal{H}_{\varepsilon}^{R}(A) = \mathcal{H}_{\varepsilon}(A).$$

Indeed, the inequality

$$\mathcal{N}_{\varepsilon}(A) \leqslant \mathcal{N}_{\varepsilon}^{R}(A)$$

appears in the statement of Theorem IV. In a centralizable space, however, each ε-covering of A by sets

$$U_1, \ldots, U_N$$

can be assigned an ε-net of the same number of points

$$x_0^{(1)}, \ldots, x_0^{(N)}$$

and hence

$$\mathcal{N}_{\varepsilon}^{R}(A) \leqslant \mathcal{N}_{\varepsilon}(A).$$

Theorem VI. *The space D^X of bounded functions on an arbitrary set X with metric*

$$\rho(f, g) = sup_{x \in X} |f(x) - g(x)|$$

is centralizable.

Suppose $U \subset D^X$ is a set of diameter

$$d = \sup_{f,g} \sup_{x \in X} |f(x) - g(x)|.$$

Putting

$$\bar{f}(x) = \sup_{f \in U} f(x), \ \underline{f}(x) = \inf_{f \in U} f(x),$$

we see that

$$d = \sup_{x \in X} (\bar{f}(x) - \underline{f}(x)).$$

Now it is easy to show that for the function

$$f_0(x) = 1/2(\bar{f}(x) - \underline{f}(x))$$

and any function $f \in U$, we have

$$|f(x) - f_0(x)| \leqslant d/2,$$

whenever $x \in X$, i.e.

$$\rho(f, f_0) \leqslant d/2 \ \ \forall f \in U,$$

which was to be proved.

Theorem VII (A.G. Vitushkin [3]). *Any completely bounded set A may be embedded in a centralizable space R.*

By the Mazur-Banach theorem [17, Chapter IX], any A (we are only interested in completely bounded sets, and hence certainly for separable A) can be isometrically embedded in the space C, and therefore in the space D^I which contains it, where I is the unit closed interval $[0,1]$; the theorem is proved.

By Theorem VI, the space D^I is centralizable, while Theorem V implies that for any set A contained in it we have

$$\mathcal{N}_\varepsilon^{D^I}(A) = \mathcal{N}_\varepsilon(A).$$

By the inequality

$$\mathcal{N}_\varepsilon(A) \leqslant \mathcal{N}_\varepsilon^R(A)$$

(Theorem IV) and Theorem VII we have

$$\mathcal{N}_\varepsilon(A) = \min_{A \subseteq R} \mathcal{N}_\varepsilon^R(A), \ \mathcal{H}_\varepsilon(A) = \min_{A \subseteq R} \mathcal{H}_\varepsilon^R(A),$$

which justifies calling $\mathcal{H}_\varepsilon(A)$ the *minimal ε-entropy*.

§2. Examples of exact calculation of the functions $\mathcal{H}_\varepsilon(A)$ and $\mathcal{E}_\varepsilon(A)$ and their estimates.

1. A is the closed interval $\Delta : \{a \leqslant x \leqslant b\}$ of length $|\Delta| = b - a$ on the line D with metric $\rho(x, x') = |x - x'|$.

In this case[4]

$$\mathcal{N}_\varepsilon(A) = \mathcal{H}_\varepsilon^D(A) = M_{2\varepsilon}(A) = \begin{cases} |\Delta|/2\varepsilon & \text{for integer } |\Delta|/2\varepsilon, \\ [|\Delta|/2\varepsilon] + 1 & \text{for non-integer } |\Delta|/2\varepsilon, \end{cases} \quad (3)$$

i.e.

$$\mathcal{N}_\varepsilon(A) = \frac{|\Delta|}{2\varepsilon} + O(1), \ \mathcal{N}_\varepsilon(A) = \frac{|\Delta|}{\varepsilon} + O(1),$$

$$\mathcal{H}_\varepsilon(A) = \frac{|\Delta|}{2\varepsilon} + O(\varepsilon), \ \mathcal{E}_\varepsilon(A) = \frac{|\Delta|}{\varepsilon} + O(\varepsilon). \quad (4)$$

Indeed it is easy to see that Δ is centralizable in the sense of §1 and therefore, in accordance to Theorem V, we have

$$\mathcal{N}_\varepsilon^D(\Delta) = \mathcal{N}_\varepsilon(\Delta). \quad (5)$$

The number of sets of an ε-covering of Δ cannot be smaller than $[|\Delta|/2\varepsilon]$ or equal to $[|\Delta|/2\varepsilon]$ in the case when $[|\Delta|/2\varepsilon]$ is not an integer, since in this case the family of all these sets would have measure no greater than $2\varepsilon[|\Delta|/2\varepsilon] < |\Delta|$.

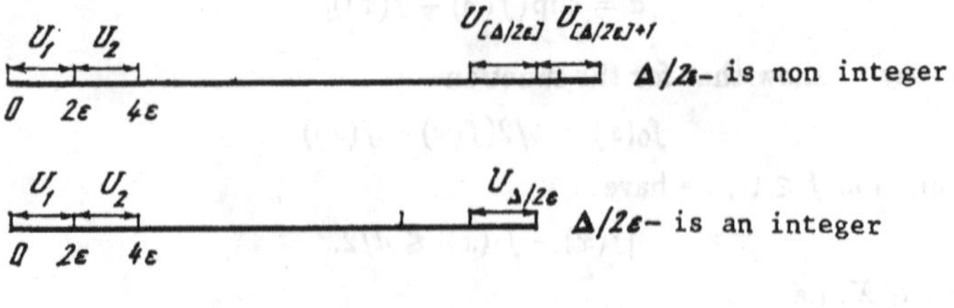

FIGURE 1

---

[4]The notation $[A]$ stands for the interger part of the number A.

However the reader will verify without difficulty that it is always possible to construct an ε-covering consisting of $|\Delta|/2\varepsilon + 1$ sets, and whenever $|\Delta|/2\varepsilon$ is an integer, it turns out to be sufficient to use $|\Delta|/2\varepsilon$ sets (Figure 1).

On the other hand, when $|\Delta|/2\varepsilon$ is non-integer, dividing Δ into $[|\Delta|/2\varepsilon]$ equal parts by the points $a = x_1, x_2, \ldots, x_{[|\Delta|/2\varepsilon]+1}$, we obtain $[|\Delta|/2\varepsilon]+1$ points forming a 2ε-distinguishable set. If $|\Delta|/2\varepsilon$ is an integer, we divide Δ in a similar way into $|\Delta|/2\varepsilon - 1$ equal parts and obtain $|\Delta|/2\varepsilon$ points constituting a 2ε-distinguishable set. The above implies that

$$\mathcal{M}_{2\varepsilon} \geqslant \mathcal{N}_\varepsilon(\Delta),$$

hence, using (5) and inequality (1), we obtain (3).

2. A is the set $F_1^\Delta(L)$ of functions $f(x)$ defined on the closed interval $\Delta = [a, b]$, satisfying the Lipschitz condition

$$|f(x) - f(x')| \leqslant L|x - x'|$$

and vanishing at the point a.

The given set may be considered in the space $\mathbf{D}^\Delta$, i.e. in the metric

$$\rho(f(x), g(x)) = \sup_{x \in \Delta} |f(x) - g(x)|.$$

Let us show that

$$\mathcal{H}_\varepsilon(A) = \mathcal{E}_{2\varepsilon}(A) = \mathcal{H}_\varepsilon^{D^\Delta}(A) = \begin{cases} \dfrac{|\Delta|L}{\varepsilon} - 1 \text{ for integer } \dfrac{|\Delta|L}{\varepsilon}, \\ [\dfrac{|\Delta|L}{\varepsilon}] \quad \text{for non-integer } \dfrac{|\Delta|L}{\varepsilon}, \end{cases} \qquad (6)$$

i.e.

$$\mathcal{H}_\varepsilon(A) = \frac{|\Delta|L}{\varepsilon} + O(1), \quad \mathcal{E}_\varepsilon(A) = \frac{2|\Delta|L}{\varepsilon} + O(1). \qquad (7)$$

Let us note first of all that this set can be mapped uniquely and isometrically, by substitution the independent variable $t = L(x - a)$, onto the set $F_1^{\Delta'}(1)$, i.e., onto the set of functions given on the closed interval $[0, \Delta']$, where $\Delta' = |\Delta|L$, satisfying the Lipchitz condition with constant equal to 1 and vanishing at the origin. It is obvious that here the values of $\mathcal{H}_\varepsilon$ and $\mathcal{E}_\varepsilon$ will not change. We shall compute the functions $\mathcal{H}_\varepsilon$ and $\mathcal{E}_\varepsilon$ for this last set.

The idea of our constructions will be to find an ε-covering of $F_1^{\Delta'}(1)$ and a 2ε-distinguishable set in $F_1^{\Delta'}(1)$ consisting of the same number of elements K_ε.

Then according to the definition of the functions $\mathcal{N}_\varepsilon$ and $\mathcal{M}_{2\varepsilon}$, we get

$$\mathcal{N}_\varepsilon(F_1^{\Delta'}(1)) \leqslant K_\varepsilon, \quad \mathcal{M}_{2\varepsilon}(F_1^{\Delta'}(1)) \geqslant K_\varepsilon, \qquad (8)$$

hence, using inequality (1) as in Subsection 1 of this section, we get

$$\mathcal{M}_{2\varepsilon}(F_1^{\Delta'}(1)) = \mathcal{N}_\varepsilon(F_1^{\Delta'}(1)) = K_\varepsilon. \qquad (9)$$

To the relations (9), let us add one more, namely

$$\mathcal{N}_\varepsilon(F_1^{\Delta'}(1)) = \mathcal{N}_\varepsilon^{D^{\Delta'}}(F_1^{\Delta'}(1)),$$

which follows from the fact that $D^{\Delta'}$ is centralized.

Suppose $\varepsilon > 0$ is such that Δ'/ε is not an integer. The number $[\Delta'/\varepsilon]$ will be denoted by n the expressions $k\varepsilon$ by t_k $(k = 1, 2, \ldots, n)$. Divide the closed interval $[0, \Delta']$ into $n + 1$ segments

$$\Delta_k = [(k-1)\varepsilon, k\varepsilon], \quad k = 1, 2, \ldots, n, \quad \Delta_{n+1} = [n\varepsilon, |\Delta'|].$$

Denote by $\varphi(t)$ the function defined on the closed inteval $[\varepsilon, |\Delta'|]$ which equals ε at the point $t = \varepsilon$ and which is linear on the segment Δ_k with angular coefficient equal to $+1$ or -1.

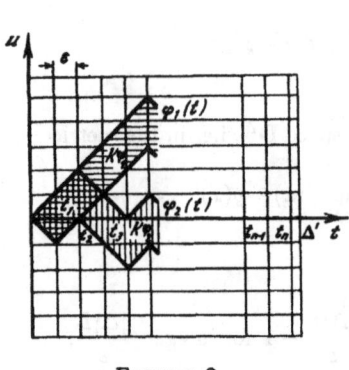

FIGURE 2

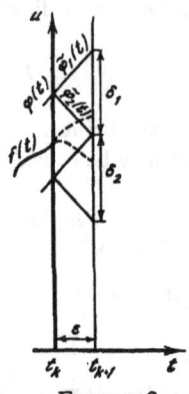

FIGURE 3

The set of points of the plane (t, u) such that

$$-t \leqslant u \leqslant t, \ t \in \Delta_1, \ \varphi(t) - 2\varepsilon \leqslant u \leqslant \varphi(t), \ t \in [\varepsilon, |\Delta'|],$$

will be called an ε-corridor (Figure 2) and denoted by K_φ.

The number of all corridors will be denoted by K_ε. For our choice of ε, as can be easily seen, the number K_ε equals $2^{[|\Delta'|/\varepsilon]}$.

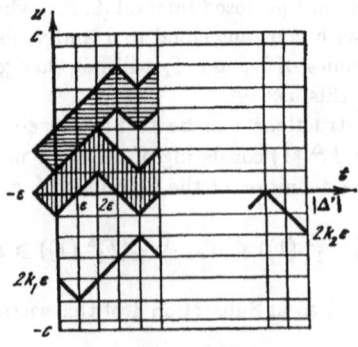

FIGURE 4

Let us show that the set of all ε-corridors constitutes an ε-covering of the set $F_1^{\Delta'}(1)$. (Further we shall say that the function $f(t)$ belongs to certain corridor K_φ if

$$\varphi(t) - 2\varepsilon \leqslant f(t) \leqslant \varphi(t).)$$

The proof of this statement will be obtained by induction.

Indeed, on the segment $\Delta_1 = [0, \varepsilon]$ all the functions of our set belong to all the ε-corridors. Now suppose by inductive assumption that for each function f from our set on the segment $[0, t_k]\, (k \leqslant n)$ there exists an ε-corridor K_φ such that

$$\varphi(t) - 2\varepsilon \leqslant f(t) \leqslant \varphi(t),\ 0 \leqslant t \leqslant t_k.$$

Then it follows from the Lipchitz condition (Figure 3) that $f(T_{k+1})$ either belongs to the segment

$$\delta_1 = [\varphi(t_k) - \varepsilon,\ \varphi(t_k) + \varepsilon],$$

or to the segment

$$\delta_2 = [\varphi(t_k) - 3\varepsilon,\ \varphi(t_k) - \varepsilon]$$

and on the segment $[0, t_{k+1}]$ the function f belongs to the corridor $K_{\tilde\varphi}$, where $\tilde\varphi$ is the function coinciding with φ on $[0, t_k)$ and which on $\Delta_k = [t_k, t_{k+1}]$ is linear with angular coefficient equal to $+1$ in the case when $f(t_{k+1}) \in \delta_1$ and to -1 if $f(t_{k+1}) \in \delta_2$. The induction is over. Now we must construct a 2ε-distinguishable set consisting of K_ε elements. Divide $[0, |\Delta'|]$ into n equal to segments $\bar\Delta_1, \ldots, \bar\Delta_n$. The length of each segment satisfies $|\bar\Delta_i| = |\Delta'|/n > \varepsilon$ (recall that $n[|\Delta'|/\varepsilon]$). Consider the set M_n of functions which vanish at the origin and which on the segment $\bar\Delta_i$ are linear with angular coefficients equal to $+1$. Two different functions from M_n differ from each other at least on $2|\Delta'|/n$, i.e. they are 2ε-distinguishable and their number is equal to $2^n = K_\varepsilon$.

If $|\Delta'|/\varepsilon$ is an integer, then the value of K_ε, equal to the number of all ε corridors, turns out to be $2^{|\Delta'|/\varepsilon - 1}$. A set similar to the set M_n considered above, where $n = |\Delta'|/\varepsilon - 1$, as can be easily shown, is 2ε-distinguishable and consists of $2^n = K_\varepsilon$ functions.

Thus we have shown that

$$\mathcal{N}_\varepsilon(A) = \mathcal{M}_{2\varepsilon(A)} = \mathcal{N}_\varepsilon^{D^\Delta}(A) = K_\varepsilon = \begin{cases} 2^{|\Delta'|/\varepsilon - 1} & \text{for integer } |\Delta'|/\varepsilon, \\ 2^{[|\Delta'|/\varepsilon - 1]} & \text{for non-integer } |\Delta'|/\varepsilon, \end{cases}$$

which implies (6) if we take logarithms, since $|\Delta'| = |\Delta|L$.

3. Now suppose A is the set $F_1^\Delta(C, L)$ of functions $f(t)$ defined on the closed interval $\Delta = [a, b]$ and satisfying the Lipschitz conditions with constant L, bounded on Δ by the constant C. This set will also be viewed as a metric space in the uniform metric on Δ.

The space $F_1^\Delta(C, L)$ differs from the space $F_1^\Delta(L)$, which we considered in the previous subsection, in that it involves the condition

$$|f(t)| \leqslant C,\ t \in \Delta,$$

instead of $f(0) = 0$.

We shall obtain the following estimates

$$\frac{|\Delta|L}{\varepsilon} + \log\frac{C}{\varepsilon} - 3 \leqslant \mathcal{E}_{2\varepsilon}(A) \leqslant \mathcal{H}_\varepsilon \leqslant \frac{|\Delta|L}{\varepsilon} + \log\frac{C}{\varepsilon} + 3 \qquad (10)$$

for $\varepsilon \leqslant \min(C/4, C^2/16|\Delta|L)$, i.e.,

$$\mathcal{H}_\varepsilon = |\Delta|L/\varepsilon + \log(C/\varepsilon) + O(1),$$
$$\mathcal{E}_{2\varepsilon} = 2|\Delta|L/\varepsilon + \log(C/\varepsilon) + O(1). \qquad (11)$$

We see that here we succeed in obtaining estimates for $\mathcal{H}_\varepsilon$ and $\mathcal{E}_\varepsilon$ from below and from above which, for ε not too large, differ from each other only by a few units. This kind of result may be viewed as quite satisfactory from the point of view of practical applications.

Let us pass to the proof of inequalities (10).

As in the previous subsection, everything reduces to the space $F_1^{\Delta'}(C, 1)$, where $\Delta' = [0, |\Delta|L]$.

The estimate from above for $\mathcal{H}_\varepsilon$ is obtained by using the results of the previous subsection. Indeed, choosing $\varepsilon > 0$, let us consider the set of points of the (t, u)-plane with coordinates $(-\varepsilon, 2k\varepsilon)$, $|k| \leqslant [C/2\varepsilon]$ (Figure 4). At each of these points, construct a set of ε-corridors as we did in the previous subsection for zero. The set of ε-corridors constructed at the point $(-\varepsilon, 2k\varepsilon)$, according to the previous considerations, will turn out to be an ε-covering for the set of functions $f \in F_1^{\Delta'}(C, L)$ satisfying $2k\varepsilon - \varepsilon \leqslant f(0) \leqslant 2k\varepsilon + \varepsilon$ and therefore the set of all ε-corridors will be a ε-covering for $F_1^{\Delta'}(C, L)$. Thus we obtain

$$\mathcal{H}_\varepsilon(F_1^{\Delta'}(C, L)) \leqslant \left(2\left[\frac{C}{2\varepsilon}\right] + 1\right)\mathcal{H}_\varepsilon(F_1^{\Delta' \cup [-\varepsilon, 0]}(L)),$$

hence taking logarithms and using (6), we get

$$\mathcal{H}_\varepsilon(A) \leqslant \frac{|\Delta'|}{\varepsilon} + \log\left(\frac{C}{\varepsilon} + 1\right) \leqslant \frac{|\Delta'|}{\varepsilon} + \log\frac{C}{\varepsilon} + 2, \qquad (12)$$

so that

$$\varepsilon \leqslant C\log(C/\varepsilon + 1) \leqslant \log(C/\varepsilon) + 1.$$

The estimate from above for $\mathcal{E}_{2\varepsilon}$ is somewhat more difficult to obtain. Denote by $2r$ the maximal even integer such that $2r\varepsilon < |\Delta'|$. Divide Δ' into $2r$ different segments $\Delta_1, \Delta_2, \ldots, \Delta_{2r}$. Construct the set $\bar{M}_{2r}$ of functions which at zero are equal to one of the values $2k\varepsilon$, $|k| \leqslant [C/2\varepsilon]$, while on the segments Δ_i, $i = 1, 2, \ldots, 2r$ they are linear with angular coefficient equal to ± 1. Functions from $\bar{M}_{2r}$ are 2ε-distinguishable and belong to $F_1^{\Delta'}(1)$. It remains to estimate the number of functions belonging to $F_1^{\Delta'}(C, 1)$, i.e., those whose absolute value is no greater than C. Further we carry out this estimate by elementary means. From the point of view of probability theory, this situation reduces to the estimation of the number of paths (corresponding to the function from $\bar{M}_{2r}$ constructed above) which do not leave the given limits for ordinary random walks with probability at each step of moving up or down equal to one half and the necessary estimate may be obtained much faster by using inequalities well-known in probability theory.

We shall carry out the required estimate in several steps.

α) Denote by $N(k_1, k_2)$ the number of functions $\varphi(t) \in \bar{M}_{2r}$ such that

$$\varphi(0) = 2k_1\varepsilon, \quad \varphi(|\Delta'|) = 2k_2\varepsilon,$$

where k_1 and k_2 are any admissible integers; suppose, to be definite, that $k_2 \geqslant k_1$; now $N(k_1, k_2)$ is equal to the number of functions φ for which n_1 angular coefficients are equal to $+1$ (the number of "uphill steps"), minus the number n_2 of coefficients equal to -1 (the number of "downhill steps"), and so is equal to $2(k_2 - k_1)$.

This gives

$$n_1 + n_2 = 2r, \; n_1 - n_2 = 2(k_2 - k_1),$$

i.e., $n_1 = r + k_2 - k_1$, while the number of functions with n_1 uphill steps is equal to $\binom{2r}{n_1}$. Thus

$$N(k_1, k_2) = \binom{2r}{k_2 - k_1 + r}. \tag{13}$$

β) Consider the set $U_1(k)$ of functions from $\overline{M}_{2r}$ satisfying the condition

$$\varphi(0) = 2k\varepsilon, \; \varphi(|\Delta'|) > 2k\varepsilon.$$

The number $N_1(k)$ of such functions satisfies the relation

$$N_1(k) = \sum_{\substack{k_1 \leqslant k \\ k_2 > k}} N(k_1, k_2) \leqslant \sum_{\substack{k_1 \leqslant k \\ k_2 > k}} \binom{2r}{k_2 - k_1 + r}. \tag{14}$$

In (14), the term $\binom{2r}{s+r}$ appears s times, so that

$$N_1(k) \leqslant \sum_{0 \leqslant s \leqslant r} s \binom{2r}{s + r}. \tag{15}$$

γ) Suppose $U_2(k)$ denotes the set of functions from $\bar{M}_{2r}$ such that

$$\varphi(0) \leqslant 2k\varepsilon, \; \max_{t \in \Delta'} \varphi(t) > 2k\varepsilon.$$

Suppose $\varphi(t)$ belongs to $U_2(k)$ but does not belong to $U_1(k)$. Consider the last of the points $s\varepsilon$ satisfying $\varphi(s\varepsilon) = (k+1)\varepsilon$. Construct a new function $\tilde{\varphi}(t)$ which coincides with $\varphi(t)$ up to the point $s\varepsilon$ and further is its mirror image with respect to the line $u = k + 1$. Obviously $\tilde{\varphi}(t) \in U_1(k)$.

To the function $\varphi \in U_1(k) \cap U_2(k)$ we assign this function itself. We have obtained a map of the set $U_2(k)$ into $U_1(k)$ for which, as can be easily verified, the inverse image of each function from $U_1(k)$ consists of no more than two functions, i.e. the number of functions $N_2(k)$ in $U_2(k)$ is no greater than $2N_1(k)$.

It is easy to understand that the number of functions from $\overline{M}_{2r}$ such that

$$|\varphi|(0) \leqslant 2k\varepsilon, \quad \max_{t \in \Delta} |\varphi(t)| > 2k\varepsilon, \tag{16}$$

is no greater than $2N_2(k) \leqslant 4N_1(k)$.

It remains only to note that the set of functions from $\bar{M}_{2r}$ which do not belong to (16) for $k = [C/2\varepsilon]$ is precisely the set whose number of function was to be estimated from the very beginning.

Thus

$$\mathcal{M}_{2\varepsilon}(F_1^{\Delta'}(C,1)) \geqslant \left(2\left[\frac{C}{2\varepsilon}\right]+1\right) 2^{2r} - 4N_1(k) \geqslant$$

$$\geqslant \left(2\left[\frac{C}{2\varepsilon}\right]+1\right) 2^{2r} - 4\sum_{s=0}^{r} s2r\binom{s+r}{.} \tag{17}$$

Let us estimate the right-hand side of (17). We have

$$\sum_{s=0}^{r} s\binom{2r}{r+s} = \left(2r\binom{2r-1}{r}\right) - \frac{r\binom{2r}{r}}{2} = \frac{r}{2}\binom{2r}{r}. \tag{18}$$

The proof of the simple relation (18) is left to the reader.

Further we use the elementary inequality

$$\binom{2r}{r} < 2^{2r}/\sqrt{2r+1} \tag{19}$$

(which is contained, for example, in [18], and can be proved without difficulty). Thus by using (18) and (19), we see that (17) implies

$$\mathcal{M}_{2\varepsilon}(F_1^{\Delta'}(C,1)) \geqslant 2^{2r}\left(2\left[\frac{C}{2\varepsilon}\right]+1-\frac{2r}{2r+1}\right) \geqslant$$

$$2^{2r}\left(\frac{C}{\varepsilon}-1-\sqrt{2r}\right). \tag{20}$$

Taking logarithms of (20) and using the fact that $|\Delta'| - 2\varepsilon \leqslant 2r\varepsilon \leqslant |\Delta'|$, we get

$$\mathcal{E}_{2\varepsilon}(A) \geqslant \frac{|\Delta'|}{\varepsilon} - 2 + \log\frac{C}{\varepsilon} + \log\left(1-\frac{\varepsilon}{C}-\sqrt{\frac{\varepsilon|\Delta'|}{C^2}}\right) \geqslant$$

$$\geqslant \frac{|\Delta'|}{\varepsilon} + \log\frac{C}{\varepsilon} - 3 \quad \text{when } \varepsilon \leqslant \min\left(\frac{C}{4}, \frac{C^2}{16|\Delta'|}\right),$$

which was to be proved.

4. A is the set $A_h(C)$ of functions f given on D which are periodic with period 2π, analytic in a strip of width $h : |\Im z| \leqslant h$, $z = x + iy$, and bounded there by the

constant C. The given set is considered in the uniform metric on D. For this set we have the following formulas

$$\mathcal{H}_\varepsilon(A) = 2\frac{(\log(1/\varepsilon))^2}{h \log e} + O\left(\log\frac{1}{\varepsilon}\log\log\frac{1}{\varepsilon}\right),$$

$$\mathcal{E}_\varepsilon(A) = 2\frac{(\log(1/\varepsilon))^2}{h \log e} + O\left(\log\frac{1}{\varepsilon}\log\log\frac{1}{\varepsilon}\right). \tag{21}$$

The relation (21) will be obtaind as in the previous subsections by estimating from above the number of sets constituting an ε-covering of $A_h(C)$ and from below the number of 2ε-distinguishable functions in $A_h(C)$.

To obtain the estimates of the expressions $\mathcal{H}_\varepsilon(A)$ from above, let us expand the function $f \in A_h(C)$ in a Fourier series

$$f(x) = \sum_{k=-\infty}^{+\infty} c_k e^{ikx}. \tag{22}$$

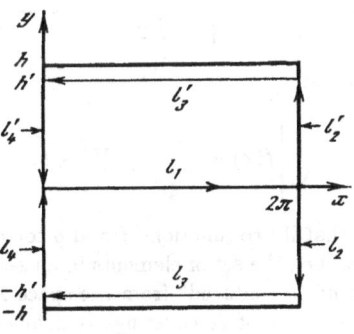

FIGURE 5

Further we shall use the following inequality for the coefficients in (22)

$$|c_k| \leqslant Ce^{-|k|h}. \tag{23}$$

Inequality (23) is obtained as follows. Suppose $k > 0$; integrate $f(z)e^{ikz}$ along the contour shown on Figure 5. Since f is periodic, we have

$$\int_{l_2} f(\xi)e^{ik\xi}d\xi = -\int_{l_4} f(\xi)e^{ik\xi}d\xi,$$

hence by the Cauchy theorem

$$|c_k| = \left|\frac{1}{2\pi}\int_0^{2\pi} f(x)e^{ikx}dx\right| = \left|\frac{1}{2\pi}\int_0^{2\pi} f(x-ih')e^{ikx}e^{-kh'}dx\right| \leqslant$$
$$\leqslant Ce^{-|k|h'} \leqslant Ce^{-|k|h},$$

so that $h' > 0$ is an arbitrary number less than h. The case $k < 0$ is similar.

Using (23), let us rewrite (22) in the form

$$f(x) = \sum_{|k| \leqslant n} c_k e^{ikx} + R_n(x),$$

$$|R_n(x)| \leqslant | \sum_{|k| > n} c_k e^{ikx} \leqslant 2C e^{-nh} \frac{e^{-h}}{1 - e^{-h}}.$$

$$(24)$$

Suppose $\varepsilon > 0$, choose the minimal n such that $|R_n(x)| < \varepsilon/2$. It follows from (24) that

$$n = \frac{\log(1/\varepsilon)}{h \log e} + O(1).$$

$$(25)$$

Let us approximate, further, each coefficient $c_k = \alpha_k + i\beta_k$ in absolute value with precision $\varepsilon/2(2n+1)$ by the expression $c'_k = \alpha'_k + i\beta'_k$, putting to this end

$$\alpha'_k = \left[\frac{2\sqrt{2}(2n+1)\alpha_k}{\varepsilon} \right] \frac{\varepsilon}{2\sqrt{2}(2n+1)} = m_k^1 \frac{\varepsilon}{2\sqrt{2}(2n+1)},$$

$$\beta'_k = \left[\frac{2\sqrt{2}(2n+1)\beta_k}{\varepsilon} \right] \frac{\varepsilon}{2\sqrt{2}(2n+1)} = m_k^2 \frac{\varepsilon}{2\sqrt{2}(2n+1)}.$$

$$(26)$$

Here obviously we shall get

$$\left| f(x) - \sum_{|k| \leqslant n} c'_k e^{ikx} \right| \leqslant \varepsilon.$$

$$(27)$$

It is clear from (27) that if two functions f and g correspond to the same family $\{c'_k\}$, then $||f - g|| \leqslant 2\varepsilon$, i.e., the set of elements in an ε-covering may be estimated by the number of all families "induced" from the space $A_h(C)$.

According to (26), each set of 2ε-coverings is defined by the following matrix with integer entries:

$$U = \begin{pmatrix} m_{-n}^1 & \cdots & m_0^1 & \cdots & m_n^1 \\ m_{-n}^2 & \cdots & m_0^2 & \cdots & m_n^2 \end{pmatrix}.$$

Now, using (23), we obtain inequalities for m_k^i:

$$|m_k^i| \leqslant \frac{2\sqrt{2}C(2n+1)e^{-|k|h}}{\varepsilon} = N_k, \qquad i = 1, 2,$$

$$(28)$$

which in turn imply that

$$\log N_k = \log(1/\varepsilon) - |k|h \log e + \log n + D,$$

$$(29)$$

where D is a bounded constant independent of k. It follows from (28) that the number of all possible matrices U is no greater than

$$N = \prod_{k=-n}^{n} N_k^2,$$

and therefore, recalling (25), we get

$$\mathcal{H}_\varepsilon(A) \leqslant \log N = 2 \sum_{|k| \leqslant n} \log N_k \leqslant 4n \log \frac{1}{\varepsilon} - 2n^2 h \log e +$$

$$+ O(n \log n) = 2 \frac{(\log(1/\varepsilon))^2}{h \log e} + O\left(\log \frac{1}{\varepsilon} \log \log \frac{1}{\varepsilon}\right).$$

Now let us carry out an estimate from below for $\mathcal{E}_{2\varepsilon}(A_h(0))$. We shall use two considerations in doing this.

α) For each $h' > h$ the inequality

$$|c_k| \leqslant C \frac{1 - e^{-(h'-h)}}{1 + e^{-(h'-h)}} e^{-|k|h'} \tag{30}$$

implies that the function

$$f(z) = \sum_{k=-\infty}^{+\infty} c_k e^{ikz} \tag{31}$$

belongs to the space $A_h(C)$, since then the series (31) uniformly converges in a strip of width h, and if $|\text{Im}z| < h$ we have

$$|f(z)| \leqslant \sum_{k=-\infty}^{+\infty} |c_k| e^{|k|h} \leqslant C \frac{1 - e^{-(h'-h)}}{1 + e^{-(h'-h)}} \sum_{k=-\infty}^{+\infty} e^{-|k|(h'-h)} = C. \tag{32}$$

β) $\|f\| = \max_{x \in D} |f(x)| \geqslant \max_k |c_k|$, where c_k are the coefficients in (31). The latter follows from the fact that

$$|c_k| = \left| \frac{1}{2\pi} \int_0^{2\pi} f(x) e^{ikx} dx \right| \leqslant \max_{x \in D} |f(x)|.$$

Choosing $\varepsilon > 0$ and putting $h' = (1 + (\log(1/\varepsilon))^{-1})$, let us choose the largest m such that

$$C \frac{1 - e^{-(h'-h)}}{1 + e^{-(h'-h)}} e^{-mh'} \geqslant 2\sqrt{2}\varepsilon,$$

which implies

$$m = \frac{\log(1/\varepsilon)}{h \log e} + O(\log \log \frac{1}{\varepsilon}). \tag{33}$$

Now construct a set $\{\Phi\}$ of polynomials of degree m

$$\varphi(z) = \sum_{|k| \leqslant m} c_k e^{ikz}, \quad c_k = (s_k^1 + is_k^2)2\varepsilon, \tag{34}$$

where s_k^j $(j = 1, 2)$ are arbitrary integers such that

$$|s_k^j| \leqslant \left[\frac{C(1 - e^{-(h'-h)})}{(1 + e^{-(h'-h)})} \frac{e^{-|k|h'}}{2\sqrt{2}\varepsilon} \right] = M_k. \tag{35}$$

The inequality (35) guarantees that inequality (30) is valid; thus $\{\Phi\} \subset A_h(C)$. Moreover, for distinct polynomials φ_1 and φ_2 from $\{\Phi\}$, we have by construction for at least one k, satisfying $|k| \leqslant m$, the inequality

$$|C_k^1 - C_k^2| \geqslant 2\varepsilon,$$

i.e. all the polynomials from $\{\Phi\}$ are 2ε-distinguishable.

According to (34), each polynomial from $\{\Phi\}$ is determined by an integer matrix

$$U = \begin{pmatrix} s^1_{-m} & \cdots & s^1_0 & s^1_m \\ s^2_{-m} & \cdots & s^2_0 & s^2_m \end{pmatrix},$$

while the number of all matrices U is no less than

$$M = \prod_{|k| \leqslant m} M_k^2.$$

It also follows from (35) that

$$\log M_k = \log \frac{1}{\varepsilon} - h|k| \log e + O\left(\log\log \frac{1}{\varepsilon}\right)$$

uniformly with respect to k, hence, using (33), we get

$$\mathcal{E}_{2\varepsilon}(A_h(C)) \geqslant \log M \geqslant 2 \sum_{|k| \leqslant m} \log M_k$$

$$\geqslant 4m \log \frac{1}{\varepsilon} - 2m^2 h \log e + O(\log \frac{1}{\varepsilon} \log\log \frac{1}{\varepsilon})$$

$$= 2\frac{(\log(1/\varepsilon))^2}{h \log e} + O\left(\log \frac{1}{\varepsilon} \log\log \frac{1}{\varepsilon}\right).$$

This concludes the proof of formula (21).

§3. Typical orders of increase of the functions $\mathcal{H}_\varepsilon(A)$ and $\mathcal{E}_\varepsilon(A)$.

The functions $f(\varepsilon)$ that we shall consider further will usually be positive and defined for all ε in the interval $0 < \varepsilon < \varepsilon_0$. Since we shall always be studying their limiting behavior as $\varepsilon \to 0$, the sign of $\varepsilon \to 0$ will often be omitted. To describe the limiting behavior of the function $f(\varepsilon)$, besides the usual notation O and o, we shall also use the following notations:

$f \sim g$ if $\lim(f/g) = 1$,

$f \overset{\scriptscriptstyle >}{\sim} g,\; g \overset{\scriptscriptstyle <}{\sim} f$ if $\overline{\lim}(g/f) \leqslant 1$,

$f \asymp g$ if $f = O(g)$ and $g = O(f)$,

$f \succeq g\; g \preceq f$ if $f = O(g)$,

$f \succ\!\succ g,\; g \prec\!\prec f$ if $f = o(g)$.

The relation $f \sim g$ is called *strong equivalence*, the relation $f \asymp g$ *weak equivalence* of the functions f and g. The examples considered in §2 give us three typical orders of growth of the functions $\mathcal{H}_\varepsilon$ and $\mathcal{E}_\varepsilon$ as $\varepsilon \to 0$.

I. If A is a bounded set in n-dimensional Euclidean space $\mathbf{D}^n$ possessing inner points, then

$$\mathcal{M}_\varepsilon(A) \asymp \mathcal{N}_\varepsilon(A) \asymp (1/\varepsilon)^n. \tag{36}$$

Formula (36) is valid in any n-dimensional Banach space (see §4). It implies that

$$\mathcal{E}_\varepsilon(A) \sim \mathcal{H}_\varepsilon(A) \sim n\log(1/\varepsilon). \tag{37}$$

Since by (37) the behaviour of $\mathcal{H}_\varepsilon$ and $\mathcal{E}_\varepsilon$ is determined first of all by the dimension n of the space $\mathbf{D}^n$, the natural idea to define the lower and upper metric dimension for an arbitrary completely bounded set A by the formulas

$$\underline{\mathrm{dm}}(A) = \varliminf \mathcal{H}_\varepsilon(A)/\log(1\varepsilon), \tag{38}$$

$$\overline{\mathrm{dm}}(A) = \varlimsup \mathcal{H}_\varepsilon/\log(1/\varepsilon). \tag{39}$$

arises. It is easy to deduce from Theorem IV in §1 that the formulas

$$\underline{\mathrm{dm}}(A) = \varliminf \mathcal{E}_\varepsilon(A)/\log(1\varepsilon),$$

$$\overline{\mathrm{dm}}(A) = \varlimsup \mathcal{E}_\varepsilon/\log(1/\varepsilon)$$

determine the same expressions $\underline{\mathrm{dm}}(A)$ and $\overline{\mathrm{dm}}(A)$. If

$$\underline{\mathrm{dm}}(A) = \overline{\mathrm{dm}}(A) = \mathrm{dm}(A),$$

then $\mathrm{dm}(A)$ is simply called the *metric dimension* of the set A. Obviously in our case the completely bounded sets A in $\mathbf{D}^n$ possessing interior points will satisfy

$$\mathrm{dm}(A) = n.$$

It is easy to prove (see Theorem XII in §4) that for a convex infinite-dimensional set A in Banach space, the metric dimension is always equal to $+\infty$, i.e. the order of growth of the functions $\mathcal{H}_\varepsilon(A)$ and $\mathcal{E}_\varepsilon(A)$ is superior to $\log(1/\varepsilon)$:

$$\mathcal{H}_\varepsilon(A) \succ\!\!\succ \log(1/\varepsilon), \ \ \mathcal{H}_\varepsilon(A) \succ\!\!\succ \log(1/\varepsilon).$$

In order to distinguish how massive sets of this type are, their metric dimension is useless.

II. Among infinite-dimensional sets important for analysis, the least mass is possessed by typical sets for absolutely bounded analytic functions, such as the set $A_h(C)$ considered in Example 3 in §2. The typical order of growth of functions $\mathcal{H}_\varepsilon$ and $\mathcal{H}_\varepsilon$ here is their growth according to the rule

$$\mathcal{E}_\varepsilon \asymp \mathcal{H}_\varepsilon \asymp (\log(1/\varepsilon))^s \tag{40}$$

with a certain finite constant $s > 1$ or intermediate orders of growth such as

$$\frac{(\log(1/\varepsilon))^s}{\log\log(1/\varepsilon)} \tag{41}$$

(the last function grows slower than $(\log(1/\varepsilon))^s$ but faster than $(\log(1/\varepsilon))^{s-\delta}$ for any $\delta > 0$). The simplest numerical characteristic of such orders of growth is[5]

$$df = \lim \frac{\log \mathcal{H}_\varepsilon}{\log \log(1/\varepsilon)}.$$

In §7 we shall see that in many important and sufficiently general cases of sets of analytic functions, $df(A)$ coincides with the number of independent variables or, geometrically, for classes of functions $f(P)$ of a point P of a complex manifold, with the complex dimension of this manifold. We shall agree to say, somewhat freely, that $df(A)$ is the *functional dimension* of the set A.

III. The examples from subsections 2 and 3 in §2 show that the growth of $\mathcal{H}_\varepsilon$ and $\mathcal{E}_\varepsilon$ for functions that only satisfy the Lipschitz condition is considerably greater than the orders of growth for the classes of analytic functions indicated above. It turns out that the *orders of growth*

$$\mathcal{E}_\varepsilon \asymp \mathcal{H}_\varepsilon \asymp (1/\varepsilon)^q \tag{42}$$

with finite positive q are typical for classes of functions which are differentiable a finite number of times p and possess derivatives of highest order p satisfying the Hölder condition of order α. Here for functions in n variables we obtain the formula

$$\mathcal{E}_\varepsilon \asymp \mathcal{H}_\varepsilon \asymp (1/\varepsilon)^{n/(p+\alpha)} \tag{43}$$

(see §5). Very roughly, formula (43) may be explained as follows: the "mass" of the set of p times differentiable functions in n variables, whose pth derivatives satisfy the Hölder condition of order α, is determined by the relation

$$q = n/(p+\alpha) \tag{44}$$

between the *number of independent variables n* and the *"smoothness index"* $p+\alpha$. Concerning the origin of these important relations (43), (44), see Appendix I. The general definition of the *metric order* of a completely bounded set A is given by the formula[6]

$$q(A) = \lim \frac{\log \mathcal{H}_\varepsilon(A)}{\log(1/\varepsilon)}. \tag{45}$$

IV. Considerably higher orders of growth of $\mathcal{H}_\varepsilon$ and $\mathcal{E}_\varepsilon$ than in (42) arise when we consider functionals of a certain smoothness given on infinite-dimensional completely bounded sets (see §9).

---

[5]The reader will have no difficulty in writing out the definition of the lower functional dimension $\underline{df}(A)$ and the upper functional dimension $\overline{df}(A)$. Using functions from $\mathcal{H}_\varepsilon$ instead of those from $\mathcal{E}_\varepsilon$, and Theorem IV from §1, we get the same $df(A), \underline{df}(A), \overline{df}(A)$.

[6]Here it is also natural to introduce, in the expected way, the notion of lower metric order $\underline{q}(A)$ and upper metric order $\bar{q}(A)$. The use of the functions $\mathcal{H}_\varepsilon$ instead of the functions $\mathcal{E}_\varepsilon$ does not change the values of $q(A), \bar{q}(A), \underline{q}(A)$.

§4. ε-entropy and ε-capacity in finite-dimensional spaces.

We shall consider n-dimensional space in the form of n-dimensional coordinate space $\mathbf{D}^n$ with points

$$x = (x_1, \ldots, x_n),$$

whose coordinates $x_1, \ldots, x_n$ are real numbers, although essentially the relations which interest us in most cases are not related to the choice of coordinates. The choice of a coordinate system involves a certain auxiliary apparatus, for example, the special norm

$$\|x\|_0 = \max_{1 \leqslant k \leqslant n} |x_k|.$$

The unit ball in this norm, i.e. the set S_0 of points $x \in \mathbf{D}^n$ such that

$$\|x\|_0 \leqslant 1,$$

is called the *unit cube*. Multiplying by $b > 0$ and shifting the unit cube by a, we obtain the cube

$$S_0(a, b) = bS_0 - a;$$

by $2b$ we denote the diameter of the cube $S_0(a, b)$.

The set of points x_i whose coordinates are of the form $(dm_1^i, \ldots, dm_n^i)$, where m_k^i $(k = 1, 2, \ldots, n)$ are all possible integers, will be called a *cubic lattice of diameter* d.

It is well known that the point x is called an *inner point* of the set $A \subset \mathbf{D}^n$ if A contains a certain cube $S_0(a, b)$ of center $a \in A$. The sets consisting only of inner points are called *open*, and this determines the usual topology in $\mathbf{D}^n$.

The set $A \subset \mathbf{D}^n$ is called *convex* if $x \in A$, $y \in A$, $\alpha + \beta = 1$, $\alpha \geqslant 0$, $\beta \geqslant 0$ imply $\alpha x + \beta y \in A$. A closed convex set possessing at least one inner point will be called a *convex body*.

Any bounded convex body symmetric with respect to the origin may be chosen to be the unit sphere. The norm $\|x\|_S$ corresponding to a convex symmetric body S is determined by the formula

$$\|x\|_S = \inf \left\{ b \mid \frac{x}{b} \in S, b > 0 \right\}.$$

In its turn, the norm $\|x\|$ can be used to define the unit ball by means of the formula

$$S = \{x \mid \|x\| \leqslant 1\}.$$

Thus we have defined an arbitrary Banach metric in $\mathbf{D}^n$. The space $\mathbf{D}^n$ with metric generated by a symmetric convex bounded body S is denoted by D_S^n. In the cases

$$\|x\|_p = \left(\sum_{k=1}^n |x_k|^p \right)^{1/p}, \quad p \geqslant 1,$$

the unit balls are denoted by S_p^n and the corresponding space by $\mathbf{D}_p^n$. The space $\mathbf{D}_2^n$ is ordinary Euclidean (or "Cartesian") coordinate space.

It is well known that any Banach metric defines the same topology in $\mathbf{D}^n$. The notion of volume and measure of the set in $\mathbf{D}^n$ also does not essentially depend on the choice of the metric. In the standard theory of Lebesgue measure μ^n, the measure $1/2S_0$ is taken to be equal to 1. If for the unit of measure we take the volume of $1/2S$ in some different metric, then the ordinary way of constructing Lebesgue measure theory will yield the measure

$$\mu_S^n(A) = \mu^n(A)/\mu^n(1/2S),$$

which differs from μ^n only by a constant factor.

The set A is Jordan measurable if it is Lebesgue measurable and its boundary has measure equal to zero. The measure $\mu^n(A)$ of such a set will be called its *"volume"* (if the measure of the boundary is not equal to zero a set will not be assigned any volume).

It is easy to understand that the question of finding minimal ε-nets in the space $\mathbf{D}_S^n$ reduces to the problem of finding the "most economical covering" of subsets in the space $\mathbf{D}^n$ by shifted solids εS, while the problem of finding maximal 2ε-distinguishable sets reduces to the problem of finding the most advantageous disposition in $\mathbf{D}^n$ of shifted non-intersecting bodies εS with centres in the given set.

There is a vast literature on this question. A fairly complete bibliography can be found by the reader in Toth's book [19][7].

In the present section we shall only deal with the most general and simplest results about estimates of the functions $\mathcal{H}_\varepsilon(A)$ and $\mathcal{E}_\varepsilon(A)$ for $A \subset \mathbf{D}_S^n$.

Theorem VIII. *For any bounded set $A \subset \mathbf{D}_S^n$ there exists a constant $\alpha > 0$ such that for all $\varepsilon \leqslant \varepsilon_1$ we have*

$$\mathcal{N}_\varepsilon(A) \leqslant \alpha(1/\varepsilon)^n. \tag{46}$$

If A has an inner point in $\mathbf{D}^n$, then there exists a constant $\beta > 0$ such that $\varepsilon \leqslant \varepsilon_2$ implies

$$\mathcal{M}_{2\varepsilon}(A) \geqslant \beta(1/\varepsilon)^n. \tag{47}$$

In the notations of §3, Theorem VIII means that for bounded set in $\mathbf{D}_S^n$ possessing an inner point we have

$$\mathcal{N}_\varepsilon(A) \asymp \mathcal{M}_\varepsilon(A) \asymp (1/\varepsilon)^n. \tag{48}$$

We have mentioned this theorem previously in §3 (see (36)).

Theorem IX. *For $\mathbf{D}_S^n$ there exist constants $\theta = \theta(n, S)$ and $\tau = \tau(n, S)$ such that for any set A of volume $\mu^n(A) > 0$ we have*

$$\mathcal{N}_\varepsilon(A) \sim \theta \frac{\mu^n(A)}{\mu^n(S)} \left(\frac{1}{\varepsilon}\right)^n, \quad \mathcal{M}_{2\varepsilon}(A) \sim \tau \frac{\mu^n(A)}{\mu^n(S)} \left(\frac{1}{\varepsilon}\right)^n. \tag{49}$$

It is natural to call the constant θ the minimal density of the covering $\mathbf{D}^n$ by the body S and the constant τ the "maximal density of disposition of the solid S in $\mathbf{D}^n$".

---

[7] Many essential facts will be found by the reader in the commentary to this book by I. M. Yaglom.

For θ and τ we have the obvious inequalities

$$\tau \leqslant 1 \leqslant \theta.$$

Proof of Theorem VIII. We choose constants b and B so as to have simultaneously

$$bS_0 \subset S \subset BS_0, \quad A \subset BS_0.$$

Suppose $\varepsilon > 0$; in order to cover the cube BS_0, we require no more than

$$(B/\varepsilon b + 2)^n \tag{50}$$

cubes $\varepsilon bS_0 + a^i = S_0(a^i, \varepsilon b)$, where a^i has coordinates $2b\varepsilon(m_1^i, m_2^i, \ldots, m_n^i)$, the m_j^i are integers, less in absolute value than $[B/2\varepsilon b] + 1$. If we include the cubes $\varepsilon bS_0 + a^i$ into solids of the form $\varepsilon S + a^i$, we obtain an ε-covering of A, and therefore

$$\mathcal{N}_\varepsilon(A) \leqslant (B/\varepsilon b + 2)^n,$$

which implies (46).

Suppose further that A has an inner point a. Let us choose a c such that $S_0(a, c) \subset A$. The cube $S_0(a, c)$ contains more than $(2[c/2\varepsilon B])^n$ cubes $S_0(a^i, \varepsilon B)$, where the a^i are vectors $2\varepsilon B(m_1^i, \ldots, m_n^i)$ and the m_j^i are integers in absolute value no greater than $[c/2\varepsilon B]$. The set a^i is 2ε-distinguishable, and this implies (47).

Proof of Theorem IX. Denote[9]

$$\lim_{\varepsilon \to 0} \mathcal{N}_\varepsilon(S_0)\mu^n(S)\left(\frac{\varepsilon}{2}\right)^n = \lim_{\varepsilon \to 0} \mathcal{N}_\varepsilon(S_0)\frac{\mu^n(S)}{\mu^n(S_0)}\varepsilon^n$$

by θ. Let us show that the following limit actually exists

$$\lim_{\varepsilon \to 0} \mathcal{N}_\varepsilon(S_0)\mu^n(S)(\varepsilon/2)^n = \theta.$$

Suppose $\delta > 0$. Choose an ε_1 such that $\varepsilon \leqslant \varepsilon_1$ implies

$$\theta - \delta \leqslant \mathcal{N}_\varepsilon(S_0)\mu^n(S)(\varepsilon/2)^n,$$

while

$$\mathcal{N}_{\varepsilon_1}(S_0)\mu^n(S)(\varepsilon_1/2)^n \leqslant \theta + \delta/2.$$

The cube S_0 may be covered by no more than

$$(1/\lambda + 2)^n$$

cubes $S_0(a^i, \lambda)$ for any λ, $0 < \lambda < 1$ (see (50)), while each of the latter is covered by $\mathcal{N}_{\varepsilon_1}$ solids $\varepsilon\lambda S$.

---

[9]Recall that S_0 is the cube $\|x_0\| \leqslant 1$; its measure is equal to 2^n.

Thus

$$\mathcal{N}_{\varepsilon_1 \lambda}(S_0) \leqslant (1/\lambda + 2)^n \mathcal{N}_{\varepsilon_1}(S_0),$$

i.e.

$$\mathcal{N}_{\varepsilon_1 \lambda}(S_0)\mu^n(S)(\frac{\varepsilon_1}{2})^n \lambda^n \leqslant \lambda^n(\frac{1}{\lambda} + 2)^n \mathcal{N}_{\varepsilon_1}(S_0)(\frac{\varepsilon_1}{2})^n \mu^n(S).$$

Choosing λ_1 so that $\lambda \leqslant \lambda_1$ implies

$$\lambda^n(1/\lambda + 2)^n(\theta + \delta/2) \leqslant \theta + \delta,$$

we obtain (for all $\lambda \leqslant \lambda_1$) the relation

$$\theta - \delta \leqslant \mathcal{N}_{\varepsilon_1 \lambda}(S_0)\mu^n(S)(\varepsilon_1 \lambda/2)^n \leqslant \theta + \delta, \tag{51}$$

i.e.,

$$\lim_{\varepsilon \to 0} \mathcal{N}_\varepsilon(S_0)\mu^n(S)(\varepsilon/2)^n = \theta.$$

Now, further, suppose A is an arbitrary solid of volume $\mu^n(A) > 0$, Π_{2d} is the cubic lattice of diameter $2d$. Choosing $\delta' > 0$, let us find a d so that the sum of measures of the cubes $S_0'(\bar{a}^i, d)$ $(i = 1, \ldots, m)$, $\bar{A}^i \in \Pi_{2d}$, possessing common points with the boundary of A will be less than δ'. Suppose $a^i(i = 1, 2, \ldots, k)$ are the centres of the cubes $S(a^i, d)$ each of which is entirely contained inside A. Then of course

$$k \geqslant (\mu^n(A) - \delta')/(2d)^n. \tag{52}$$

Choosing an arbitrary $d' < d$, let us consider the set of cubes $S_0(A^i, d')$ (Figure 6).

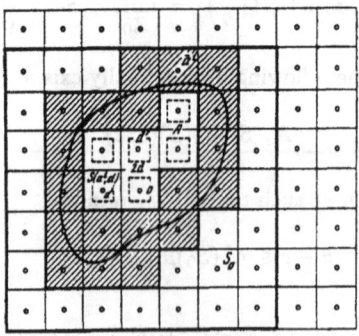

FIGURE 6

According to (51), for $\varepsilon \leqslant \varepsilon_1 \lambda_1$, we have

$$\theta - \delta \leqslant \mathcal{N}_{\varepsilon d'}(d'S_0)\mu^n(S)(\varepsilon/2)^n \leqslant \theta + \delta.$$

Let us choose ε_2 so small as to have $2\varepsilon d' < d - d'$ as soon as $\varepsilon \leqslant \varepsilon_2$. Then obviously

$$\mathcal{N}_{\varepsilon d'}(A) \geqslant k\mathcal{N}_{\varepsilon d'}(d'S_0),$$

since the coverings of adjacent cubes do not intersect, and for $\varepsilon \leqslant \varepsilon_2$, we have

$$\mathcal{N}_{\varepsilon d'}(A)(\frac{\varepsilon}{2})^n \mu^n(S) \geqslant \frac{\mu^n(A) - \delta'}{(2d)^n}(\theta - \delta),$$

so that

$$\lim_{\varepsilon \to 0} \mathcal{N}_\varepsilon(A)\varepsilon^n \mu^n(S) \geqslant \theta\mu^n(A)$$

since δ' and δ were arbitrary and $d' < d$. On the other hand,

$$\mathcal{N}_{\varepsilon d}(A) \leqslant (k+m)\mathcal{N}_{\varepsilon d}(dS_0) \leqslant \frac{\mu^n(A) + \delta'}{(2d)^n}(\theta + \delta)\frac{1}{\mu^n(S)}\left(\frac{2}{\varepsilon}\right)^n,$$

i.e.,

$$\overline{\lim_{\varepsilon \to 0}} \, \varepsilon^n \mu^n(S)\mathcal{N}_\varepsilon(A) \leqslant \theta\mu^n(A), \tag{53}$$

which was to be proved.

The reader is asked to carry out the proof of the theorem for $\mathcal{M}_{2\varepsilon}$ on his own.

Besides Theorems VIII and IX, let us note the following very rough inequality, which estimates θ and τ from above and from below.

Theorem X. *For any* $\mathbf{D}_S^n$ *we have*

$$\theta/\tau \leqslant 2^n, \tag{54}$$

and therefore

$$2^{-n} \leqslant \tau \leqslant 1 \leqslant \theta \leqslant 2^n. \tag{55}$$

Choosing $\delta > 0$, let us find an ε_0 so that for a certain A of volume $\mu^n(A) > 0$ we simultaneously have the inequalities

$$\varepsilon^n \mu^n(S)\mathcal{N}_\varepsilon(A) \geqslant (\theta - \delta)\mu^n(A),$$
$$\varepsilon^n \mu^n(S)\mathcal{M}_{2\varepsilon}(A) \leqslant (\tau + \delta)\mu^n(A),$$

when $\varepsilon \geqslant \varepsilon_0$. Further, using the inequality $\mathcal{N}_\varepsilon(A) \leqslant \mathcal{M}_\varepsilon(A)$ from Theorem IV in §1, we obtain

$$\theta - \delta \leqslant (\tau + \delta)2^n,$$

which implies Theorem X.

Note that while the *existence* of the constants τ and θ (Theorem IX) is very easy to prove, their computation or even their estimate for various specific spaces often turn out to be very difficult problems.

A trivial case is the one of the space $\mathbf{D}_\infty^n$, since the unit sphere in this space is the unit cube S_0 and the entire space may be decomposed into unit cubes without overlapping, so that the distance between the centres of the cubes is equal to the diameter of the cubes (the space is "decomposable"). In any decomposable space, as can be easily seen, we have

$$\theta(n, S) = \tau(n, S) = 1. \tag{56}$$

Let us indicate two more examples of decomposable spaces: the space $\mathbf{D}_1^2$ and the space $\mathbf{D}_S^2$, where S is the regular hexagon. In the first space the sphere is a square and in the second the hexagon can be used to construct a paving of the entire plane.

(Note that this decomposability effect appeared in an infinite-dimensional space – the space $F_1^\Delta(L)$; §1, Subsection 2.)

The values of θ and τ were calculated in the case of the space $\mathbf{D}_2^2$ [19]. It turns out that the best covering of the plane by disks can be carried out as follows. The plane is covered by regular hexagons and then a circle is circumscribed to each of them. The best disposition of circles on the plane which do not intersect each other is obtained if a circle is inscribed in each hexagon.

It is easy to calculate that we then have

$$\tau = \frac{\pi}{\sqrt{12}} = 0,9069\dots, \quad \theta = \frac{2\pi}{\sqrt{27}} = 1,2092\dots$$

Already in the case of three-dimensional space, the problem of finding the densest filling of the space by balls and the related problem of computing the constants τ and θ for the space $\mathbf{D}_2^3$ has not been entirely solved yet. And of course the values of τ_n and θ_n have not been computed (this is the notation for τ and θ in the space $\mathbf{D}_2^n$). The best estimates for θ_n at the present time were obtained by Davenport [20] and Watson [21].

Before we state this result, let us introduce a definition. Let R be a lattice in n-dimensional space such that the spheres $S_i = S(a^i, 1)$, $a^i \in R$ cover the entire space $\mathbf{D}^n$. Suppose $\sum_\lambda$ denotes the sum over those i for which $a^i \in R \cap \lambda S$. The expression

$$\lim_{\lambda \to \infty} \left(\sum_\lambda \mu^n(S_i)/\mu^n(\lambda S) \right),$$

in the case when the limit exists, is naturally called the density of the covering of the space $\mathbf{D}^n$ by the solid S with centres in R. Denote this expression by $\theta_R(n, S)$. The value of $\theta_R(n, S)$, as can be easily seen, estimates that of $\theta(n, S)$ from above:

$$\theta(n, S) \leqslant \theta_R(n, S). \tag{57}$$

Davenport and Watson constructed a lattice R in D^n such that

$$\theta_R(N, S_2^n) \leqslant (1,07)^n, \tag{58}$$

and therefore

$$\theta_n(1,07)^n. \tag{59}$$

The theorems of Davenport and Watson are based on delicate construction in n-dimensional space; we do not present them here, referring the reader to the corresponding literature.

Estimates of the value of τ_n may be obtained from an older paper by Blichfeldt [22].

Its basic ingredient is the following simple

Lemma I. *(Blichfeldt) Suppose we are given m points in Euclidean space whose pairwise distance is no less than 2. Then the sum of squares of their distances to any point of space is no less than $2(m-1)$.*

Proof. Obviously it suffices to consider the case when the given point of space is the origin. Choose an orthonormed basis, denoting the coordinates of the i-th point by $(x_1^i, \ldots, x_n^i)$. By assumption,

$$\sum_{k=1}^{n}(x_k^i - x_k^j)^2 \geqslant 4, \; i \neq j. \tag{60}$$

Adding (60) for $i \neq j$, we obtain

$$m\sum_{i=1}^{m}((x_1^i)^2 + \cdots + (x_n^i)^2) - (x_1^1 + \cdots + x_1^m)^2 -$$
$$- (x_n^1 + \cdots + x_n^m)^2 \geqslant 2m(m-1),$$

hence

$$\sum_{k=1}^{n}\sum_{i=1}^{m}(x_k^i)^2 \geqslant 2(m-1),$$

which was to be proved.

The lemma proved above immediately implies that if $n+1$ points of Euclidean space have pairwise distances no less than 2, then the radius of the ball containing all these points is no less than $\sqrt{2n/(n+1)}$.

Suppose we are given an arbitrary $\varepsilon\sqrt{2n/(n+1)}$-net. No more than $n+1$ points with pairwise distance 2ε can be placed inside each sphere. Hence for every solid A it is easy to obtain

$$\mathcal{M}_{2\varepsilon}(A) \leqslant (n+1)\mathcal{N}_{\varepsilon\sqrt{2n/(n+1)}}(A).$$

Now (59) and (55) immediately imply that

$$2^{-n} \leqslant \tau_n \leqslant (n+1)\theta_n\left(\frac{n+1}{2n}\right)^{n/2} \leqslant (n+1)\sqrt{e}2^{-n/2}(1,07)^n \leqslant$$
$$\leqslant (n+1)\sqrt{e}\left(\frac{11}{14}\right)^n. \tag{61}$$

Concluding this section, let us show that its main results do not admit meaningful generalizations to infinite-dimensional Banach spaces. Bounded sets in infinite-dimensional space are no longer necessarily (unlike the case of D^n) completely bounded. But even for completely bounded sets in infinite-dimensional space, it is impossible to obtain any universal (valid for all completely bounded sets in the given infinite-dimensional space E) estimate from above for the growth of $\mathcal{H}_\varepsilon$ and $\mathcal{E}_\varepsilon$ (in D^n we have seen that they are always $\preceq \varepsilon^{-n}$). This is obvious from the following theorem:

Theorem XI. *Suppose $\varphi(\varepsilon)$, being monotone increasing, tends to $+\infty$ as $\varepsilon \to 0$ and the Banach space E is infinite-dimensional. Then it is possible to find a compact set K in E for which*

$$\mathcal{N}_\varepsilon(K) \succeq \varphi(\varepsilon), \quad \mathcal{M}_\varepsilon(K) \succeq \varphi(\varepsilon).$$

Obviously, by Theorem IV, §1 it suffices to prove Theorem XI for $\mathcal{M}_\varepsilon$. Consider a sequence of linear subspaces in E

$$E_1 \subset E_2 \subset \cdots \subset E_n \subset \ldots, \quad \dim E_n = n.$$

Let us show that there exists a sequence of points

$$x_n \in E_n$$

with the property

$$\|x_n\| = 1, \quad \rho(x_n, E_{n-1}) = 1.$$

To do that, let us choose a point y_n not belonging to E_{n-1} in each E_n. Let us find (this is possible, [23, p. 16]) in E_n one of the "nearest" points z_n to y_n, i.e., a point z_n such that

$$\|y_n - z_n\| = \rho(y_n, E_{n-1}) = \rho_n,$$

and put

$$x_n = \rho_n^{-1}(y_n - z_n).$$

It is easy to see that the x_n possess the required property.

Now let us introduce the function $\psi(n)$ inverse to the function $\varphi(n)$. Construct the set K from the point O and the points

$$\xi_n = 2\psi(n)x_n, \quad n = 1, 2, \ldots$$

The fact that K is compact is obvious since

$$\|\xi_n\| = 2\psi(n) \to 0 \text{ as } n \to \infty.$$

For any ε, taking

$$n_1 = [\varphi(\varepsilon)]$$

we obtain

$$\psi(n_1) \geqslant \varepsilon,$$

which implies that the set of points

$$0, \xi_1, \ldots, \xi_{n_1}$$

is ε-distinguishable in E. Since the number of elements of this set satisfies

$$n_1 + 1 \geqslant \varphi(\varepsilon),$$

we have

$$\mathcal{M}_\varepsilon(K) \succeq \varphi(\varepsilon).$$

The set A contained in the Banach space E is said to be *n-dimensional*, if it can be placed in a n-dimensional linear subspace but cannot be placed in a linear subspace of lesser dimension. By Theorem VIII, any convex n-dimensional bounded set in any Banach space has metric dimension n and, moreover, for it we have

$$\mathcal{N}_\varepsilon(A) \asymp \mathcal{M}_\varepsilon(A) \asymp \varepsilon^{-n}.$$

The set A in a Banach space is called *infinite-dimensional*, if cannot be placed in any linear subspace of finite dimension. We have the following

Theorem XII. *If the set A in Banach space is infinite-dimensional and convex, then we have*

$$\mathcal{N}_\varepsilon(A) \gg \left(\frac{1}{\varepsilon}\right)^n, \quad \mathcal{M}_\varepsilon(A) \gg \left(\frac{1}{\varepsilon}\right)^n$$

for any n.

Proof. For any $n' > n$, in A there exists $n' + 1$ affinely independent points. The n'-dimensional simplex that they generate, because of the convexity of A, is entirely contained in A, but for such a simplex by Theorem VIII the functions $\mathcal{N}_\varepsilon$ and $\mathcal{M}_\varepsilon$ are of order of growth ε^{-n}.

§5. ε-entropy and ε-capacity of functions of finite smoothness.

Formula (43) may be established without serious additional efforts for functions given on an arbitrary connected compact set K possessing a finite metric dimension

$$\mathrm{dm}(K) = n$$

and contained in a finite-dimensional Banach space. To be more definite, let us suppose that this Banach space is realized coordinate-wise as D^n with norm $\|x\|_0$ (as described in §4). Obviously, we do not always have $n \leqslant N$, and since we assumed that K is compact, we always have $n \geqslant 1$. Actually the most interesting is the simplest case when K is a parallelepipedon of dimension n in D^n.

As usual, in the space D^K of bounded real functions on K, the norm is equal to

$$\|f\| = \sup_{x \in K} |f(x)|. \tag{62}$$

A function in this space is of smoothness $q > 0$ ($q = p + \alpha$, p is an integer, $0 < \alpha \leqslant 1$), if for any vectors $x \in K$ and $x + h \in K$ we have

$$f(x + h) = \sum_{k=0}^{p} \frac{1}{k!} B_k(h, x) + R(h, x), \tag{63}$$

where $B_k(h, x)$ is a homogeneous form with respect to h of degree k and

$$|R(h, x)| \leqslant C\|h\|_0^q, \tag{64}$$

where C is a certain constant.

All functions f satisfying (62) and (63) with a given constant C constitute the class $F_q^K(C)$, while the functions which satisfy (62) and (63) for some constant C (depending on f) form the class F_q^K. If K is an n-dimensional parallelepipedon, then F_q^K is simply the class of functions possessing partial derivatives of all orders $k \leqslant p$ for which the partial derivatives of order p satisfy the Hölder condition of order α.

Theorem XIII. *For any bounded set $A \subset F_q^K(C)$ in the sense of the metric (62), we have the estimates*

$$\mathcal{H}_\varepsilon(A) \preceq \left(\frac{1}{\varepsilon}\right)^{n/q}, \quad \mathcal{E}_\varepsilon(A) \preceq \left(\frac{1}{\varepsilon}\right)^{n/q}, \tag{65}$$

which cannot be improved in the sense that there exists a bounded set A satisfying the relation

$$\mathcal{E}_\varepsilon(A) \asymp \mathcal{H}_\varepsilon(A) \asymp \left(\frac{1}{\varepsilon}\right)^{n/q}. \tag{66}$$

The first part of the theorem already includes the statement (which can be easily proved independently) that a bounded $A \subset F_q^K(C)$ is completely bounded (i.e. has a compact closure in D^K). It is possible, further, to prove that for functions f from a bounded set $A \subset F_q^K$, the forms $B_k(h, x)$ are uniformly bounded

$$|B_k(h, x)| \leqslant C_k \|h\|_0^k, \quad k = 0, 1, \ldots, p, \tag{67}$$

where the constants C_k depend only on A and not on $f \in A$. If we take into consideration this circumstance, it is clear that for the proof of Theorem XIII it suffices to consider the sets

$$F^K(C_0, \ldots, C_q, C),$$

consisting of all functions f satisfying conditions (63), (64), (67) with given constants $C_0, \ldots, C_p, C$. Therefore Theorem XIII may be deduced[9] from the following statement.

Theorem XIV. *If all the constants $C_0, \ldots, C_p, C$ are positive, then*

$$\mathcal{E}_\varepsilon(F_q^K(C_0, \ldots, C_p, C)) \asymp \mathcal{H}_\varepsilon(F_q^K(C_0, \ldots, C_p, C)) \asymp \left(\frac{1}{\varepsilon}\right)^{n/q}. \tag{68}$$

In the one-dimensional case (K is the closed interval $[a, b]$ of the numerical line) Theorem XIV can be stated as follows:

Theorem XV. *Suppose*

$$A = F_q^K(C_0, \ldots, C_p, C)$$

is a set of functions $f(x)$ defined for $x \in [a, b]$ and satisfying the conditions

$$|f^{(k)}(x)| \leqslant C_k, \quad k = 0, \ldots, p, \tag{69}$$

$$|f^{(p)}(x + h) - f^{(p)}(x)| \leqslant C|h|^\alpha, \quad 0 < \alpha \leqslant 1, \tag{70}$$

If the constants $C_0, \ldots, C_p, C$ are positive, then

$$\mathcal{H}_\varepsilon(A) \asymp \mathcal{E}_\varepsilon(A) \asymp \varepsilon^{-1/q}, \quad q = p + \alpha. \tag{71}$$

We present a detailed proof of Theorem XV and indicate what modifications must be carried out in the proof in order to obtain Theorem XIV. By Theorem IV

---

[9] We do not present proofs (not difficult in principle, but rather cumbersome) of the properties of bounded sets $A \subset F_q^K(C)$ used above. The reader who does not succeed in recovering them, must assume proved, among the results of this section, only Theorem XIV, which is the only one used in the sequel.

from §1 it is clear that the conclusion of Theorem XV is equivalent to the following system of asymptotic inequalities

$$\mathcal{H}_{\varepsilon}(A) \preceq \varepsilon^{-1/q}, \tag{72}$$

$$\mathcal{E}_{\varepsilon}(A) \succeq \varepsilon^{-1/q}. \tag{73}$$

Proof of inequality (72). Put

$$\Delta = (\varepsilon/2C)^{1/q} \tag{74}$$

and choose points

$$x_r = x_0 + r\Delta, \ r = 0, \ldots, s,$$

so that the segment $[a, b]$ is contained in the segment $[x_0, x_s]$. It is easy to see that the number s may be chosen of order

$$s \asymp \Delta^{-1} \asymp \varepsilon^{-1/q}. \tag{75}$$

Putting

$$\varepsilon_k = \varepsilon/e\Delta^k \asymp \Delta^{q-k} \asymp \varepsilon^{1-k/q}, \tag{76}$$

$$\beta_r^{(k)}(f) = [f^{(k)}(x_r)/\varepsilon_k], \tag{77}$$

let us assign to each function $f \in A$ the matrix

$$\beta = \|\beta_r^{(k)}(f)\|, \ k = 0, \ldots, p; \ r = 0, \ldots, s,$$

consisting of integers $\beta_r^{(k)}(f)$. The set of functions $f \in A$ with fixed matrix β is denoted by U_β. Let us now prove that the diameter of each set U_β is no greater than 2ε. Indeed, putting

$$g = f_1 - f_2, \ f_1 \in U_\beta, \ f_2 \in U_\beta,$$

we obtain, according to (77),

$$|g^{(k)}(x_r)| = |f_1^{(k)}(x_r) - f_2^{(k)}(x_r)| \leqslant \varepsilon_k, \tag{78}$$

and by (70) and (74) we get

$$|g^{(p)}(x+h) - g^{(p)}(x)| \leqslant 2C|h|^{\alpha}. \tag{79}$$

For any point $x \in [a, b]$ let us find x_r, $|h| = |x - x_r| \leqslant \Delta$ and write the Taylor expansion

$$g(x) = \sum_{k=0}^{p} \frac{1}{k!} g^{(k)}\left(x_r\right)h^k + \frac{h^p}{p!}(g^{(p)}(\xi) - g^{(p)}(x_r)), \ |xi - x_r| \leqslant \Delta. \tag{80}$$

From (78)-(80), (76), (74) we obtain

$$|g(x)| \leqslant \sum_{k=0}^{p} \frac{1}{k!} \varepsilon_k \Delta^k + 2C\Delta^{p+\alpha} \leqslant \frac{\varepsilon}{e}\left(1 + \frac{1}{1!} + \cdots + \frac{1}{p!}\right) + \varepsilon < 2\varepsilon.$$

Thus the sets U_β constitute a 2ε-covering of the set A. It remains to calculate the number of N of non-empty sets U_β. To do this, note that the number of possible values of the index $\beta_0^{(k)}$ by (69) and (77) will be of order ε_k^{-1}, while the number N' of various forms of the first row

$$\beta_0^{(0)}, \ldots, \beta_0^{(p)}$$

of the matrix β is

$$N' \asymp (\varepsilon_0 \cdot \ldots \cdot \varepsilon_s)^{-1} \asymp \varepsilon^{-\omega}, \quad \omega = \sum_{k=0}^{p}(1 - \frac{k}{p}) > 0. \tag{81}$$

Now assume that the first r rows of the matrix β are

$$\beta_0^{(0)} \ldots \beta_0^{(p)}$$
$$\ldots$$
$$\beta_r^{(0)} \ldots \beta_r^{(p)}$$

and are fixed, and consider the number of various possible (for non-empty sets U_β) values of the index $\beta_{r+1}^{(k)}$. Since

$$f^{(k)}(x_{r+1}) = f^{(k)}(x_r) + f^{(k+1)}(x_r)\Delta + \cdots + \frac{1}{(p-k)!}f^{(p)}(x_r)\Delta^{p-k} + R_k,$$
$$R_k = \frac{1}{(p-k)!}\Delta^{p-k}(f^{(p)}(\xi) - f^{(p)}(x_r)), \quad |\xi - x_r| \leqslant \Delta, \tag{82}$$

we have by (76)

$$\frac{f^{(k)}(x_{r+1})}{\varepsilon_k} = \frac{f^{(k)}(x_r)}{\varepsilon_k} + \frac{f^{(k+1)}(x_r)}{\varepsilon_{k+1}} + \cdots + \frac{1}{(p-k)!}\frac{f^{(p)}(x_r)}{\varepsilon_p} + \frac{R_k}{\varepsilon_k},$$

where

$$\frac{f^{(m)}(x_r)}{\varepsilon_m} = \beta_r^{(m)} + \tau_r^{(m)}, \quad |\tau_r^{(m)}| \leqslant 1,$$

hence

$$\beta_{r+1}^{(k)} = \left[\frac{f^{(k)}(x_{r+1})}{\varepsilon_k}\right] = \left[\beta_r^{(k)} + \frac{\beta_r^{(k+1)}}{1!} + \cdots + \frac{\beta_r^{(p)}}{(p-k)!} + D\right],$$

so that by (70), 82), (74), (76) we have

$$|D| \leqslant |\tau_r^{(k)}| + \frac{1}{1!}|\tau_r^{(k+1)}| + \cdots + \frac{1}{(p-k)!}|\tau_r^{(p)}| + \frac{|R_k|}{\varepsilon_k} \leqslant$$

$$\leqslant 1 + \frac{1}{1!} + \cdots + \frac{1}{(p-k)!} + \frac{\Delta^{p-k}C\Delta^\alpha}{\varepsilon_k} < 2e.$$

Hence the number of different possible values of β_{r+1}^k is no greater than $4e + 2$, while the number of various possible $p+1$ rows of the matrix β is no greater than

$$\gamma = (4e + 2)^{p+1}.$$

Now it is easy to understand that the total number N of non-empty sets U_β has the estimate

$$N \leqslant N'\gamma^s,$$

from which, by (75) and (81), we obtain

$$\mathcal{H}_\varepsilon(A) \leqslant \log N = \log N' + s \log \gamma \preceq \log \frac{1}{\varepsilon} + \left(\frac{1}{\varepsilon}\right)^{1/q} \asymp \left(\frac{1}{\varepsilon}\right)^{1/q}.$$

i.e. the inequality (72).

Proof of inequality (73). Put

$$\varphi(y) = \begin{cases} a(1+y)^q(1-y)^q, & |y| \leqslant 1, \\ 0, & |y| \geqslant 1. \end{cases}$$

(Figure 7). It is easy to check that $\varphi^{(p)}(y)$ satisfies the Hölder condition of degree α. We can choose $\alpha > 0$ so that this constant will be $\leqslant C$. Take

$$\Delta = (\varepsilon/a)^{1/q}$$

and in the closed interval $[a, b]$ choose the points

$$x_r = x_0 + 2r\Delta, \quad r = 0, 1, \ldots, s.$$

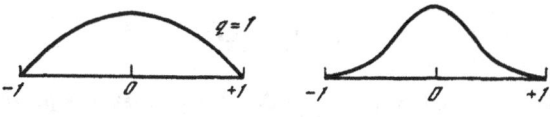

FIGURE 7

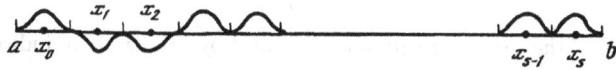

FIGURE 8

It is easy to see that their number may be taken to be of order

$$s \asymp \Delta^{-1} \times \varepsilon^{-1/q}.$$

Consider the set U of functions of the form

$$g(x) = \sum_{r=0}^{s} \gamma_s \Delta^q \varphi\left(\frac{x - x_r}{\Delta}\right), \quad \gamma_r = \pm 1$$

(an example of such a function is shown on Figure 8). It is easy to check that for a sufficiently small ε the function $g(x)$ from U belongs to A. At some point x_r two different functions of this kind have opposite signs and are equal to $a\Delta^q = \varepsilon$ in absolute value. Therefore their distance in the metric (62) is greater than ε, i.e. the set U is ε-distinguishable. The number of elements in the set U is equal to

$$N = 2^{s+1}, \quad \log N = s + 1 \times \varepsilon^{-1/q},$$

which proves inequality (73).

Theorem XV is proved.

The proof of Theorem XIV may be carried out along the same lines. For the set

$$A = F_q^K(C_0, C_1, \ldots, C_p, C)$$

it is sufficient to establish the estimates

$$\mathcal{H}_\varepsilon(A) \preceq \varepsilon^{-n/q}, \tag{83}$$

$$\mathcal{E}_\varepsilon(A) \succeq \varepsilon^{-n/q}. \tag{84}$$

Proof of inequality (83). As in the one-dimensional case, let us set

$$\Delta = (\varepsilon/2C)^{1/q}$$

and in the set K choose points

$$x^0, x^1, \ldots, x^s,$$

constituting a Δ connected Δ-net, i.e. a Δ-net such that it is possible to join any of its points to any other one of its points by a chain of points which are pairwise at a distance of no greater than Δ. Since the given compact set is connected, such will be, as can be easily seen, each $(\Delta/2)$-net. Clearly, s can be chosen of order

$$s \asymp \Delta^{-n} \times \varepsilon^{-n/q}.$$

The role of the matrix β will now be played by the family of indices

$$\beta_r^{(k_1, \ldots, k_n)}, \quad k_1 + k_2 + \cdots + k_n = k,$$

which are the integer parts obtained by dividing ε_k by the coefficients

$$f^{(k_1,\ldots,k_n)}(x^r)$$

appearing in the forms

$$B^{(k)}(x^r, h) = \frac{k_1! k_2! \ldots k_n!}{k!} f^{(k_1,\ldots,k_n)}(x_r) h_1^{k_1} \ldots h_n^{k_n}.$$

The ε_k themselves must be taken equal to

$$\varepsilon_k = \varepsilon / n^k \varepsilon \Delta^k,$$

which does not change their order $\varepsilon_k \asymp \Delta^{q-k}$.

The proof of the fact that the diameter of the sets U_β is no greater than 2ε is quite similar to the one in the one-dimensional case.

To estimate the number N of non-empty set U_β, we must number the points

$$x^0, x^1, \ldots, x^s$$

so that for a point x^{r+1} there always exists, among the points $x^0, \ldots, x^r$, a point x^l satisfying the relation

$$\|x^l - x^{r+1}\| \leqslant \Delta.$$

Under this condition, as before, one proves that the index $\beta_{r+1}^{(k_1,\ldots,k_n)}$ for a fixed set of indices $\beta_l^{(k_1,\ldots,k_n)}$ can assume only a bounded number of values, the bound being given by a constant γ. Denoting by P the number of indices β with the given r, we obtain the estimates[10]

$$N \leqslant N'(\gamma^P)^s, \quad N' \asymp \varepsilon^{-\omega}, \quad \log N \asymp s \asymp \varepsilon^{-n/q},$$

which directly leads to (83).

Proof of (84). Just as in the one-dimensional case, put

$$\varphi(y) = \begin{cases} a \displaystyle\prod_{l=1}^{n}(1+y_l)^q(1-y_l)^q, & \text{if } |y_l| \leqslant 1, \ l = 1,\ldots,m, \\ 0 \quad \text{otherwise.} \end{cases}$$

The function φ possesses partial derivatives of all orders up to p and its partial derivatives of order p satisfy the Hölder condition of degree q. The constant a may be chosen so that the function φ belongs to the class $F_q^K(C)$.

In the compact set K, choose a 2Δ-distinguishable set of points

$$x^0, x^1, \ldots, x^s,$$

_____

[10] N' here is the number of different possible groups of indices $\beta_0^{(k_1,\ldots,k_m)}$.

and, as in the one-dimensional case, put

$$\Delta = (\varepsilon/a)^{1/q}.$$

Obviously s may be taken of order

$$s \asymp \Delta^{-n} \asymp \varepsilon^{-n/q}.$$

Now consider the set U of functions of the form

$$g(x) = \sum_{r=0}^{s} \gamma_r \Delta^q \varphi \left(\frac{x - x^r}{\Delta} \right), \quad \gamma_r = \pm 1.$$

The number of elements in these functions is equal to

$$N = 2^{s+1}, \quad \log N \succeq s \asymp \varepsilon^{-n/q}.$$

As in the one-dimensional case, one proves that U is an ε-distinguishable set in A (if ε is sufficiently small).

Remark 1. For the questions treated in §8, it is of interest to make the estimates in Theorem XV uniform with respect to the length $T = b - a$ of the segment $[a, b]$. Analyzing the proof of Theorem XV, it is easy to notice that for fixed $C_0, C_1, \ldots, C_p, C$ and for $T \geqslant T_0 > 0$, we have uniformly

$$\mathcal{H}_\varepsilon(A) \asymp \mathcal{E}_\varepsilon(A) \asymp T(1/\varepsilon)^{1/q}.$$

Remark 2. In the proof of Theorem XIV, we used the fact that the compact set K is connected only in the proof of inequality (83). Without assuming that K is connected, we must estimate for each r the number of various possible indices $\beta_r^{(k_1, \ldots, k_n)}$ as

$$N' \asymp (1/\varepsilon)^\omega.$$

For N we obtain

$$s \log N' \asymp (1/\varepsilon)^{n/q} \log(1/\varepsilon)$$

and instead of (83) we get

$$\mathcal{H}_\varepsilon(A) \preceq (1/\varepsilon)^{n/q} \log(1/\varepsilon).$$

Instead of Theorem XIV we then get (without assuming that K is connected)

$$\left(\frac{1}{\varepsilon} \right)^{n/q} \preceq \mathcal{E}_\varepsilon(A) \asymp \mathcal{H}_\varepsilon(A) \preceq \left(\frac{1}{\varepsilon} \right)^{n/q} \log(1/\varepsilon).$$

Remark 3. The set F_q^K may be viewed as a linear space. The natural topology in it is determined by the norm

$$\|f\| = \sum_{k=0}^{p} \sup_{x \in K} \|B_f^{(k)}(x)\| + \sup_{x \in K} \|R_f(x)\|,$$

where

$$\|B_f^{(k)}(x)\| = \sup_{x+h\in K} \frac{|B_k(h,x)|}{\|h\|^k},$$

$$\|R_f(x)\| = \sup_{x+h\in K} \frac{|R(h,x)|}{\|h\|^q}.$$

It is easy to obtain the following corollary from Theorem XIV: *for any bounded subset A of F_q^K possessing at least one inner point, the functions $\mathcal{H}_\varepsilon(A)$ and $\mathcal{E}_\varepsilon(A)$, computed in the metric (62) (!) are of order*

$$\mathcal{H}_\varepsilon(A) \asymp \mathcal{E}_\varepsilon(A) \asymp \varepsilon^{-n/q}.$$

§6. ε-entropy of the class of differentiable functions in the metric L^2.

The present section deals with the same topics as the previous one, although the methods of proof are essentially a continuation of §4. In writing it, the authors used research results due to V. I. Arnold.

Denote by $P_\alpha^{[0,2\pi]}(C)$, $\alpha > 0$, or simply by $P_\alpha(C)$, the class of real functions $f(x)$ periodic with period 2π which are equal to zero in the mean and possess a derivative of order α in L^2 (in the sense of Weyl) uniformly bounded in the mean by the constant C. The norm in this class will be taken as in L^2:

$$\|f\| = \frac{1}{\pi} \int_0^{2\pi} |f(x)|^2 dx.$$

Recall that the function f^α is called a *derivative of order α in the sense of Weyl* of a function $f \in L^2$

$$f(x) \sim \frac{a_0}{2} + \sum_{k=1}^{\infty}(a_k \cos kx + b_k \sin kx),$$

if the corresponding Fourier series

$$f^{(\alpha)}(x) = \sum_{k=1}^{\infty} k^\alpha((a_k \cos \frac{\pi}{2}\alpha + b_k \sin \frac{\pi}{2}\alpha) \cos kx +$$
$$+ (b_k \cos \frac{\pi}{2}\alpha - a_k \sin \frac{\pi}{2}\alpha) \sin kx),$$

converges in the mean[11]. Uniform convergence in the mean of $f^{(\alpha)}$ for $f \in P_\alpha(C)$ means that

$$\|f^{(\alpha)}\| = (\sum_{k=1}^{\infty} k^{2\alpha}(a_k^2 + b_k^2))^{1/2} \leqslant C. \tag{85}$$

The spaces L^2 and l^2, as is known, are isometric. Consider the orthonormed basis in l^2 consisting of vectors $\{e_1, \ldots, e_n, \ldots\}$ such that e_1 corresponds to $\sqrt{2}/2$, $e_{2k} \leftrightarrow$

---

[11]For integers $\alpha = n$, the derivative $f^{(n)}(x)$ in the sense of Weyl coincides with the ordinary derivative $d^n f/dx^n$.

$\sin kx$, $e_{2k+1} \leftrightarrow \cos kx$, $k \geqslant 1$ for some isometric map of L^2 into l^2. It is easy to see that in this basis, to the class $P_\alpha(C)$ will correspond the set of vectors $x = (x_1, \ldots, x_n, \ldots)$ whose coordinates satisfy the inequality

$$x_1 = 0, \quad \sum_{k=1}^{\infty} \left[\frac{k}{2}\right]^{2\alpha} x_k^2 \leqslant C^2. \tag{86}$$

This set, which will also be denoted by $P_\alpha(C)$, is an ellipsoid with semi-axes $a_k = C/[k/2]^{12}$.

Further we shall consider the particular case of ellipsoids in l^2 whose semi-axes a_k are equal to B/k^α ($k = 1, 2, \ldots$). We shall denote them by Σ. In view of the fact that

$$\frac{C}{(k/2)^\alpha} \leqslant \frac{C}{[k/2]^\alpha} \leqslant \frac{C}{((k-1)/2)^\alpha}, \quad k = 2, 3, \ldots,$$

we obtain

$$\Sigma'_\alpha(C2^\alpha) \subset P_\alpha(C) \subset \Sigma_\alpha(C2^\alpha), \tag{87}$$

where $\Sigma'_\alpha(C2^\alpha)$ is an ellipsoid for which $a_1 = 0$. The inclusion (87) allows to estimate $\mathcal{H}_\varepsilon(P_\alpha)$ and $\mathcal{E}_\varepsilon(P_\alpha)$ in terms of $\mathcal{H}_\varepsilon(\Sigma_\alpha)$ and $\mathcal{E}_\varepsilon(\Sigma_\alpha)$. For the latter, we shall show that

$$\mathcal{E}_{2\varepsilon}(\Sigma_\alpha(B)) \gtrsim \left(\frac{B}{2\varepsilon}\right)^{1/\alpha} \alpha \log e, \tag{88}$$

$$\mathcal{H}_\varepsilon(\Sigma_\alpha(B)) \lesssim \left(\frac{\sqrt{2}B}{\varepsilon}\right)^{1/\alpha} \alpha \log e \left(1 + \frac{\gamma}{\alpha}\right), \tag{89}$$

where γ is a bounded constant not depending on α. Then (87) implies

Theorem XVI.

$$\left(\frac{C}{2\varepsilon}\right)^{1/\alpha} 2\alpha \log e \lesssim \mathcal{E}_{2\varepsilon}(P_\alpha(C)) \leqslant \mathcal{H}_\varepsilon(P_\alpha(C)) \lesssim$$

$$\lesssim \left(\frac{\sqrt{2}C}{\varepsilon}\right)^{1/\alpha} 2\alpha \log e \left(1 + \frac{\gamma}{\alpha}\right), \tag{90}$$

where γ is a bounded constant not depending on α.

For $\alpha \geqslant \alpha_0$, relation (90) can be rewritten as

$$\left(\frac{C}{\varepsilon}\right) 2\alpha \log e \left(1 + \frac{\gamma_1}{\alpha}\right) \lesssim \mathcal{E}_{2\varepsilon}(P_\alpha(C)) \leqslant \mathcal{H}_{2\varepsilon}(P_\alpha(C)) \lesssim$$

$$\lesssim (C/\varepsilon)^{1/\alpha} 2\alpha \log e \left(1 + \frac{\gamma_2}{\alpha}\right), \tag{91}$$

---

[12]An ellipsoid in l^2 is a set x for which

$$\sum_{k=1}^{\infty} x_k^2/a_k^2 \leqslant 1;$$

the a_k are called the *semi-axes* of this ellipsoid.

where γ_1 and γ_2 are bounded constants.

The result that we have obtained in Theorem XVI is interesting in that, for the space of differentiable functions in the metric L^2, it is possible to find sufficiently narrow inequalities (90) for $\mathcal{E}_{2\varepsilon}$ and $\mathcal{H}_\varepsilon$ asymptotically converging to each other as α (the order of smoothness) tends to infinity (see (91)). This is all the more interesting if we have in mind that a similar result in the metric C has not been obtained (see §5).

Thus it remains to prove (88) and (89).

Suppose $\Sigma_\alpha^n(B)$ denotes $\Sigma_\alpha(B) \cap D_2^n$, where D_2^n is the n-dimensional Euclidean space containing the first n axes of the ellipsoid $\Sigma_\alpha(B)$. Assume that $\mathcal{M}_{2\varepsilon}$ is the maximal number of non-intersecting n-dimensional balls S_2^n of radius ε whose centres are in $\Sigma_\alpha^n(B)$; then we get

$$\mathcal{M}_{2\varepsilon} \geqslant \frac{\mu_n(\Sigma_\alpha^n(B))}{\mu^n(2\varepsilon S_2^n)} = \frac{\mu_n(\Sigma_\alpha^n(B))}{(2\varepsilon)^n \mu^n(S_2^n)}, \tag{92}$$

since by the maximality assumption concerning the number $\mathcal{M}_{2\varepsilon}$, spheres of radius 2ε must cover the entire set $\Sigma_\alpha^n(B)$. Now put

$$B/n^\alpha < 2\varepsilon \leqslant B/(n-1)^\alpha,$$

i.e.,

$$n = (B/2\varepsilon)^{1/\alpha} + O(1), \tag{93}$$

and using the relations

$$1) \mu^n(\Sigma_\alpha^n(B)) = \mu^n(S_2^n) B^n (n!)^{-\alpha},$$
$$2) \log n! < (n+1/2) \log n - n \log e$$

(Stirling's formula), we obtain the following relation from (92) and (93):

$$\mathcal{E}_{2\varepsilon}(\Sigma_\alpha(B)) \geqslant \log \mathcal{M}_{2\varepsilon}$$
$$\geqslant n \log \frac{B}{2\varepsilon} - \alpha n \log n - \frac{\alpha}{2} \log n + \alpha n \log e$$
$$\geqslant \left(\frac{B}{2\varepsilon}\right)^{1/\alpha} \alpha \log e + O\left(\log \frac{1}{\varepsilon}\right) \overset{\sim}{\sim} \left(\frac{B}{2\varepsilon}\right)^{1/\alpha} \alpha \log e.$$

To get (89), let us choose m so as to have

$$\frac{B}{m^\alpha} \leqslant \frac{\sqrt{2}}{2}\varepsilon < \frac{B}{(m-1)^\alpha},$$

i.e.,

$$m = (\sqrt{2}B/\varepsilon)^{1/\alpha} + O(1). \tag{94}$$

Consider the ellipsoid $\Sigma_\alpha^m(B)$. It is easy to see that the distance from any point of $\Sigma_\alpha(B)$ to $\Sigma_\alpha^m(B)$ is less than or equal to $(\sqrt{2}/2)\varepsilon$, and therefore any $(\sqrt{2}/2)\varepsilon$-net for $\Sigma_\alpha^m(B)$ is also ε-net for $\Sigma_\alpha(B)$.

In D_2^m construct a cubic lattice of diameter $\varepsilon/\sqrt{2m}$. (For the definition of a lattice see §4.)

Suppose N is the number of cubes of diameter $\varepsilon/\sqrt{2m}$[13] (with centres at the points of the lattice) intersecting $\Sigma_\alpha^m(B)$. These cubes obviously lie in an $(\varepsilon\sqrt{2}/2)$-neighbourhood U of the ellipsoid $\Sigma_\alpha^m(B)$. U is contained inside an ellipsoid homothetic to $\Sigma_\alpha^m(B)$ with least axis equal to $B/m^\alpha+(\sqrt{2}/2)\varepsilon$, or a homothety coefficient less than 2. Thus the volume of all the cubes of the lattice are no greater than the volume of this ellipsoid. Hence we obtain the inequality

$$N \leqslant \frac{\mu(S_2^m)B^m}{(m!)^\alpha}2^m\left(\frac{\sqrt{2m}}{\varepsilon}\right)^m. \tag{95}$$

Now let us use the following facts:

1)$\mu^m(S_2^m) = \pi^{m/2}/\Gamma(m/2+1)$ is the volume of the sphere ([24, volume 3, p. 474])

2)$\log\Gamma(a) = (a-1/2)\log a - a\log e + O(1)$ ([24, Vol. 2, p.819]),
then from (94) and (95) we get

$$\mathcal{H}_\varepsilon(\Sigma_\alpha(B)) \leqslant \log N \leqslant m\log\frac{\sqrt{2}B}{\varepsilon} - \alpha m\log m + m(\alpha\log e + O(1))$$

$$+ O(\log m) \leqslant \left(\frac{\sqrt{2}B}{\varepsilon}\right)^{1/\alpha}\alpha\log e\left(1+\frac{\gamma}{\alpha}\right),$$

which was to be proved.

In his study of the estimate of the expression $\mathcal{H}_\varepsilon$ from above, Arnold used results of Davenport and Watson, which were already mentioned in §4. Naturally, the author obtained more precise estimates than ours, since the lattice constructed by Davenport and Watson is much more economical than the cubic one. Let us present without proof the inequalities obtained by Arnold

$$\mathcal{H}_\varepsilon(\Sigma_\alpha(B)) \leqslant \left(\frac{\sqrt{2}B}{\varepsilon}\right)^{1/\alpha}\alpha\log e\left(1+\frac{1,2}{\alpha}\right) + \alpha\log e.$$

§7. ε-entropy of classes of analytic functions.

Besides the notes by the authors [6] and [7] containing estimates of an ε-entropy type of weak equivalence, this section is based on the works of A. G. Vitushkin [3] and V. D. Yerokhin [4]. The proofs of Lemmas II and III stated below contain in general form methods developed by A. G. Vitushkin; Theorem XVIII is especially constructed for the system of functions introduced by V. D. Yerokhin. In part, the results obtained by Yerokhin were simultaneously proved by different methods by K. I. Babenko [1]. Theorem XX is due to A. G. Vitushkin [3], while Theorem XXI belongs to V. M. Tikhomirov.

1. We shall assume that in a linear normed space F we are given a system of elements

$$\varphi_{k_1k_2\ldots k_s}, \quad k_r = 0, 1, 2, \ldots, \quad r = 1, 2, \ldots, s,$$

---

[13]In this case the diameter of the cube in the Euclidean metric is equal to $\varepsilon\sqrt{2}/2$.

and the corresponding system of linear functionals

$$c_{k_1 k_2 \ldots k_s}$$

such that each element f can be represented in the form[14]

$$f = \sum_{k \in K} c_k \varphi_k,$$

and there exists a constant C_1 such that

$$\sup_{k \in K} |c_k| \leqslant \|f\| \leqslant C_1 \sum_{k \in K} |c_k|. \tag{96}$$

The following two basic lemmas hold.

Lemma II. *Suppose $A \subset F$ and there exists a constant C_2 and a positive vector $h = (h_1, \ldots, h_s)$ such that $f \in A$ implies*

$$|c_k| \leqslant C_2 e^{-(h,k)}. \tag{97}$$

Then

$$\mathcal{H}_\varepsilon(A) \leqslant \frac{2(\log(1/\varepsilon))^{s+1}}{(s+1)!(\log e)^s h^\theta} + O\left(\left(\log \frac{1}{\varepsilon}\right)^s \log \log \frac{1}{\varepsilon}\right) \tag{98}$$

Lemma III. *Suppose that for given C_3 and h the existence of a positive vector Δ satisfying*

$$|c_k| \leqslant C_3 \Delta^\theta e^{-(h+\Delta, k)} \tag{99}$$

implies that the element

$$f = \sum_{k \in K} c_k \varphi_k$$

belongs to $A \subset F$. Then

$$\mathcal{E}_{2\varepsilon}(A) \geqslant \frac{2(\log(1/\varepsilon))^{s+1}}{(s+1)!(\log e)^s h^\theta} + O\left(\left(\log \frac{1}{\varepsilon}\right)^s \log \log \frac{1}{\varepsilon}\right). \tag{100}$$

We shall say that *the system of conditions* (I) *holds for the set $A \subset F$* if the three conditions (96), (97) and (99) stated above hold for this set.

Inequalities (98) and (100) taken together obviously yield the following theorem.

---

[14]In this section k denotes an integer non-negative vector $k = (k_1, \ldots, k_s)$. By K we denote the set of all such vectors. The vector $h = (h_1, \ldots, h_s)$ will be called *positive* if all its components are strictly greater than zero. The notation h^k means that for the s-dimensional vector h and $k \in K$ we have the relation $h^k = h_1^{k_1} \ldots h_s^{k_s} = \prod_{i=1}^s h_i^{k_i}$. The scalar product (h, k) has the ordinary meaning $\sum_{i=1}^s h_i k_i$; θ denotes the unit vector $(1, \ldots, 1)$, so that $h^\theta = h_1 \ldots h_s = \prod_{i=1}^s h_i$.

Theorem XVII. *If the system of conditions (I) holds for the set A, then*

$$\mathcal{H}_{2\varepsilon}(A) = \frac{2}{(s+1)!(\log e)^s h^\theta}\left(\log\frac{1}{\varepsilon}\right)^{s+1} + O\left(\left(\log\frac{1}{\varepsilon}\right)^s \log\log\frac{1}{\varepsilon}\right),$$

$$\mathcal{E}_{2\varepsilon}(A) = \frac{2}{(s+1)!(\log e)^s h^\theta}\left(\log\frac{1}{\varepsilon}\right)^{s+1} + O\left(\left(\log\frac{1}{\varepsilon}\right)^s \log\log\frac{1}{\varepsilon}\right). \tag{101}$$

We shall say that *the system of conditions* (II) *holds for the set* $A \subset F$, if for any positive vector α there exist constants $C_1(\alpha), C_2(\alpha)$ and $C_3(\alpha)$ such that

1)
$$\sup_{k\in K}|c_k| \leqslant \|f\| \leqslant C_1(\alpha)\sum_{k\in K}|c_k|, \tag{102}$$

2) $f \in A$ implies

$$|c_k| \leqslant C_2(\alpha)e^{-(h-\alpha,k)},$$

3) the inequalities

$$|c_k| \leqslant C_3(\alpha)e^{-(h+\alpha,k)}$$

imply that the element

$$f = \sum_{k\in K}c_k\varphi_k$$

belongs to the set $A \subset F$.

It is easy to deduce from inequalities (98) and (100) the following

Theorem XVIII. *If the system of conditions (II) holds for the set* $A \subset F$, *then*

$$\mathcal{E}_\varepsilon(A) \sim \mathcal{H}_\varepsilon(A) \sim \frac{2(\log(1/\varepsilon))^{s+1}}{(s+1)!(\log e)^s h^\theta}. \tag{103}$$

Indeed, if we fix α, inequality (98) yields

$$\varlimsup_{\varepsilon\to 0}\frac{\mathcal{H}_\varepsilon(A)}{(\log(1/\varepsilon))^{s+1}} \leqslant 2(s+1)!(\log e)^s(h-\alpha)^\theta. \tag{104}$$

On the basis of inequality (100), we see that

$$\lim_{\varepsilon\to 0}\frac{\mathcal{E}_{2\varepsilon}(A)}{(\log(1/\varepsilon))^{s+1}} \geqslant \frac{2}{(s+1)!(\log e)^s(h+\alpha)^\theta}. \tag{105}$$

Taken together, inequalities (104) and (105) give us (103); since α is arbitrary, we get

$$\mathcal{E}_{2\varepsilon}(A) \leqslant \mathcal{H}_\varepsilon(A) \leqslant \mathcal{E}_\varepsilon(A).$$

Before we pass to the proofs of Lemmas II and III, let us make the following

Remark. In Theorem XVII, the set K consisted of non-negative vectors k. In our subsequent exposition the reader will easily understand what changes must be made in the proof in the case when we are dealing with arbitrary vectors $k = (k_1,\ldots,k_s)$, $-\infty < k_i < \infty$. Let us state the final results. Let $|k|$ denote the vector with coordinates $(|k_1|,\ldots,|k_s|)$ and $\check{K}$ the set of all integer vectors.

Theorem XIX. *Suppose that the system of conditions (I') for the set K holds i.e.,*

$$\sup_{k \in \check{K}} |c_k| \leqslant \|f\| \leqslant C_1 \sum_{k \in \check{K}} |c_k|, \tag{106}$$

2) $f \in A$ implies

$$|c_k| \leqslant C_2 e^{-(h,|k|)},$$

3) for a given constant C_3 and for any positive vector Δ the inequalities

$$|c_k| \leqslant C_3 \Delta^\theta e^{-(h+\Delta,|k|)} \tag{107}$$

imply that the element

$$f = \sum_{k \in \check{K}} c_k \varphi_k$$

belongs to the set $A \subset F$. Then

$$\mathcal{H}_\varepsilon(A) = \frac{2^{s+1}}{(s+1)!(\log e)^s h^\theta} \left(\log \frac{1}{\varepsilon} \right)^{s+1} + O\left(\left(\log \frac{1}{\varepsilon} \right)^s \log \log \frac{1}{\varepsilon} \right),$$

$$\mathcal{E}_\varepsilon(A) = \frac{2^{s+1}}{(s+1)!(\log e)^s h^\theta} \left(\log \frac{1}{\varepsilon} \right)^{s+1} + O\left(\left(\log \frac{1}{\varepsilon} \right)^s \log \log \frac{1}{\varepsilon} \right). \tag{108}$$

Theorem XIX is especially worded to work with Fourier series.
2. Thus it remains to prove Lemmas II and III.
Consider the set S_n of integer vectors $k \in K$ such that

$$(h, k) \leqslant n. \tag{109}$$

By $\bar{S}_n$ denote $K \setminus S_n$ (Figure 9).
Condition (96) implies that

$$\|f\| \leqslant C_1 \sum_{k \in S_n} |c_k| + C_1 \sum_{k \in \bar{S}_n} |c_k| = C_1 \sum_{k \in S_n} |c_k| + R_n. \tag{110}$$

The inequalities (97) yields

$$|R_n| \leqslant C_1 C_2 \sum_{k \in \bar{S}_n} e^{-(h,k)}. \tag{111}$$

Assuming further that $n \geqslant (\max_i h_i / \min_i h_i) + 1 = \lambda$, let us show that

$$\sum_{k \in \bar{S}_n} e^{-(h,k)} \leqslant \int_{\substack{(h,x) \geqslant n-\lambda \\ x \geqslant 0}} e^{-(h,x)} dx. \tag{112}$$

Assume that B_k is the unit cube

$$B_k = \{x | k_r - 1 \leqslant x_r \leqslant k_r, \ r = 1, \ldots, x\}.$$

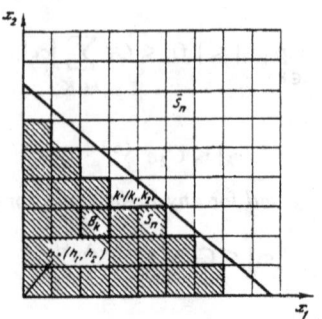

FIGURE 9

We shall call k the maximal vertex of the cube B_k. Obviously the difference $k - x$ is a non-negative vector for any point of the cube x. It is easy to understand that the union of all B_k for $k \in K$ covers the set of all non-negative x. The function $e^{-(h,x)}$ in the cube B_k has a minimum at the point k. Indeed, if we construct the hyperplane $(h, x) = (h, x_0)$ through an arbitrary point $x_0 \in B_k$, we see that

$$e^{-(h,x)} = e^{-(h,k)}e^{(h,k-x)} \geqslant e^{-(h,k)} \text{ since } (h, k - x) \geqslant 0,$$

because h is a positive vector, while k is the maximum vertex.

Now choose a minimal λ so that the hyperplane $(h, x) = n - \lambda$ has no common points with B_k whenever $k \in S_n$. Then from the above it follows that

$$\sum_{k \in S_n} e^{-(h,k)} \leqslant \int\limits_{\substack{(h,x) \geqslant n-\lambda \\ x \geqslant 0}} e^{-(h,x)}\, dx.$$

It is easy to calculate that it suffices to take the number $(\max_i h_i / \min_i h_i) + 1$ for λ. It is only the boundedness of λ that is important for us.

The integral (112) can be calculated in an elementary way[15]:

$$\int\limits_{\substack{(h,x) \geqslant n-\lambda \\ x \geqslant 0}} e^{-(h,x)}\, dx = \frac{1}{h^\theta} \int_{n-\lambda}^{\infty} e^{-\xi} \frac{\xi^{s-1} d\xi}{(s-1)!} \leqslant B e^{-n} n^s,$$

where B is a bounded constant. Now let us take the minimal n for which we have

$$C_1 C_2 B e^{-n} n^s \leqslant \varepsilon/2. \tag{113}$$

Hence

$$n = \frac{\log(1/\varepsilon)}{\log e} + O\left(\log\log \frac{1}{\varepsilon}\right). \tag{114}$$

---

[15]Recall that $h^\theta = \prod_{i=1}^{s} h_i$.

The estimate (113) means that $|R_n|$ in (110) is no greater than $\varepsilon/2$. Now approximate the remaining finite-dimensional part

$$\sum_{k \in S_n} c_k \varphi_k.$$

Denote by P_n the number of vectors $k \in S_n$. It is easy to understand from the above that the P_n satisfy the inequalities

$$\frac{(n-\lambda)^s}{s! h^\theta} = \int\limits_{\substack{(h,x)\leqslant n-\lambda \\ x \geqslant 0}} dx \leqslant P_n \leqslant \int\limits_{\substack{(h,x)\geqslant n \\ x \geqslant 0}} dx = \frac{n^s}{s! h^\theta},$$

hence

$$P_n = \frac{n^s}{s! h^\theta} \left(1 + \frac{O(1)}{n}\right). \tag{115}$$

Further let us approximate the coordinate $c_k = \alpha_k + i\beta_k$ with precision $\varepsilon/P_n C_1$ by the expression $\alpha'_k + i\beta'_k$, setting

$$\begin{aligned}
\alpha'_k &= \left[\frac{\sqrt{2}C_1 P_n \alpha_k}{\varepsilon}\right] \frac{\varepsilon}{\sqrt{2}P_n C_1} = m_k^1 \frac{\varepsilon}{\sqrt{2}P_n C_1}, \\
\beta'_k &= \left[\frac{\sqrt{2}C_1 P_n \beta_k}{\varepsilon}\right] \frac{\varepsilon}{\sqrt{2}P_n C_1} = m_k^2 \frac{\varepsilon}{\sqrt{2}P_n C_1}.
\end{aligned} \tag{116}$$

To each element f we have therefore assigned a family of indices

$$U = \{m_k^i\}, \ k \in S_n, \ i = 1, 2.$$

For two elements f_1 and f_2 from A corresponding to the same family U, we have

$$\|f_1 - f_2\| \leqslant C_1 \sum_{k \in S_n} |c_k^1 - c_k^2| + 2R_n \leqslant \sum_{k \in S_n} \frac{\varepsilon}{P_n} + \varepsilon = 2\varepsilon$$

(see the relations (110) and (116)), i.e., the set of elements of an ε-covering of A is no greater than the number of all families U corresponding to A. It only remains to estimate the number N of these families. Since, according to (97), we have $\max\{\alpha_k, \beta_k\} \leqslant C_2 e^{-(h,k)}$, it follows from (116) that

$$|m_i^k| \leqslant \frac{\sqrt{2}C_1 C_2 P_n}{\varepsilon} e^{-(h,k)} = N_k, \ i = 1, 2. \tag{117}$$

This means

$$\log N = \log(1/\varepsilon) - (h, k)\log e + \log P_n + D_1;$$

here D_1 is a bounded constant not depending on k. The number N is therefore estimated by the expression

$$\prod_{k \in S_n} (2N_k + 1)^2,$$

where

$$\mathcal{H}_\varepsilon(A) \leqslant \log N \leqslant 2 \sum_{k \in S_n} \log(2N_k + 1) =$$

$$= 2P_n \log \frac{1}{\varepsilon} = 2 \sum_{k \in S_n} (k,h) \log e + 2P_n \log P_n + O(P_n).$$

It follows from considerations similar to the one used to estimate R_n that the expression $\sum_{k \in S_n}(k,h)$ will not be larger than the integral

$$\int_{\substack{(h,x) \leqslant n-\lambda \\ x \geqslant 0}} (h,x)dx = \frac{s(n-\lambda)^{s+1}}{(s+1)!h^\theta} = \frac{sn^{s+1}}{(s+1)!h^\theta} = \frac{sn^{s+1}}{(s+1)!h^\theta}\left(1 + \frac{O(1)}{n}\right),$$

so that finally, using (114) and (115), we obtain

$$\mathcal{H}_\varepsilon(A) \leqslant \frac{2}{(s+1)!(\log e)^s h^\theta}\left(\log \frac{1}{\varepsilon}\right)^{s+1} + O\left(\left(\log \frac{1}{\varepsilon}\right)^{s+1} \log \log \frac{1}{\varepsilon}\right).$$

Proof of Lemma III. Suppose $\varepsilon > 0$. Put

$$h' = h + \Delta,$$

where the vector Δ has the coordinates

$$\Delta_i = \frac{h_i}{\log(1/\varepsilon)}, \quad i = 1, 2, \ldots, s. \tag{118}$$

Choose the maximal m so that

$$C_3 \Delta^\theta e^{-m} = \frac{C_3 h^\theta}{(\log(1/\varepsilon))^s} e^{-m} \geqslant 2\sqrt{2}\varepsilon, \tag{119}$$

where ε is taken from the conditions of Lemma III.
    Denote by S_m the set of vectors k from K such that

$$(h', k) \leqslant m.$$

Consider the set $\{\Psi\}$ of functions $\Psi = \sum_{k \in K} c_k \varphi_k$ for which

α) $c_k = 0, \ k \notin S_m$;

β) $c_k = (s_k^1 + is_k^2)2\varepsilon, \ k \in S_m$; $\tag{120}$

γ) $s_k^i \ (i = 1, 2)$ are any integers such that

$$|s_k^i| \leqslant \frac{C_3 \Delta^{\theta_\varepsilon - (h',k)}}{2\sqrt{2}\varepsilon} = M_k. \tag{121}$$

The inequality (121) means, according to condition (97) of Lemma III, that the set $\{\Psi\}$ is contained in A.

By the construction of $\{\Psi\}$, it follows from (96) that the set $\{\Psi\}$ is 2ε–distinguishable. According to (120), each element of $\{\Psi\}$ is determined by a family of indices

$$U = \{s_k^i\}, \ k \in S_m, \ i = 1, 2,$$

while the number of all U according to (121) is no greater than

$$M = \prod_{k \in S_m} M_k^2.$$

From (121) we get

$$\log M_k = \log(1/\varepsilon) - (h', k) \log e + \log C_3 \Delta^\theta + D_2, \tag{122}$$

where D_2 is a bounded constant not depending on k. Hence, recalling (118), we can transform (122) to

$$\log M_k = \log \frac{1}{\varepsilon} - (h', k) \log e + O \left(\log \log \frac{1}{\varepsilon} \right).$$

From (119) we get

$$m = \frac{\log(1/\varepsilon)}{\log e} + O \left(\log \log \frac{1}{\varepsilon} \right),$$

and finally[16]

$$\mathcal{E}_{2\varepsilon}(A) \geqslant \log M = 2 P_m \log \frac{1}{\varepsilon} - 2 \sum_{k \in S_m} (h', k) \log e +$$

$$+ O \left(P_m \log \log \frac{1}{\varepsilon} \right) = \frac{2}{(s+1)!(\log e)^s h'^\theta} \left(\log \frac{1}{\varepsilon} \right)^{s+1} +$$

$$+ O \left(\left(\log \frac{1}{\varepsilon} \right)^s \log \log \frac{1}{\varepsilon} \right) = \frac{2}{(s+1)!(\log e)^s h^\theta} \left(\log \frac{1}{\varepsilon} \right)^{s+1} +$$

$$+ O \left(\left(\log \frac{1}{\varepsilon} \right)^s \log \log \frac{1}{\varepsilon} \right).$$

This concludes the proof of Lemma III.

3. We now pass to the applications of the main theorems XVII and XVIII of this section.

Everywhere in the sequel we shall consider the class $\mathcal{A}_G^K$ of analytic functions $f(z_1, \ldots, z_s)$ given on a bounded continuum K possessing an analytic continuation to the domain $G \supset K$ of s-dimensional complex space. Here all the functions $f \in \mathcal{A}_G^K(C)$ are assumed uniformly bounded in G by the constant C.

---

[16] P_m has the same meaning as in Lemma II. For it we have relation (115).

The classes A_G^K will be viewed as *metric spaces* in the uniform metric

$$\|f\| = \max\{|f(z_1,\ldots,z_s)|, \ (z_1,\ldots,z_s) \in K\}.$$

First we consider classes of analytic functions of one complex variable.

1) The class $A_{K_R}^{\bar{K}_r}(C)$, where $\bar{K}_r$ is the closed disk $|z| \leqslant r$, while K_R is the open disc $|z| < R, \ R > r$.

Functions of the class $A_{K_R}^{\bar{K}_r}(C)$ can be expanded into Taylor series

$$f(z) = a_0 + a_1 z + \cdots + a_n z^n + \cdots = c_0 + \frac{c_1}{r} z + \cdots + \frac{c_n}{r^n} z^n + \ldots, \qquad (123)$$

where $c_k = a_k r^k$. The set K here consists of integers $k = 0, 1, \ldots$. Put $e^h = R/r$ or $h = \log(R/r)/\log e$. From the Cauchy inequality, we get

$$|c_k| \leqslant C(r/R)^k = Ce^{-hk}. \qquad (124)$$

It follows directly from (123) that if

$$|c_k| \leqslant C'\Delta e^{-(h+\Delta)k} = C'\Delta(r/R)^k e^{-\Delta k}, \qquad (125)$$

then the series (123) uniformly converges in K_R and

$$|f(z)| \leqslant C'\Delta \sum_{k=0}^{\infty} e^{-\Delta k} = \frac{C'\Delta}{1 - e^{-\Delta}}.$$

Choose a C' so as to have

$$C'\Delta/(1 - e^{-\Delta}) \leqslant C \qquad (126)$$

for all $\Delta \leqslant \Delta_0$. Then

$$f(z) \in A_{K_R}^{\bar{K}_r}(C).$$

Directly from (123) we get

$$\|f\| = \max_{z \in \bar{K}_r} |f(z)| \leqslant \sum_{k=0}^{\infty} |a_k| r^k = \sum_{k=0}^{\infty} |c_k|. \qquad (127)$$

Moreover

$$|a_k| = \left|\frac{1}{2\pi i} \int_{Z_r} \frac{f(\xi)}{\xi^{k+1}} d\xi\right| \leqslant \frac{\max_{\xi \in Z_r} |f(\xi)|}{r^k} = \frac{\|f\|}{r^k}, \qquad (128)$$

where Z_r is a circle of radius r. Thus from relations (124), (125), (127), (128) it follows that the system of conditions (I) holds for $A_{K_R}^{\bar{K}_r}(C)$, where $C_1 = 1, C_2 = C, C_3 = C'$ is taken from (126) and therefore[17]

$$\mathcal{H}_\varepsilon(A_{K_R}^{\bar{K}_r}(C)) = \frac{(\log(1/\varepsilon))^2}{\log(R/r)} + O\left(\log\frac{1}{\varepsilon} \log\log\frac{1}{\varepsilon}\right). \qquad (129)$$

---

[17]All the results are written only for the function $\mathcal{H}_\varepsilon$. For $\mathcal{E}_\varepsilon$ all the formulas are similar.

2) The class $\mathcal{A}_h(C)$ of functions periodic on D with period equal to 2π analytic and bounded by the constant C in the strip $|\mathrm{Im}z| < h$. This class was studied in detail in §2. For it we have conditions (I') and therefore, according to formula (108),

$$\mathcal{H}_\varepsilon(\mathcal{A}_h(C)) = \frac{2}{h\log e}\left(\log\frac{1}{\varepsilon}\right)^2 + O\left(\log\frac{1}{\varepsilon}\log\log\frac{1}{\varepsilon}\right). \tag{130}$$

3) The class $A_{\Sigma_\lambda}^\Delta(C)$. Here by Δ we mean $[-1,1]$ and by Σ_λ the ellipse with sum of semi-axes equal to λ and foci at the points ± 1.

Functions of this class can be developed into series of Chebyshev polynomials

$$f(z) = \sum_{k=0}^{\infty} a_k T_k(z).$$

The fact that Theorem XVII is applicable can be proved independently but we shall obtain it as a particular case of our consideration of the following class of functions $A_{C_R(K)}^K(C)$.

For the class $A_{\Sigma_\lambda}^\Delta(C)$, we have $h = (\log\lambda^{-1})/\log e$, i.e.,

$$\mathcal{H}_\varepsilon(A_{\Sigma_\lambda}^\Delta(C)) = \frac{(\log(1/\varepsilon))^2}{\log\lambda^{-1}} + O\left(\log\frac{1}{\varepsilon}\log\log\frac{1}{\varepsilon}\right). \tag{131}$$

4) Suppose K denotes an arbitrary bounded continuum containing more than one point. Suppose G_∞ is the one of its adjacent domains which contains the point $z = \infty$. This is a simply connected domain of the extended plane whose boundary Γ is part of the continuum K. Let us map G_∞ on the exterior of the unit disk of the plane ω. The inverse image $|\omega| = R$, $R > 1$ under this map will be called the R-level line of the continuum K and denoted by $C_R(K)$. Consider the class $A_{C_R(K)}^K(C)$.

It turns out (see [25, p. 414-423]), that there exists a special class of series, called series of Faber polynomials, using which one can find a unique expansion of the function belonging to the class $A_{C_R(K)}^K$:

$$f(z) = \sum_{k=0}^{\infty} c_k \Phi(k)(z). \tag{132}$$

Here the polynomials $\Phi_k(z)$ and coefficients c_k in (132) satisfy the following conditions:

$$\alpha)\,|a_r| \leqslant M_r/r^k,$$

where

$$M_r = \max_{z \in C_r(K)} |f(z)|.$$

Hence as $r \to R$ we get

$$|a_k| \leqslant C/R^k, \tag{133}$$

while as $r \to 1$ we obtain

$$|a_k| \leqslant \max_{z \in K} |f(z)| = \|f\|. \tag{134}$$

$$\beta) \, |\Phi_k(z)| \leqslant DR^{k'}, \, z \in C_R(K), \, R' > R, \tag{135}$$

where D is a certain constant.

$$\gamma) \, \max_{z \in K} |\Phi_k(z)| \leqslant B(\delta)(1 + \delta)^k. \tag{136}$$

The relations (132)-(136), as can be easily seen, imply that the system of conditions (II) holds for the class $A^K_{C_R(K)}(C)$. Here $h = \log R / \log e$. Thus according to Theorem XVIII, we have

$$\mathcal{H}_\varepsilon(A^K_{C_R(K)}(C)) \sim \frac{(\log(1/\varepsilon))^2}{\log R}. \tag{137}$$

The Chebyshev polynomials, a particular case of the Faber polynomials when K is the segment $\Delta = [-1, 1]$, unlike the general Faber polynomials, are bounded on Δ:

$$|T_k(x)| \leqslant 1, \, x \in \Delta, \tag{138}$$

and the relations (133)-(135), (138) already allow the application of Theorem XVIII. It is possible to prove that the results of Theorem XVII remain valid for sufficiently a smooth continuum K, but we shall not dwell on this.

5) After the publication of Vitushkin's paper [3] in which he obtained formulas (129)-(131) (and their generalizations to the space of many complex variables, which we shall consider below), A. N. Kolmogorov expressed the conjecture that in the case of the class $A^K_G(C)$, where G is an arbitrary domain and C is a simply connected continuum contained in it, $K \subset G$, there exists a constant $\tau = \tau(G, K)$ such that

$$\mathcal{H}_\varepsilon(A^K_G(C)) \sim \tau(G, K) \left(\log \frac{1}{\varepsilon}\right)^2. \tag{139}$$

This conjecture was proved by V. D. Yerokhin [4] and K. I. Babenko [1] almost simultaneously. Note that the results of Yerokhin are somewhat more general.

The methods of Yerokhin and Babenko are essentially different from each other.

The method used by Yerokhin to obtain formula (139) fits neatly into the outline (based on applying Theorem XVIII) presented above.

Yerokhin showed that, in the domain G, there exists a sequence of functions $\{f_k(z)\}_{k \geqslant 1}$ analytic in G such that

$$f(z) = c_0 + \sum_{k=1}^{\infty} c_k f_k(z)$$

for any function $f(z)$ analytic in G, and in this situation we have the following properties of the functions $f_k(z)$ and coefficients c_k:

$$1^0)\, \|f_k\| \leqslant C(\delta)(1+\delta)^k,$$

$$2^0)\, \sup_{z \in G} |f_k(z)| \leq C(\delta)(1+\delta)^k R^k,$$

$$3^0)\, |c_k| \leqslant C(\delta)(1+\delta)^k / R^k,$$

$$4^0)\, |c_k| \leqslant \|f\|,$$

where $C(\delta)$ is a constant not depending on k while $R = R(G, K)$ is the radius of the exterior circle in the map of a doubly connected domain bounded by K and by the boundary of G on the annulus

$$1 < |\omega| < R;$$

R is sometimes called the *conformal radius* of the doubly connected domain GK. It is easy to see that properties 1)-4) imply that Theorem XVIII is applicable and that we have $h = \log R / \log e$.

Thus we have shown that

$$\mathcal{H}_\varepsilon(A_G^K(C)) \sim \frac{2}{\log R} \left(\log \frac{1}{\varepsilon} \right)^2. \tag{140}$$

Note that the paper [4] contains a more general result than the one stated here.

We shall not develop the construction of the functions f_k in this paper. Let us pass to the class of functions $f(z_1, \ldots, z_s)$ of several complex variables.

6) The class $A_{K_R}^{\bar{K}_r}(C)$, $R = (R_1, \ldots, R_s)$, $r = (r_1, \ldots, r_s)$, where $\bar{K}_r = \bar{K}_{r_1} \times \cdots \times \bar{K}_{r_s}^s$ is a closed polycylinder, i.e. the Cartesian product of the disks

$$\bar{K}_{r_i}^i = \{z_i : |z_i| \leqslant r_i\},\ i = 1, 2, \ldots, s;$$

$K_R = K_{R_1}^1 \times \cdots \times K_{R_s}^s$ is a polycylinder, the Cartesian product of the disks

$$K_{R_i}^i = \{z_i : |z_i| < R_i\},\ i = 1, 2, \ldots, s.$$

Functions of this class can be expanded into a Taylor series

$$f(z) = \sum_K a_k z^k,\ z = (z_1, \ldots, z_s).$$

The applicability of Theorem XVII is easy to establish. The vector h in this case is of the form $h = (h_1, \ldots, h_s)$, where $h_i = \log(R_i/r_i)/\log e$, $i = 1, 2, \ldots, s$ so that

$$\mathcal{H}_\varepsilon(A_{K_R}^{\bar{K}_r}(C))$$

$$= \frac{2}{(s+1)! \prod_{i=1}^s \log(R_i/r_i)} \left(\log \frac{1}{\varepsilon} \right)^{s+1} + O\left(\left(\log \frac{1}{\varepsilon} \right)^s \log\log \frac{1}{\varepsilon} \right). \tag{141}$$

7) The class $A_h(C)$, $h = (h_1, \ldots, h_s)$ of functions $f(z_1, \ldots, z_s)$ periodic in each variable with period 2π and bounded in the Cartesian product

$$\Pi_{h_1}^1 \times \cdots \times \Pi_{h_s}^s, \quad \Pi_{h_i}^i = \{z_i : |\Im z_i| < h_i\}.$$

In this case the functions are expandable in a *Fourier series*

$$f(z) = \sum_{k \in \check{K}} c_k e^{i(k,z)}, \quad z = (z_1, \ldots, z_s).$$

The fact that the assumptions of Theorem XIX (see the Remark in subsection 2) hold is easily verified, so that

$$\mathcal{H}_\varepsilon(A_h(C)) = \frac{2^{s+1}}{(s+1)!(\log e)^s h^\theta} \left(\log \frac{1}{\varepsilon} \right)^{s+1} +$$
$$+ O\left(\left(\log \frac{1}{\varepsilon} \right)^s \log \log \frac{1}{\varepsilon} \right). \tag{142}$$

Further let us present the generalization of formula (140) (also see [4]):

$$\mathcal{H}_\varepsilon(A_{G^s}^{K^s}(C)) \sim \frac{2}{(s+1)! \prod_{i=1}^s \log R_i} \left(\log \frac{1}{\varepsilon} \right)^{s+1}. \tag{143}$$

In formula (143), $G^s = G_1 \times \cdots \times G_s$ is the Cartesian product of simply connected domains G_i while $K^s = K_1 \times \cdots \times K_s$ is the Cartesian product of simply connected continua $K_1 \subset C_i$, where the R_i are the conformal radii of the domains $G_i K_i$ defined in our proof of formula (140).

4. Let us conclude this section by considering two classes of entire analytic functions.

In this subsection let us denote by z the s-dimensional complex vector $z = (z_1, \ldots, z_s)$; by $|z|$ we mean the real s-dimensional vector

$$|z| = (|z_1|, \ldots, |z_s|).$$

We shall say that the entire function f is a function of finite order if there exists a positive vector $\bar{p} = (\bar{p}_1, \ldots, \bar{p}_s)$ such that the inequality

$$|f(z)| < e^{(\theta, |z|^p)} = e^{|z_1|^{p_1} + |z_s|^{p_s}} \tag{144}$$

is satisfied for all z such that all their coordinates have a modulus greater than a certain $r(\bar{p})$.

The greater lower bound of the numbers $\bar{p}_1, \ldots \bar{p}_s$ for which we have inequality (144) is called the *order of the entire function* f. The order of a function will usually be written as a vector $p = (p_1, \ldots, p_s)$.

The greater lower bound of the positive numbers $\bar{\sigma}_1, \ldots, \bar{\sigma}_s$ for which we have asymptotically

$$|f(z)| < e^{(\bar{\sigma}, |z|^p)} = e^{\bar{\sigma}_1 |z_1|^{p_1} + |z_s|^{p_s}}, \tag{145}$$

will be called the *type of the entire function f* of order p.

Types of functions will be generally denoted by vectors $\sigma = (\sigma_1, \ldots, \sigma_s)$.

We shall study the ε-entropy of the following classes of entire functions of finite order and type:

1. The class $\Phi_{p\sigma}^s(C)$ (here σ and p are arbitrary positive vectors) of functions f satisfying the inequality

$$|f(z)| \leqslant Ce^{(\sigma, |z|^p)}. \tag{146}$$

The class $\Phi_{p\sigma}^s(C)$ will be considered in the metric

$$\|f\| = \max_{z \in \bar{K}_\theta^s} |f(z)|,$$

where $\bar{K}_\theta^s = \bar{K}_1^1 \times \cdots \times \bar{K}_1^s$ is the polycylinder, i.e. the product of $\bar{K}_1^i = \{z_i : |z_i| \leqslant 1\}$.

2. The class $F_{p\theta,\sigma}^s$, where σ is an arbitrary positive vector and $p\theta$ is the vector with components $(p, p, \ldots, p)$, consisting of entire functions satisfying the inequality

$$|f(z)| \leqslant Ce^{(\sigma|\mathrm{Im}z|^{p\theta})}, \ p > 1, \tag{147}$$

and periodic with period 2π in each variable.

The class $F_{p\theta,\sigma}^s$ will be viewed in the uniform metric on D^s.

We have the following theorems.

Theorem XX.

$$\mathcal{E}_\varepsilon(\Phi_{p\sigma}^s(C)) \sim \mathcal{H}_\varepsilon(\Phi_{p\sigma}^s(C)) \sim \frac{2}{(s+1)!} p^\theta \frac{(\log(1/\varepsilon))^{s+1}}{\log\log(1/\varepsilon)}. \tag{148}$$

Theorem XXI.

$$\mathcal{E}_\varepsilon(F_{p\theta,\sigma}^s(C)) \sim \mathcal{H}_\varepsilon(F_{p\theta,\sigma}^s(C)) \sim \Delta_{s,p,\sigma} \left(\log \frac{1}{\varepsilon}\right)^{s(p-1)/p+1}, \tag{149}$$

where

$$\Delta_{s,p,\sigma} = 2^{s+1} \frac{\sigma^{\theta/p} p(p-1)^{s/p}(\Gamma(1-1/p))^s}{(sp-s+p)(\log e)^{s(1-1/p)}\Gamma(s-s/p+1)}.$$

Note the particular case of formula (149), when there is only one variable, which relates to the next section:

$$\mathcal{E}_\varepsilon(F_{p,\sigma}^1) \sim \mathcal{H}_\varepsilon(F_{p,\sigma}^1) \sim \frac{4\sigma^{1/p} p^2 (\log(1/\varepsilon))^{2-1/p}}{(2p-1)((p-1)\log e)^{1-1/p}}. \tag{150}$$

5. It can be shown [4] that the result of Theorem XX is not changed if the polycylinder $K_1^1 \times \cdots \times K_1^s$ is replaced by the Cartesian product

$$K^1 \times \cdots \times K^s,$$

where K^i are arbitrary continua in the planes z_i.

Functions of class $\Phi^s_{p\sigma}(C)$ can be developed into *Taylor series*, while functions of the class $F^s_{p\theta,\sigma}(C)$ are developed into *Fourier series* converging for all finite z.

Let us deduce inequalities for the coefficients of these expansions.

In the first case

$$f(z) = \sum_{k \in K} a_k z^k,$$

where k and K have the same meaning as in Theorem XVII of this section. Using the Cauchy inequalities and relations (146), we get

$$|a_k| \leqslant \frac{\max_{z \in \mathring{K}^1_{R_1} \times \cdots \times z \in \mathring{K}^s_{R_s}} |f(z)|}{R^k} \leqslant \frac{Ce^{(\sigma, R^p)}}{R^k}$$

for any $R = (R_1, \ldots, R_s)$, and therefore the inequality remains valid for the minimum of the right-hand side with respect to R, which can be easily computed in an elementary way.

As the result, we obtain

$$|a_k| \leqslant C \prod_{i=1}^{s} \left(\frac{\sigma_i p_i e}{k_i} \right)^{k_i/p_i} =$$

$$= C \exp \left(-\frac{1}{\log e} \sum_{i=1}^{s} \left(\frac{k_i \log k_i}{p_i} - \frac{k_i B_i}{p_i} \right) \right), \tag{151}$$

where by B_i we denote $\log(\sigma_i p_i e)$.

In the second case

$$f(z) = \sum_{k \in \check{K}} c_k e^{i(z,k)},$$

where $\check{K}$ has the meaning given to it in Remark 14 in Subsection 1. In this case we use (147) and the inequalities for c_k which can be easily deduced in a way similar to the one used in subsection 3 in §2:

$$|c_k| \leqslant \max_{\substack{|Im z_i| \leqslant h_i \\ i=1,\ldots,s}} |f(z)| e^{-(h,|k|)}.$$

Thus we obtain

$$|c_k| \leqslant Ce^{-((\sigma, h^p) - (h,|k|))}, \tag{152}$$

where $h = (h_1, \ldots, h_s)$ is any positive vector. The inequalities (152) remains valid also for the minimum of the right-hand side with respect to h; hence, after elementary transformations, we get

$$|c_k| \leqslant Ce^{-(A,|k|^{p/(p-1)})}, \tag{153}$$

where

$$A = (A_1, \ldots, A_s),$$

$$A_i = \left(\frac{1}{\sigma_i p} \right)^{1/(p-1)} \left(\frac{p-1}{p} \right).$$

The inequalities (151) and (153) will play the same role as condition (97) in Theorem XVIII.

Now let us state some inequalities which play the role of conditions (99) in Theorem XVII.

For any positive vector $\delta = (\delta_1, \ldots, \delta_s)$, there exists a constant $C'(\delta)$ such that:

α) if for any $k \in K$ we have the inequalities

$$|a_k| \leqslant C'(\delta) \exp\left(-\frac{1}{\log e} \sum_{i=1}^{n} \left(\frac{k_i \log k_i}{p_i} - \frac{k_i B_i}{p_i} + k_i \delta_i\right)\right), \qquad (154)$$

where B_i is taken from (151), then the function

$$f(z) = \sum_{k \in K} a_k z^k$$

belongs to the class $\Phi_{p\sigma}^s(C)$;

β) if for any $k \in \tilde{K}$ we have the inequalities

$$|c_k| \leqslant C'(\delta) \exp(-\sum_{i=1}^{s}(A_i|k_i|^{p/(p-1)} + |k_i|\delta_i), \qquad (155)$$

where A_i is taken from (153), then the function

$$f(z) = \sum_{k \in \tilde{K}} c_k e^{i(z,k)}$$

belongs to the class $F_{p\theta,\sigma}^s(C)$.

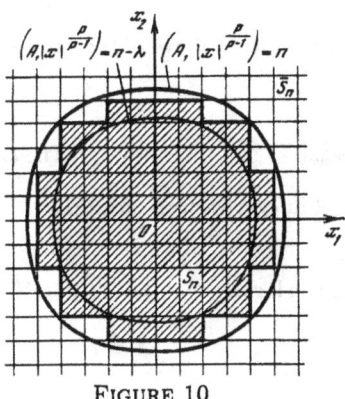

FIGURE 10

The inequalities (154) and (155) have similar proofs. Let us give the one for inequality (155). We have

$$|f(z)| \leqslant \sum_{k \in \tilde{K}} |c_k| e^{(|\Im z|,|k|)} \leqslant C'(\delta) e^{(\sigma,|\Im z|^p)} \sum_{k \in \tilde{K}} e^{-(\delta,k)} \leqslant C e^{(\sigma,|\Im z|^p)},$$

if we put

$$C'(\delta) = \frac{C}{\sum_{k \in \check{K}} e^{-(\sigma, |k|)}}.$$

Inequality (96) turns out to be valid (as can be easily checked) for our classes, namely

$$\sup_{k \in K} |a_k| \leqslant \|f\| \leqslant \sum_{k \in K} |a_k| \qquad (156)$$

for the class $\Phi_{p\sigma}^s(C)$ and

$$\sup_{k \in \check{K}} |c_k| \leqslant \|f\| \leqslant \sum_{k \in \check{K}} |c_k| \qquad (157)$$

for the class $F_{p\theta, \sigma}^s$.

6. The proof of Theorems XX and XXI is similar in its structure to the proof of Theorem XVII carried out in Subsection 2. Therefore we shall develop their proofs in parallel, with certain comments.

To reduce notation, we shall introduce the following symbols: k/p and x/p (where k, p and x are vectors) denote the vectors with components $(k_1/p_1, \ldots, k_s/p_s)$ and respectively $(x_1/p_1, \ldots, x_s/p_s)$. The notation $\log k$ for the vector k, where $k_i \geqslant 1$, will denote the vector with components $(\log k_1, \ldots, \log x_s)$.

The volume of the set $(A, |x|^{p/(p-1)}) = 1$ appearing below will be denoted further by $\Delta(A)$. It is equal to (see [24, volume 3, p.477]) the expression

$$\Delta(A) = \frac{((1 - 1/p)\Gamma(1 - 1/p))^s 2^s}{\Gamma(s - s/p + 1) A^{(1 - 1/p)\theta}}.$$

Let us begin the proof.

Estimate from above of the expression $\mathcal{H}_\varepsilon$

The class $\Phi_{p\sigma}^s(C)$ | The class $F_{p\theta, \sigma}^s(C)$ (see Fig. 10)

Consider the set S_n of integer vectors such that

$$(k/p, \log k - B) \leqslant n \log n. \qquad | \qquad (A, |k|^{p/(p-1)}) \leqslant n^{p/(p-1)}. \qquad (158)$$

By $\bar{S}_n$ denote

$$k \setminus S_n. \qquad | \qquad \check{K} \setminus S_n.$$

We have according to (156) and (157)

$$\|f\| \leqslant \sum_{k \in S_n} |a_k| + \sum_{k \in \bar{S}_n} |a_k| = \qquad | \qquad \|f\| \leqslant \sum_{k \in S_n} |c_k| + \sum_{k \in \bar{S}_n} |c_k| =$$

$$\sum_{k \in S_n} |a_k| + R_n. \qquad\qquad \sum_{k \in S_n} |c_k| + R_n. \qquad (159)$$

The inequalities (151) and (153) yields

$$|R_n| \leqslant C \sum_{k \in S_n} e^{-1/\log e(k/p, \log k - B)} \qquad |R_n| \leqslant C \sum_{k \in S_n} e^{(A, |k|^{p/(p-1)})}. \qquad (160)$$

The functions $-(x/p, \log x - B)\, x \geqslant 0$ and $-(A, x^{p/(p-1)})x \geqslant 0$ have a minimum on the cube B_k (see Figure 9) at the point k[18].

The proof of this fact will be left to the reader. Thus we have the possibility of estimating the sum (160) by the integrals

$$C \int_{\substack{(x/p, \log x - B) \geqslant (n-\lambda)\log(n-\lambda) \\ x \geqslant 0}} \exp\left(-\frac{1}{\log e}\left(\frac{x}{p}, \log x - B\right)\right) dx,$$

$$\qquad (161)$$

$$C \int_{(A, |x|^{p/(p-1)}) \geqslant (n-\lambda)^{p/(p-1)}} \exp(-(A, |x|^{p/(p-1)}))dx,$$

where λ is bounded. The integrals (161) are estimated by the relations

$$\int_{(x/p, \log x - B) \geqslant R \log R} \exp\left(-\frac{1}{\log e} \times \right.$$

$$\left. \times \left(\frac{x}{p}, \log x - B\right)\right) dx =$$

$$= e^{-R \log R} R^s \left(1 + \frac{O(1)}{\log R}\right).$$

$$\int_{(A, |x|^{p/(p-1)}) \geqslant R^{p/(p-1)}} \exp(-(A,$$

$$|x|^{p/(p-1)}))dx = O(Re^{s - R^{p/(p-1)}}),$$

hence for a minimal n satisfying $|R_n| \leqslant \varepsilon/2$, we get

$$n = \frac{\log(1/\varepsilon)}{\log\log(1/\varepsilon)} + O\left(\log\log\frac{1}{\varepsilon}\right). \qquad\qquad n = \left(\frac{\log(1/\varepsilon)}{\log e}\right)^{1-1/p}$$

$$+ O\left(\log\log\frac{1}{\varepsilon}\right). \qquad (162)$$

Denote by P_n the number of vectors k belonging to S_n. For P_n we have the inequalities

$$\int_{\substack{(x/p, \log x - B) \leqslant (n-\lambda)\log(n-\lambda) \\ x \geqslant 0}} dx \leqslant P_n \leqslant \qquad \int_{(A, |x|^{p/(p-1)}) \leqslant (n-\lambda)^{p/(p-1)}} dx \leqslant P_n \leqslant$$

---

[18]Except for a finite number of k in the case $\Phi_{p\sigma}^s(C)$, namely k for which we have $(k/p, \log k - B) \leqslant 0$. This set will be neglected.

$$\leqslant \int_{\substack{(x/p,\log x - B)\leqslant(n-\lambda)\log(n-\lambda) \\ x\geqslant 0}} dx \quad \Big| \quad \int_{(A,|x|^{p/(p-1)})\leqslant(n+\lambda)^{p/(p-1)}} dx. \quad (163)$$

The integrals written out $\quad|\quad$ The integrals written out

satisfy the estimate $\quad\quad|\quad$ are equal $\Delta(A)(n \pm \lambda)^s$, hence

$$P_n = \frac{n^s p^\theta}{s!}\left(1 + \frac{O(1)}{\log n}\right). \quad \Big| \quad P_n = \Delta(A)n^s\left(1 + \frac{O(1)}{n}\right). \quad (164)$$

Further let us approximate the coefficient $a_k = \alpha_k + i\beta_k$ (respectively $c_k = \alpha_k + i\beta_k$) by the expression $\alpha'_k + i\beta'_k$ with precision up to ε/P_n by setting

$$\alpha'_k = \left[\frac{\sqrt{2}P_n\alpha_k}{\varepsilon}\right]\frac{\varepsilon}{\sqrt{2}P_n} = m_k^1 \frac{\varepsilon}{\sqrt{2}P_n},$$

$$\beta'_k = \left[\frac{\sqrt{2}P_n\beta_k}{\varepsilon}\right]\frac{\varepsilon}{\sqrt{2}P_n} = m_k^2 \frac{\varepsilon}{\sqrt{2}P_n}. \quad (165)$$

For m_k^i, $i = 1, 2$, we obtain the inequalities

$$|m_k^i| \leqslant \quad\quad\quad | \quad\quad\quad |m_k^i| \leqslant$$

$$\left|\frac{C'P_n \exp(-1/\log e(\frac{k}{p}, \log k - B))}{\varepsilon}\right| = N_k \left|\left|\frac{C'P_n \exp(-(A, k|^{p/(p-1)}))}{\varepsilon}\right|\right| = N_k, \quad (166)$$

where C' is bounded. Hence

$$\log N_k = \log\frac{1}{\varepsilon} - \left(\frac{k}{p}, \log k-\right) \quad\Big|\quad \log N_k = \log\frac{1}{\varepsilon} - (A, |k|^{p/(p-1)})\times$$

$$-B) + \log P_n + D_1, \quad\quad\quad|\quad\quad \times\log e + \log P_n + D_2,$$

where D_1 and D_2 are bounded constants not depending on k. The expressions

$$\sum_{k\in S_n}\left(\frac{k}{p}, \log k - B\right) \quad\quad\Big|\quad\quad \sum_{k\in S_n}(A, |k|^{p/(p-1)})$$

are estimated by the integrals

$$\int_{(x/p,\log x - B)\leqslant(n-\lambda)\log(n-\lambda)} (\frac{x}{p}, \log x - B))\,dx,$$

$$\int_{(A,|x|^{p/(p-1)})\leqslant(n-\lambda)^{p/(p-1)}} (A, |x|^{p/(p-1)})dx,$$

whose principal term equals

$$\frac{sp^{\theta}n^{s}n\log n}{(s+1)!} + O(n^{s}).$$

which are equal (see [24,vol. 3, p. 479])

$$\frac{s\Delta(A)n^{s+p/(p-1)}}{s+p/(p-1)}. \qquad (167)$$

Bringing together all the results obtained, we shall have, using (162),

$$\mathcal{H}_{\varepsilon}(\Phi_{p\sigma}^{s}(C)) \underset{\sim}{\prec} \frac{2n^{s}p^{\theta}}{s!}\log\frac{1}{\varepsilon} -$$

$$\mathcal{H}_{\varepsilon}(F_{p\sigma}^{s}(C)) \underset{\sim}{\prec} 2\Delta(A)n^{s}\log 1\varepsilon -$$

$$- \frac{2sn^{s}p^{\theta}}{(s+1)!}n\log n \sim$$

$$- 2\Delta(A)\frac{s}{s+p/(p-1)}n^{s+p/(p-1)} \times$$

$$\sim \frac{2p^{\theta}}{(s+1)!}\frac{(\log(1/\varepsilon))^{s+1}}{(\log\log(1/\varepsilon))}.$$

$$\times \log e \sim \Delta_{sp\sigma}\left(\log\frac{1}{\varepsilon}\right)^{s(p-1)/p+1} \qquad (168).$$

Estimate from below of the expression E_{ε}

Choosing an arbitrary $\delta > 0$, let us find the largest m satisfying

$$C'(\delta)m^{-m} \geqslant 2\sqrt{2}\varepsilon, \qquad\qquad C'(\delta)e^{-m^{p/(p-1)}} \geqslant 2\sqrt{2}\varepsilon, \qquad (169)$$

where $C'(\delta)$ is the constant from (154) and (155).

Consider the set S_{m} of integer vectors introduced above for m satisfying (169). Suppose Ψ denotes the set of functions

$$f(z) = \sum_{k\in K} a_{k}z^{k}$$

$$f(z) = \sum_{k\in \tilde{K}} c_{k}e^{i(k,z)}$$

such that

$\alpha)$ $a_{k} = 0$, $k \notin S_{m}$;

$\alpha)$ $c_{k} = 0$, $k \notin S_{m}$;

$\beta)$ $a_{k} = (s_{k}^{1} + is_{k}^{2})2\varepsilon$, $k \in S_{m}$;

$\beta)$ $c_{k} = (s_{k}^{1} + is_{k}^{2})2\varepsilon$, $k \in S_{m}$;

$\gamma)$ s_{k}^{i}, $i = 1, 2$, $-$integers and

$\gamma)$ s_{k}^{i}, $i = 1, 2$, $-$integers and

$$|s_{k}^{i}| \leqslant \frac{C'(\delta)}{2\sqrt{2}\varepsilon}\exp\left(-\frac{1}{\log e}\left(\frac{p}{k},\right.\right.$$

$$|s_{k}^{i}| \leqslant \frac{C'(\delta)}{2\sqrt{2}\varepsilon}\exp(-A,|k|^{p/(p-1)}) -$$

$$\log k - B + p\delta)) = M_{k}.$$

$$-(\delta, |k|)) = M_{k}. \qquad (170)$$

Thus according to (154) and (155)

$$\{\Psi\} \subset \Phi^s_{p\sigma}(C). \qquad \qquad \{\Psi\} \subset F^s_{p\theta,\sigma}(C).$$

By using (168) and (170), we get

$$m = \frac{\log(1/\varepsilon)}{\log\log(1/\varepsilon)} + O(1), \qquad \qquad m = \left(\frac{\log(1/\varepsilon)}{\log e}\right)^{1-1/p} + O(1),$$

$$\log M_k = \log\frac{1}{\varepsilon} - (k/p,\ \log k - \qquad \log M_k = \log\frac{1}{\varepsilon} - (A, |k|^{p/(p-1)})\times$$

$$-B + p\delta) + \log C'(\delta) + D'_1, \qquad |\times \log e - (\delta, |k|) \log e + \log C'(\delta) + D'_2, (171)$$

where D'_1 and D'_2 are bounded constants. Hence computations similar to the ones for the estimate above finally give

$$\mathcal{E}_{2\varepsilon}(\Phi^s_{p\sigma}(C)) \underset{\sim}{\succ} \frac{2p^\theta}{(s+1)!} \times \qquad \mathcal{E}_{2\varepsilon}(F^s_{p\theta,\sigma}(C)) \underset{\sim}{\succ} \Delta_{sp\sigma} \times$$

$$\times \frac{(\log(1/\varepsilon))^{s+1}}{(\log\log(1/\varepsilon))^s}. \qquad \times \left(\log\frac{1}{\varepsilon}\right)^{s((p-1)/p)+1}. \qquad (172)$$

Now relations (168) and (172) imply (148) and (149), which was to be proved.

§8. The ε-entropy of classes of analytical functions bounded on the real axes[19].

1. The results of the present section are due to V. M. Tikhomirov. They were published in part in the note [7].

The subject matter of the present section appeared in connection with certain ideas of information theory. In communication theory, an important role is played by the fact that signals with bounded spectrum placed in a strip of frequency of width 2σ are determined by a discrete family of their values taken at equidistant points[20].

In the paper [26], V. A. Kotelnikov indicated an important application of this property to the theory of communications. He noted that the amount of information contained in the definition on a segment of length T of a function with spectrum bounded in the frequency range of width 2σ, for large T, is equivalent to the amount of information used in defining $2\sigma T/\pi$ real numbers. In the literature, this statement is often called the *Kotelnikov theorem*. This same idea in a somewhat different form was expressed also by C. Shannon [27, Appendix 7].

---

[19] An abridged version is published here, but the numbers of the formulas, theorems and lemmas are the original ones.

[20] In the application to functions whose Fourier transform (spectrum) vanishes outside of a certain segment of length 2σ, this idea of recovering functions from the discrete family of their values taken in an arithmetical progression with step π/σ was first established by Vallée-Poussin.

The class $\mathcal{B}_\sigma(C)$ which we shall meet in this section is the closure in the uniform metric on a segment of length T (for any $T > 0$) of the class $B_\sigma(C)$ of finite sums

$$f(t) = \sum_{k=-n}^{n} c_k e^{i\lambda k^t},$$

subjected on the real axis to the condition $|f| \leqslant C$ with frequencies λ_k from the segment $-\sigma \leqslant \lambda \leqslant \sigma$ (see [23, p. 160-164]). Hence Theorem XXII formulated above may be viewed as one of the versions of establishing the Kotelnikov theorem.

In the present section we consider the following classes of functions of one complex variable:

$\mathcal{B}_\sigma(C)$ – the class of entire analytic functions satisfying the relation

$$|f(z)| \leqslant C e^{\sigma|\Im z|}.$$

The notation $\mathcal{B}_\sigma$ is taken from [23].

$\mathcal{F}_{p\sigma}(C)$ – the class of entire functions of order p and type σ satisfying the relations

$$|f(z)| \leqslant C e^{\sigma|\Im z|^p}.$$

$\mathcal{A}_h(C)$ – the class of analytic functions $f(z)$ bounded in the strip $|\Im z| < h$ by the constant C.

All the classes listed here, as can be easily seen, are bounded on the real axis D.

Note that the class $\mathcal{B}_\sigma$ coincides with the class $\mathcal{F}_{1\sigma}$, but we have listed it separately in view of its important role in the sequel[21].

The classes of functions listed above will be regarded as lying in a normed space by introducing the norms and the metric by the formulas

$$\|f\|_T = \max_{t \in \Delta_T} |f(t)|, \quad \rho_T(f_1, f_2) = \|f_1 - f_2\|_T, \tag{173}$$

where Δ_T is the closed interval $-T \leqslant t \leqslant T$ of the real line.

Everywhere further in this section the ε-entropy and the ε-capacity of the class $\mathcal{F}$ of functions defined on D with the norm (173) will be denoted respectively by $\mathcal{H}_\varepsilon^T(\mathcal{F})$ and $\mathcal{E}_\varepsilon^T(\mathcal{F})$.

To formulate the results which will be proved in this section, we shall require the definition of the upper and lower ε-entropy and ε-capacity per unit length.

The *upper ε-entropy per unit length* for the class $\mathcal{F}$ defined on D is the function

$$\bar{\mathcal{H}}_\varepsilon^B(\mathcal{F}) = \lim_{T \to \infty} \sup (2T)^{-1} \mathcal{H}_\varepsilon^T(\mathcal{F}).$$

The *lower ε-entropy per unit length* is by definition the function

$$\bar{\mathcal{H}}_\varepsilon^H(\mathcal{F}) = \lim_{T \to \infty} \inf (2T)^{-1} \mathcal{H}_\varepsilon^T(\mathcal{F}).$$

In a similar way one defines the upper and lower ε-capacity for lower length $\bar{\mathcal{E}}_\varepsilon^B(\mathcal{F})$ and $\bar{\mathcal{E}}_\varepsilon^H(\mathcal{F})$.

We have the following theorems.

---

[21] As usual, we assume that $f \in \mathcal{B}_\sigma$, $\mathcal{F}_{p\sigma}$ or $\mathcal{A}_h$ if f belongs respectively to $\mathcal{B}_\sigma(C)$, $\mathcal{F}_{p\sigma}(C)$ or $\mathcal{A}_h(C)$ with a certain constant C.

Theorem XXII.

$$\bar{\mathcal{H}}_\varepsilon^B(\mathcal{B}_\sigma(C)) \sim \bar{\mathcal{H}}_\varepsilon^H(\mathcal{B}_\sigma(C)) \sim \frac{2\sigma}{\pi} \log \frac{1}{\varepsilon}. \tag{174}$$

Theorem XXIII.

$$\bar{\mathcal{H}}_\varepsilon^B(A_h(C)) \sim \bar{\mathcal{H}}_\varepsilon^H(A_h(C)) \sim \frac{1}{\pi h \log e} \left(\log \frac{1}{\varepsilon}\right)^2. \tag{175}$$

Theorem XXIV.

$$\bar{\mathcal{H}}_\varepsilon^B(\mathcal{F}_{p\sigma}(C)) \sim \bar{\mathcal{H}}_\varepsilon^H(\mathcal{F}_{p\sigma}(C)) \sim \frac{2\sigma^{1/p} p^2 (\log(1/\varepsilon))^{2-1/p}}{\pi(2p-1)((p-1)\log e)^{1-1/p}} \tag{176}$$

Relations for $\bar{\mathcal{E}}_\varepsilon^B$ and $\bar{\mathcal{E}}_\varepsilon^H$ are of the same form.

We suggest that the reader compare the formulas (175) and (176) written above with formulas (130) and (149) for ordinary ε-entropy of the classes $F_{p\sigma}^1$ and A_h. (Recall that these classes consist of periodic functions with period 2π belonging to the classes $F_{p\sigma}(C)$ and $A_h(C)$). The similarity of these formulas is supplemented by the similarity of their methods of proof that we shall observe in the sequel.

Also note the relation

$$\mathcal{H}_\varepsilon(\mathcal{B}_\sigma(C)) \sim (4[\sigma]+2)\log(1/\varepsilon) \tag{177}$$

for the class of functions from $\mathcal{B}_\sigma$ periodic with period 2π; it is natural to compare this relation with (174). The equivalence of (177) follows from the fact that the class $\mathcal{B}_\sigma(C)$ consists of trigonomeric polynomials of degree no greater than $|\sigma|$ (see [23, p. 190]).

2. We pass to the proof of Theorem XXII. Following the style of the previous section, we shall distinguish separately into two lemmas the properties of the class $\mathcal{B}_\sigma$ that are used further in the proof.

In Lemmas IV and V, α is an arbitrary positive number (in Lemma V less than σ) while $C_1(\alpha)$ and $C_2(\alpha)$ are positive constants depending only on α.

Lemma IV. *The function $f(z) \in \mathcal{B}_\sigma(C)$ may be represented in the form of a series*

$$f(z) = \sum_{k=-\infty}^{+\infty} c_k f_k(z). \tag{178}$$

Here we have:
a) $|c_k| \leqslant C$.
b) There exists a converging series with positive terms

$$\sum_{k=-\infty}^{+\infty} a_k = A \tag{179}$$

such that for f representd by the series (178) we have

$$|f(t)| \leqslant C_1(\alpha) \sum_{k=-\infty}^{+\infty} a_k |C_{[t(\sigma+\alpha)/\pi]+k}|. \tag{180}$$

From Lemma IV we obtain

$$\limsup_{\varepsilon \to 0} \frac{\bar{\mathcal{H}}_\varepsilon^B(\mathcal{B}_\sigma(C))}{\log(1/\varepsilon)} \leqslant \frac{2\sigma}{\pi}. \tag{181}$$

Lemma V. *There exist functions $\varphi_k(z)$, $k = 0, \pm 1, \ldots$, such that the inequality*

$$|c_k| \leqslant C_2(\alpha)C,$$

implies that the function

$$f(z) = \sum_{|k| \leqslant [T(\sigma - \alpha)/\pi]} c_k \varphi_k(z) \tag{182}$$

belongs to $\mathcal{B}_\sigma(C)$ and

$$\|f\|_T \geqslant \max_{|k| \leqslant [T(\sigma - \alpha)/\pi]} |c_k|. \tag{183}$$

Lemma V yields

$$\liminf_{\varepsilon \to 0} \frac{\bar{\mathcal{E}}_{2\varepsilon}^H(\mathcal{B}_\sigma(C))}{\log(1/\varepsilon)} \geqslant \frac{2\sigma}{\pi}. \tag{184}$$

In view of the fact that

$$\bar{\mathcal{E}}_{2\varepsilon}^H \leqslant \bar{\mathcal{E}}_{2\varepsilon}^B \leqslant \bar{\mathcal{H}}_\varepsilon^H \leqslant \bar{\mathcal{H}}_\varepsilon^B,$$

relations (181) and (184) taken together give us Theorem XXII.

Let us deduce inequalities (181) and (184) from the corresponding lemmas.

Fix α. Choose n so as to have

$$R_n < \varepsilon/2C_1(\alpha)C, \tag{185}$$

where $R_n = \sum_{|k| > n} a_k$ is the remainder of the series (179).

Put $n_T = [(\sigma + \alpha)T/\pi] + n + 1$. The coefficients $c_k = \alpha_k + i\beta_k$, $|k| \leqslant n_T$ appearing in the series (178) can be approximated with precision up to $\varepsilon/2AC_1(\alpha)$ (here A is the sum of the series (179)) by putting

$$c'_k = \left[\frac{\alpha_k 2\sqrt{2}AC_1(\alpha)}{\varepsilon}\right] \frac{\varepsilon}{2\sqrt{2}AC_1(\alpha)} + i\left[\frac{\beta_k 2\sqrt{2}AC_1(\alpha)}{\varepsilon}\right] \frac{\varepsilon}{2\sqrt{2}AC_1(\alpha)}.$$

Suppose

$$\tilde{f}(z) = \sum_{|k| \leqslant n_T} c'_k f_k(z).$$

For $|t| \leqslant T$ we have

$$|f(t) - \tilde{f}(t)| \leqslant C_1(\alpha) \sum_{k=-n}^{n} a_k |C_{[t(\sigma+\alpha)/\pi]+k} - C'_{[t(\sigma+\alpha)/\pi]+k}| +$$

$$+ C_1(\alpha) \sum_{|k| > n} A_k |C_{[t(\sigma+\alpha)/\pi]+k}| =$$

$$= \frac{\varepsilon}{2A} \sum_{|k| \leqslant n} a_k + C_1(\alpha)C \sum_{|k| > n} a_k \leqslant \varepsilon. \tag{186}$$

In the calculation of (186) we use the fact that for all $s = [t(\sigma + \alpha)/\pi] + k$, $|t| \leqslant T$, $|k| \leqslant n$, the constants c_s are approximated with precision up to $\varepsilon/2AC_1(\alpha)$ and apply relations (185).

The relation (186) means that $\|f - \tilde{f}\|_T \leqslant \varepsilon$, i.e., the set of all functions $\{\tilde{f}\}$ constitutes a ε-net for the class $\mathcal{B}_\sigma(C)$ in the metric ρ_T. It is easy to calculate that the number $N(T)$ of elements in this set $\{f\}$ satisfies the inequality[22]

$$N(T) \leqslant \left(2 \left[\frac{2\sqrt{2}AC_1(\alpha)C}{\varepsilon} \right] + 1 \right)^{4n_T + 2} = $$
$$= \left(2 \left[\frac{C'C}{\varepsilon} \right] + 1 \right)^{4[T(\sigma+\alpha)/\pi] + 4n + 6} . \tag{187}$$

Taking logarithms in (187) and using the fact that $\mathcal{H}_\varepsilon^T(\mathcal{B}_\sigma(C)) \leqslant \log N(T)$, and also the fact that α is arbitrary, we obtain (181).

Inequality (184) can be obtained in a simpler way. Fixing α, let us put $m_T = [T(\sigma - \alpha/\pi)]$. Consider the matrix of integer vectors

$$U = \begin{pmatrix} s_m^1 & \cdots & s_0^1 & \cdots & s_m^1 \\ s_{-m}^2 & \cdots & s_0^2 & \cdots & s_m^2 \end{pmatrix} .$$

Here we shall require that

$$|s_k^i| \leqslant CC_2(\alpha)/\sqrt{2}\varepsilon. \tag{188}$$

To each matrix U from the given set, we assign the function $f_U(t)$:

$$f_U(t) = 2\varepsilon \sum_{|k| \leqslant m_T} (s_k^1 + i s_k^2)\varphi_k(z).$$

It follows from (183) that the set $\{f_U(t)\}$ is 2ε-distinguishable in the metric ρ_T.

The number $M(T)$ of functions contained in this set, according to (188), for sufficiently small ε satisfies the inequality

$$M(T) \geqslant \left(2 \left[\frac{CC_2(\alpha)}{2\sqrt{2}\varepsilon} \right] \right)^{4m_T} \geqslant \left(2 \left[\frac{CC_2(\alpha)}{2\sqrt{2}\varepsilon} \right] \right)^{2[T(\sigma-\alpha)/\pi]} . \tag{189}$$

Taking logarithms in (187) and using the fact that $C_{2\varepsilon}^T(\mathcal{B}_\sigma(C)) \geqslant \log M(T)$, as well as the fact that α is arbitrary, we obtain (184).

3. To conclude the proof of Theorem XXII, we must establish Lemmas IV and V from subsection 2.

Let us use the Cartwright interpolation formula. Suppose $g(t) \in \mathcal{B}_{\sigma'}(C)$, $\sigma' < \pi$, $0 < \omega < \pi - \sigma'$. Then we have

$$g(t) = \frac{\sin \pi t}{\pi \omega} \sum_{k=-\infty}^{+\infty} (-1)^k \frac{g(k) \sin \omega(t - k)}{(t - k)^2} . \tag{190}$$

---

[22]Here by C' we denote the constant $2AC_1(\alpha)$.

The proof of formula (190) is given in [29, p. 269]. It is easy to understand that if $f \in \mathcal{B}_\sigma(C)$, then $g(t) = f(\sigma' t / \sigma) \in \mathcal{B}_{\sigma'}(C)$. Using this, let us rewrite formula (190) for the functions $f \in \mathcal{B}_\sigma$ and an arbitrary σ:

$$f(t) = \frac{\sin \pi \sigma t/\sigma'}{\pi \omega} \sum_{k=-\infty}^{+\infty} (-1)^k \frac{f(\sigma' k/\sigma) \sin \omega(\sigma t/\sigma' - k)}{(\sigma t/\sigma' - k)^2}. \tag{191}$$

Note that formula (191) expresses the possibility of recovering the function from its discrete sequence in the case of the class $\mathcal{B}_\sigma$ that was mentioned at the beginning of subsection 1.

Now for an arbitrary α put

$$\sigma' = \frac{\pi \sigma}{\sigma + \alpha}, \quad \text{or,} \quad \frac{\sigma}{\sigma'} = \frac{\sigma + \alpha}{\pi}, \quad \omega = \frac{\pi - \sigma'}{2},$$

$$f_k(z) = \frac{(-1)^{k+1} \sin(\sigma + \alpha) t \sin \omega(k - (\sigma + \alpha)t/\pi)}{\pi \omega (k - t(\sigma + \alpha)/\pi)^2},$$

$$c_k = f(\pi k/(\sigma + \alpha)).$$

Let

$$f(z) = \sum_{k=-\infty}^{+\infty} c_k f_k(z). \tag{192}$$

Let us prove that the series (192) satisfies the conditions set in Lemma IV.

Condition a) is satisfied in view of the fact that $f \in \mathcal{B}_\sigma(C)$. Further for a fixed t in the sum (192) let us distinguish separately the summands with numbers

$$k_0 = [t(\sigma + \alpha)/\pi], \quad k_0 + 1 \text{ and } k_0 - 1.$$

For the other k, i.e. the numbers k given by the formula

$$k = \pm s + k_0, \quad s = 2, 3, \ldots,$$

let us use the fact that

$$\frac{1}{(k - t(\sigma + \alpha)/\pi)^2} \leqslant \frac{1}{(|s| - 1)^2},$$

obtaining

$$|f(t)| \leqslant \sum_{s=-1}^{+1} |c_{k_0+s}| |f_{k_0+s}(t)| + \sum_{|s| \geqslant 2} |c_{k_0+s}| \left(\frac{1}{|s| - 1} \right)^2.$$

Further, denoting by $a_{-1} = a_0 = a_1$ the maximum of the function

$$f_0(t) = \frac{\sin(\sigma + \alpha) t \sin(\omega(\sigma + \alpha)t/\pi)}{P \pi \omega t^2}$$

and by $a_{\pm s}$ the expression $(|s| - 1)^{-2}$, $s = 2, 3, \ldots$, we see that condition b) of Lemma IV is satisfied.

For the function φ_k in Lemma V, let us choose the function

$$\frac{\sin(\sigma - \alpha)(z - k\pi/(\sigma - \alpha))\sin\alpha(z - k\pi/(\sigma - \alpha))}{\alpha(\sigma - \alpha)(z - k\pi/(\sigma - \alpha))^2}.$$

Lemma V is a consequence of the following obvious facts:

1) $\varphi_k \in B_\sigma$ and therefore

$$\sum_{k=-m}^{m} c_k \varphi_k \in B_\sigma;$$

2) for uniformly bounded $|c_k|$, the sums

$$\sum_{k=-m}^{m} c_k \varphi_k$$

are uniformly bounded on D.

3)
$$\sum_{k=-m}^{m} c_k \varphi_k \left(\frac{l\pi}{\sigma - \alpha}\right) = c_l, \ |l| \leqslant m.$$

This concludes the proofs of Lemmas IV and V and therefore that of Theorem XXII.

4. We shall now proceed to prove theorems XXIII and XXIV. The method we will use is, as it will be seen later, a synthesis of methods that we used in § 7 and for proving theorem XXII of this section.

This can be easily seen even from the formulations of lemmas VI and VII below. In lemma VI, by the class $B_{\Lambda,\sigma}(C)$ we understand the set of functions

$$f(t) = e^{i\Lambda t}\tilde{f}(t),$$

where

$$\tilde{f}(t) \in B_\sigma(C).$$

It is easy to see that all the relations obtained for the class $B_\sigma(C)$ are also completely valid for the class $B_{\Lambda,\sigma}(C)$ for any Λ. In lemmas VI and VII, α and σ are arbitrary positive numbers.

Lemma VI. *Any function* $f(t)$ *belonging to the class*
1) $A_n(C)$,
2) $\mathcal{F}_{p\sigma}(C)$, *can be represented in the form of a series*

$$f(z) = \sum_{n=-\infty}^{+\infty} f_n(z), \tag{193}$$

where, for the former case,

$$f_n(z) \in B_{n,\frac{1}{2}+\alpha}(C_1(\alpha)e^{-|n|h}) \quad (n = 0, \pm 1, \ldots), \tag{194}$$

and for the latter case,

$$f_n(z) \in B_{n,\frac{1}{2}+\alpha}\left(C_1(\alpha)e^{-|n|^{\frac{p}{p-1}}(\frac{1}{\sigma p})^{\frac{1}{p-1}}(\frac{p-1}{p})}\right) \quad (n = 0, \pm 1, \ldots), \tag{195}$$

$C_1(\alpha)$ *is a constant dependent only on* α.

Lemma VII. *There exist functions $\phi_{kn}(z)$ $(k = 0, \pm 1, \ldots; n = 0, \pm 1m \ldots)$ (depending on α) such that if*

$$1) \quad |c_{kn}| \leqslant C_2(\alpha, h') e^{-|n|h'}, \quad h' > h \tag{196}$$

$$2) \quad |c_{kn}| \leqslant C_2(\alpha, h') e^{-\{|n|^{\frac{p}{p-1}}(\frac{1}{\sigma p})^{\frac{1}{p-1}}(\frac{p-1}{p})+|n|\delta\}}, \quad \delta > 0, \tag{197}$$

then, in the first case,

$$f(z) = \sum_{|k| \leqslant [\frac{T}{2\pi}]} \sum_{n=-\infty}^{+\infty} c_{kn} \phi_{kn}(z) \tag{198}$$

belongs to $\mathcal{A}_n(C)$, and in the second case, to $\mathcal{F}_{p\sigma}(C)$. In addition, for any $\lambda > 0$, there exists a $\tau(\lambda)$ such that if

$$|C_{kn}| \geqslant \lambda,$$

then

$$\|f\|_{T+\tau} \geqslant C_3(\alpha)\lambda, \tag{199}$$

C_2, C_3, *and τ are positive constants.*

Note that lemma VI can be reformulated in such a way that its relation with lemmas IV and VII will be more striking. By using the result of lemma IV applied to the classes $\mathcal{B}_{n,\frac{1}{2}\alpha}$, we obtain the following lemma.

Lemma VI'. *Any function belonging to the class*
1) $\mathcal{A}_h(C)$,
2) $\mathcal{F}_{p\sigma}(C)$ *can be represented in the form of a series*

$$f(z) = \sum_{h=-\infty}^{+\infty} \sum_{n=-\infty}^{+\infty} c_{kn} f_{kn}.$$

Moreover,
 a1) $|c_{kn}| \leqslant (C_1(\alpha)) e^{-|n|h}$,
 a2) $|c_{kn}| \leqslant (C_1(\alpha)) e^{-\{|n|^{\frac{p}{p-1}}(\frac{1}{\sigma p})^{\frac{1}{p-1}}(\frac{p-1}{p})\}}$,
 b) $|F(t)| \leqslant C_1(\alpha) \sum_{h=-\infty}^{+\infty} \sum_{n=-\infty}^{+\infty} a_k |C_{[\frac{t(\frac{1}{2}+\alpha)}{\pi}]+k,n}|$, where $\sum_{k=-\infty}^{+\infty} a_k = a$ is the series
from lemma IV.

From lemma VI we obtain the inequalities:

$$\limsup_{\varepsilon \to 0} \frac{\bar{\mathcal{H}}_\varepsilon^B(\mathcal{A}_h(C))}{(\log \frac{1}{\varepsilon})^2} \leqslant \frac{1}{\pi h \log e}, \tag{200}$$

$$\limsup_{\varepsilon \to 0} \frac{\bar{\mathcal{H}}_\varepsilon^B(\mathcal{F}_{p\sigma}(C))}{(\log \frac{1}{\varepsilon})^{2-\frac{1}{s}}} \leqslant \frac{2\sigma^{\frac{1}{p}} p^2}{\pi(2p-1)\{(p-1)\log e\}^{1-\frac{1}{p}}}. \tag{201}$$

From lemma VII the following inequalities are obtained:

$$\lim_{\varepsilon \to 0} \sup \frac{\tilde{C}_\varepsilon^H(A_h(C))}{(\log \frac{1}{\varepsilon})^2} \geqslant \frac{1}{\pi h \log e}, \tag{202}$$

$$\lim_{\varepsilon \to 0} \inf \frac{\tilde{C}_{2\varepsilon}^H(F_h(C))}{(\log \frac{1}{\varepsilon})^{2-\frac{1}{s}}} \geqslant \frac{2\sigma^{\frac{1}{p}} p^2}{\pi(2p-1)\{(p-1)\log e\}^{1-\frac{1}{s}}}. \tag{203}$$

Inequalities (200) and (202) present, together, theorem XXIII, and inequalities (201) and (203) theorem XXIV.

We deduce inequalities (200) and (202 from lemmas VI and VII. Fix an $\alpha > 0$. Then, according to lemma VI, we have:

$$f(t) = \sum_{n=-\infty}^{+\infty} f_n(t),$$

where

$$f_n(t) \in B_{n,\frac{1}{2}+\alpha}(C_1(\alpha)e^{-|n|h});$$

the latter has the meaning that

$$|f_n(t)| \leqslant C_1(\alpha)e^{-|n|h}. \tag{204}$$

Hence, according to (204), we get

$$\|f\|_T \leqslant \|f\|_\infty \leqslant \max_{t \in D}\left\{ |\sum_{n \leqslant m} f_n(t)| + |R_m(t)| \right\},$$

where

$$|R_m(t)| \leqslant \sum_{|n|>m} |f_n(t)| \leqslant C_1(\alpha) \sum_{|n|\geqslant m} e^{-|n|h}.$$

Now, choose a maximal m so that the last sum become smaller than ε. Then it is obvious that

$$m = \frac{1}{h \log e} \log \frac{1}{\varepsilon} + O(1).$$

Now, we shall construct ε-nets in the classes $B_{n,\frac{1}{2}+\alpha}(C_1(\alpha)e^{-|v|h})$ for $|n| \leqslant m$ in the same manner as it was done in the proof of theorem XXII. According to inequality (187), we obtain an estimate of the number $N_k(T)$ of elements in this ε-net:

$$N_k(T) \leqslant \left(2\left[\frac{C'C_1(\alpha)e^{-|k|h}}{\varepsilon}\right]\right)^{4\left[\frac{T(\frac{1}{2}+2\alpha)}{\pi}+n+1\right]+2},$$

whence

$$\mathcal{H}_\varepsilon^T(A_h(C)) \leqslant \sum_{|k|\leqslant m} \log N_k(t) \leqslant$$

$$\leqslant 4\left(\left[\frac{T(\frac{1}{2}+2\alpha)}{\pi}\right] + n + 1\right)\left(m \log, \frac{1}{\varepsilon} + O(\log \log \frac{1}{\varepsilon})\right)$$

and, finally,

$$\limsup_{\varepsilon \to 0} \frac{\tilde{\mathcal{H}}_\varepsilon^B(\mathcal{A}_h(C))}{(\log \frac{1}{\varepsilon})^2} \leqslant \frac{2(\frac{1}{2} + 2\alpha)}{\pi h \log e} \leqslant \frac{1}{\pi h \log e}$$

in view of the arbitrariness of α.

We now prove (202). Specify $h' > h$, $\alpha > 0$, $\varepsilon > 0$ and choose a maximal r so that

$$C_2(\alpha, h')e^{-|r|h'} \geqslant 2\varepsilon.$$

Then we obtain that

$$e = \frac{\log 1\varepsilon}{h' \log e} + O(1). \tag{205}$$

Put

$$\lambda = \frac{2\varepsilon}{C_3(\alpha)}.$$

Denote

$$K_T = \frac{T}{2\pi}.$$

Consider the set of functions

$$f_U(t) = \lambda \sum_{|k| \leqslant K_T} \sum_{|n| \leqslant r} s_{kn} \phi_{kn}(t), \tag{206}$$

depending on integral matrices

$$U = \|s_{kn}^i\| \quad |k| \leqslant K_T, \quad |n| \leqslant r, \quad i = 1, 2,$$

which, consequently, have $4K_T + 2$ rows and $2r + 1$ columns. The function ϕ_{kn} is taken from lemma VII.

In (206), we take all possible integers s_{kn}^i that satisfy the inequality[23]

$$|s_{kn}^i| \leqslant \left| \frac{C_2(\alpha, h')e^{-|n|h'}}{\lambda} \right| = M_n \quad (i = 1, 2). \tag{207}$$

According to lemma VII, all the functions $f_U(t)$ belong to $\mathcal{A}_h$ and, with respect to the metric ρ_{T+r}, they are 2ε-distinguishable. The number M_T of all these matrices is, obviously, not smaller than

$$\prod_{n=-r}^{+r} M_n^{4K_T+2}. \tag{208}$$

From (207) follows

$$\log M_n = \log \frac{1}{\varepsilon} - |n|h' \log e + O(1),$$

---

[23]The constants τ, C_1, and C_2 are taken from lemma VII.

whence, taking the logarithm of (208), we obtain that

$$C_{2\epsilon}^{T+\tau}(A_h(C)) \geqslant (4K_T + 2)\left(\frac{\log \frac{1}{\epsilon}}{h' \log e} + O\left(\log \log, \frac{1}{\epsilon}\right)\right). \tag{209}$$

From (209) due to the arbitrariness of $h' > h$ and α, and taking into account that $K_T = \frac{T}{2\pi}$, we get

$$\liminf_{\epsilon \to 0} \frac{\bar{C}_{2\epsilon}^H}{(\log \frac{1}{\epsilon})^2} \geqslant \frac{2 \cdot \frac{1}{2}}{\pi h' \log e} \geqslant \frac{1}{\pi h \log e},$$

as required.

The derivation of inequalities (201) and (203) is left to the reader.

5. In order complete the proof of the theorem from this section, it remains only to establish lemmas VI and VII.

Let $f(z) \in A_h$. By the Schwartz integral over a strip,

$$f(z) = \frac{1}{4h} \int_{-\infty}^{+\infty} \frac{(u_+(t) + u_-(t))dt}{\mathrm{ch}\,\frac{\pi}{2h}(t - z)} - \frac{i\mathrm{sh}\,\frac{\pi z}{2h}}{4h} \int_{-\infty}^{+\infty} \frac{(u_+(t) - u_-(t))dt}{\mathrm{ch}\,\pi t 2h \mathrm{ch} \cdot \frac{\pi}{2h}(t - z)}, \tag{210}$$

where $u_+(t) = \mathrm{Re}\, f(t + ih)$, $u_-(t) = \mathrm{Re}\, f(t - ih)$ (see [29], p. 212).

Consider functions $\phi_1(z)$ and $\phi_2(z)$, analytic in the region $\|\mathrm{Im}\, z\| < h$, such that

$$\mathrm{Re}\,\phi_1(t - ih) = \mathrm{Re}\,\phi_1(t + ih) = u_+(t) + u_-(t) = u(t),$$
$$\mathrm{Re}\,\{i\phi_2(t - ih) = -\mathrm{Re}\,\{i\phi_2(t + ih) = \mathrm{Im}\, f(t + ih) + \mathrm{Im}\, f(t - ih) =$$
$$= v_+(t) + v_-(t) = -v(t).$$

It is easy to establish from (210) that

$$f(z) = \phi_1(z) + \phi_2(z),$$

and the function $i\phi_2(z)$ belongs to A_h which is real in D. From (210) we obtain

$$f(z) = \frac{1}{4h} \int_{-\infty}^{+\infty} \frac{u(t) + iv(t)}{\mathrm{ch}\,\frac{\pi(t-z)}{2h}} dt = \frac{1}{4h} \int_{-\infty}^{+\infty} \frac{H_f(t)dt}{\mathrm{ch}\,\frac{\pi(t-z)}{2h}} \tag{211}$$

for any

$$|H_f(t)| = |u(t) + iv(t)| \leqslant 2C$$

and for all $t \in D$.

Consider the function $\frac{f(z) - f(0)}{z} = \Phi(z)$. It is easy to see that for it the function from (211) $H_\Phi(t) \in L^2(-\infty, +\infty)$. Applying the convolution formula for Fourier transforms (see [23], p. 119), we obtain:

$$f(z) = f(0) + \frac{z}{\sqrt{2\pi}} \int_{-\infty}^{+\infty} \frac{\Psi(u)}{\mathrm{ch}\, hu} e^{int} du, \tag{212}$$

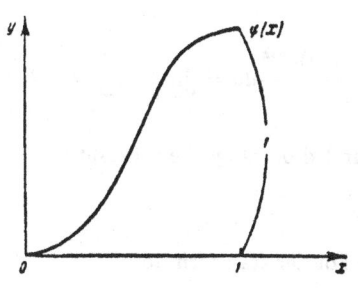 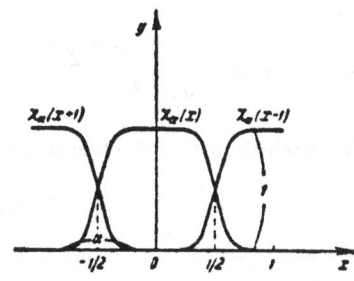

FIGURE 11

where $\Psi(u)$ is the Fourier transform of the function $H_\Phi(t)$. In obtaining (212) we used the fact that the Fourier transform of $\left(\text{ch}\,\frac{\pi t}{2h}\right)^{-1}$ is equal to $(\text{ch}\,hu)^{-1}$.

Choose an arbitrary α, and consider the function $\chi_\alpha(u)$ (Fig. 11). The function $\chi_\alpha(u)$ is equal to identity on the interval $[-\frac{1}{2}+\alpha, \frac{1}{2}-\alpha]$, and on the intervals $[-\alpha-\frac{1}{2}, -\frac{1}{2}\alpha]$ and $[\frac{1}{2}-\alpha, \frac{1}{2}+\alpha]$ it is equal to $\phi\left(\frac{x+\alpha-\frac{1}{2}}{2\alpha}\right)$ and $1-\phi\left(\frac{x-\alpha+\frac{1}{2}}{2\alpha}\right)$, respectively; here the function $\phi(x)$ is defined on the interval $[0,1]$, and, moreover, $0 = \phi(0) = \phi'(0) = \phi'(1)$, and the second derivative is of the form:

$$\phi''(x) = \begin{cases} 0, & x < 0, \\ 2, & 0 < x \leqslant \frac{1}{4}, \\ -2, & \frac{1}{4} < x \leqslant frac34, \\ 2, & \frac{3}{4} < x \leqslant 1, \\ 0, & x > 1. \end{cases} \tag{213}$$

It is easy to calculate that $\phi(1) = 1$. Put

$$f_0(z) = f(0) + \frac{z}{\sqrt{2\pi}} \int_{-\infty}^{+\infty} \frac{\Psi(u)\chi_\alpha(u)e^{iuz}}{\text{ch}\,hu}\,du,$$

$$f_n(z) = \frac{z}{\sqrt{2\pi}} \int_{-\infty}^{+\infty} \frac{\Psi(u)\chi_\alpha(u+n)e^{iuz}}{\text{ch}\,hu}\,du \tag{214}$$

$$(n = \pm 1, \pm 2, \ldots).$$

Later, we shall show that the function

$$f_n(z) \in B_{n, \frac{1}{2}+\alpha}(C_2(\alpha)e^{-|n|h}). \tag{215}$$

Then

$$|f_n| \leqslant C_2(\alpha)e^{-|n|h}.$$

It is easy to see that

$$\sum_{n=-infty}^{+\infty} \chi_\alpha(u+n) \equiv 1.$$

On account of (215), the series $\sum_{n=-\infty}^{+\infty} f_n(t)$ converges uniformly on D, whence

$$f_n(z) = f(0) + \frac{z}{\sqrt{2\pi}} \int_{-\infty}^{+\infty} \frac{\sum_{n=-\infty}^{+\infty} \Psi(u)\chi_\alpha(u+n)e^{iuz}}{\operatorname{ch} hu} du = \lim_{N \to \infty} \sum_{n=-N}^{N} f_n(z).$$

We shall show that the function $f_n(z)$ is bounded on D by the constant

$$C_2(\alpha)e^{-|n|h}.$$

By applying the convolution formula for the Fourier transform to

$$\phi_n(t) = \frac{1}{\sqrt{2\pi}} \int_{-\infty}^{+\infty} \frac{\Psi(u)\chi_\alpha(u+n)e^{iut}}{\operatorname{ch} hu} du,$$

we obtain

$$\phi_n(t) = \frac{1}{\sqrt{2\pi}} \int_{-\infty}^{+\infty} H_\Phi(s)\widetilde{\chi}_{\alpha n}(s-t)ds, \qquad (216)$$

where (as it not difficult to show by using the form of the function Φ^{24})

$$|H_\Phi(s)| \leqslant \begin{cases} CD_1, & |s| \leqslant 1, \\ \frac{CD_1}{s}, & |s| < 1, \end{cases} \qquad (217)$$

and $\widetilde{\chi}_{\alpha n}(s)$ is a Fourier transformation of the function $\frac{\chi_\alpha(u+n)}{\operatorname{ch} hu} = \chi_{\alpha n}(u)$.

The function $\chi_{\alpha n}(u)$, due to the boundedness of the second derivative of $\chi_\alpha(u)$, is twice differentiable, and its second derivative satisfies the inequality

$$|\chi''_{\alpha n}(u)| \leqslant D_2 e^{-|n|h}. \qquad (218)$$

Moreover, the function $\chi''_{\alpha n}(u) \in L(-\infty, +\infty)$, since outside the interval $1 + 4\alpha$, it is equal to zero.

Thus, we obtain that

$$|s^2 \widetilde{\chi}_{\alpha n}(s)| = \left| \frac{1}{\sqrt{2\pi}} \int_{-\infty}^{+\infty} \chi''_{\alpha n} e^{-ius} du \right| \leqslant D_3 e^{-|n|h},$$
$$|\widetilde{\chi}_{\alpha n}(s)| \leqslant D_3 e^{-|n|h}. \qquad (219)$$

As a result, using (216) and (219), and for definiteness, assuming that $t > 2$, we obtain:

$$|\phi_n(t)| \leqslant \frac{1}{\sqrt{2\pi}} C \cdot D_1 \cdot D_3 e^{-|n|h} \left\{ \int_{-\infty}^{-1} \frac{ds}{s(t-s)^2} + \int_{-1}^{+1} \frac{ds}{(t-s)^2} + \right.$$
$$\left. + \int_{+1}^{t-1} \frac{ds}{(t-s)^2 s} + \int_{t-1}^{t+1} \frac{ds}{s} + \int_{t+1}^{+\infty} \frac{ds}{s(t-s)^2} \right\} = I_1 + I_2 + I_3 + I_4 + I_5.$$

---

[24] Here and to the end of this section, D_i are some constants, inessential to us, and C is the constant entering in the definition of the classes $\mathcal{H}_h(C)$, $\mathcal{F}_{p\sigma}(C)$.

The fact that I_1, I_2, I_4, I_5 are $O(t)$, is obvious right away; I_3 is calculated in a simple way and is equal to $\frac{O(\ln t)}{t^2} = O(t)$. Thus, for $|t| > 2$ (the case of $t < -2$ is completely similar), we obtain

$$|t\phi_n(t)| \leqslant C \cdot D_5 e^{-|n|h}. \tag{220}$$

Immediately from (216), for $|t| \leqslant 2$, there follows that

$$|\phi_n(t)| \leqslant C \cdot D_5 e^{-|n|h}. \tag{221}$$

Estimates (220) and (221) together imply the boundedness of the function $f_n(t) = t\phi_n(t)$ on D by the constant $C \cdot D e^{-|n|h}$. But

$$f_n(z) = \frac{z \cdot e^{int}}{\sqrt{2\pi}} \int_{-\frac{1}{2}-\alpha}^{\frac{1}{2}+\alpha} \frac{\Psi(u-n)\chi_\alpha(u)e^{iuz}}{\operatorname{ch} h(u-n)} du.$$

The integrand above is in $L(-\infty, +\infty)$. Such integrals (see [23], p. 151) belong to $\mathcal{B}_{\frac{1}{2}+\alpha}$, whence, in virtue of (220) and (221),

$$f_n(z) \in \mathcal{B}_{\frac{1}{2}+\alpha}(C \cdot D \cdot e^{-|n|h}).$$

Each function $f(z) \in \mathcal{F}_{p\sigma}(C)$, on the basis of what has been proved, can be represented in the form of a series

$$f(z) = f(0) + \sum_{n=-\infty}^{+\infty} f_n(z),$$

where

$$f_n(z) \in \mathcal{B}_{n, \frac{1}{2}+\alpha}(C(h)D \cdot e^{-|n|h}),$$
$$C(h) \leqslant C \cdot e^{\sigma h^\bullet}$$

for any h and, consequently,

$$f_n(z) \in \mathcal{B}_{n, \frac{1}{2}+\alpha}\left(C \cdot D \cdot e^{-(|n|\sigma)^{\frac{p}{p-1}}(\frac{1}{\sigma p})^{\frac{1}{p-1}}(\frac{p-1}{p})}\right).$$

The proof of lemma VI is complete.

6. The proof of lemma VII. Let

$$\Psi_{kn}(u) = e^{-2\pi iku}\widehat{\chi}_\alpha(u + n(1 + 2\alpha)), \tag{222}$$

where $\widehat{\chi}_\alpha(u)$ is the function drawn in Fig. 12.

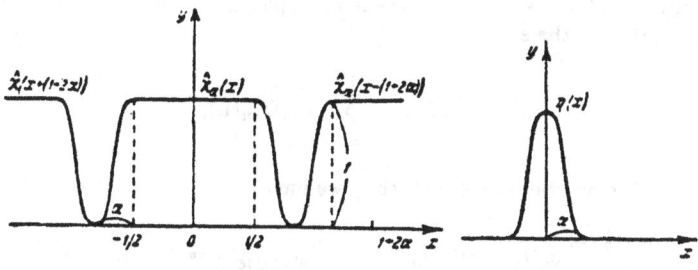

FIGURE 12

We shall show that the functions

$$\phi_{kn}(z) = \int_{-\infty}^{+\infty} \Psi_{kn}(u)e^{iuz}du \tag{223}$$

satisfy the conditions of lemma VII.
From (222) and (223) it follows that

$$\phi_{00}(t) = \int_{-(\frac{1}{2}+\alpha)}^{\frac{1}{2}+\alpha} \widehat{\chi}_\alpha(u)e^{iut}du. \tag{224}$$

By a substitution of variables: $u + n(1 + 2\alpha) = v$, we get

$$\phi_{kn}(t) = \int_{-\infty}^{+\infty} \widehat{\chi}_\alpha(u + n(1 + 2\alpha))e^{iu(t-2\pi k)}du =$$

$$= e^{-int(1+2\alpha)}\gamma_n \int_{-\infty}^{+\infty} \widehat{\chi}_\alpha(v)e^{iv(t-2\pi k)}dv, \tag{225}$$

where $\gamma_n = e^{2\pi i\alpha kn}$, i.e., $|\gamma_{kn}| = 1$. Using the boundedness of the second derivative of the function $\widehat{\chi}_\alpha(u)$ it is easy from (223) to derive the inequality $|t^2\phi_{00}(t)| \leqslant D_1(\alpha) = D_1$. Then from (225) we get that if $|c_{kn}| \leqslant C_n$ $(k = 0, \pm 1, \ldots)$, then

$$|f_n(t)| = \Big| \sum_k c_{kn}\phi_{kn}(t) \Big| \leqslant |C_n| \sum_{h=-\infty}^{+\infty} |\phi_{00}(t - 2\pi k)| \leqslant |C_n|D_1. \tag{226}$$

From (225) it also follows that

$$\phi_{kn}(t) \in \mathcal{B}_{n(1+2\alpha),\frac{1}{2}+\alpha},$$

whence, due to the boundedness of $|f_n|$ by the constant $|C_n|D_1$, we obtain that

$$f_n(z) \in \mathcal{B}_{n(1+2\alpha),\frac{1}{2}+\alpha}(D_1|C_n|); \tag{227}$$

from this the validity of inequalities (196) and (197), because of (227), implies on D a uniform convergence of the series $\sum_{n=-\infty}^{+\infty} f_n(t)$ to the function $f(t)$ from the classes $\mathcal{A}_h(c)$ or $\mathcal{F}_{p\sigma}(c)$, respectively.

We shall verify that condition (225) holds for the lemma.

According to the assumption of the lemma, $|c_{kn}| \geqslant \lambda$, $|k| \leqslant K$, $(-\infty \leqslant n \leqslant +\infty)$. By $\Psi_n(u)$, denote the sum

$$\Psi_n(u) = \sum_{s=-K}^{+K} c_{sn}\Psi_{sn}(u),$$

and by λ_n, the number $n(1 + 2\alpha)$; then we have

$$\int_{\lambda_n-(\frac{1}{2}+\alpha)}^{\lambda_n+(\frac{1}{2}+\alpha)} \Psi_n(u)e^{2\pi iku}du = \int_{\lambda-\frac{1}{2}}^{\lambda+\frac{1}{2}} \Psi_n(u)e^{2\pi iku}du + R_1 + R_2, \tag{228}$$

where

$$R_1 = \int_{\lambda_n - (\frac{1}{2} + \alpha)}^{\lambda - \frac{1}{2}} \Psi_n(u) e^{2\pi i k u} du, \quad R_2 = \int_{\lambda + \frac{1}{2}}^{\lambda_n + (\frac{1}{2} + \alpha)} \Psi_n(u) e^{2\pi i k u} du.$$

In view of the fact that $\hat{\chi}_\alpha(u + \lambda_n) = 1$ for $u \in [\lambda_n - \frac{1}{2}, \lambda_n + \frac{1}{2}]$, we obtain that

$$\int_{\lambda - \frac{1}{2}}^{\lambda + \frac{1}{2}} \Psi_n(u) e^{2\pi i k u} du = c_{kn}.$$

Two cases may arise:

$$1) \quad |R_1 + R_2| < \frac{\lambda}{2},$$

$$2) \quad |R_1 + R_2| \geqslant \frac{\lambda}{2}.$$

In the first case, because $\Psi_{s,n'}(u) = 0$ for $n' \neq n$, any s, and $u \in [\lambda_n \frac{1}{2}, \lambda_n + \frac{1}{2}]$, we have:

$$\frac{\lambda}{2} \leqslant \left| \int_{-\infty}^{+\infty} \hat{\chi}_\alpha(u + n(1 + 2\alpha)) \left(\sum_{s=-K}^{+K} \sum_{m=-\infty}^{+\infty} c_{sm} \Psi_{sm}(u) e^{2\pi i k u} \right) du \right| =$$

$$= \left| \int_{-\infty}^{+\infty} \hat{\chi}_{\alpha n}(v) f(2\pi k - v) dv \right| \tag{229}$$

(the latter according to the convolution formula). Similarly to the previous, for $\hat{\chi}_{\alpha n}(u)$ (the Fourier transform of the function $\hat{\chi}(u + n(1 + 2\alpha))$), the inequality $|\tilde{\chi}_{\alpha n}(\tau) \cdot \tau^2| \leqslant D_2$ holds. We shall choose $\tau(\lambda)$ so that

$$\frac{D_2}{\tau^2} \leqslant \frac{1}{\lambda}.$$

Then, because of (229),

$$\max |f(t)| \geqslant D_3 \cdot \lambda, \quad t \in [-2\pi k - \tau, -2\pi k + \tau]. \tag{230}$$

In the second case, we consider the function:

$$\eta_\alpha(u) = \begin{cases} \phi\left(\frac{1}{\alpha}\left(u - \lambda_n + \frac{1}{2} + \alpha\right)\right), & u \in \left[\lambda_n - \left(\frac{1}{2} + \alpha\right), \lambda_n - \frac{1}{2}\right], \\ 1 - \phi\left(\frac{1}{\alpha}\left(u - \lambda_n + \frac{1}{2} + \alpha\right)\right), & u \in \left[\lambda_n - \frac{1}{2}, \lambda_n - \frac{1}{2} + \alpha\right], \end{cases}$$

where $\phi(x)$ is the function introduced below the Fig. 11 (see also Fig. 11 itself).
Then we obtain

$$\left| \int_{-\infty}^{+\infty} \Psi_n(u) \eta_\alpha(u) e^{2\pi i k u} du \right| = |R_1 + R_2| \geqslant \frac{\lambda}{2}$$

or

$$\frac{\lambda}{2} \leqslant \left| \int_{-\infty}^{+\infty} \eta_\alpha(u) e^{2\pi i k u} \left(\sum_{s=-K}^{+K} \sum_{n=-\infty}^{+\infty} c_{sn} \Psi_{sn}(u) \right) du \right| =$$

$$= \left| \int_{-\infty}^{+\infty} \tilde{\eta}_\alpha(v) f(2\pi k - v) dv \right|,$$

where, by $\tilde{\eta}_\alpha$, we denote the Fourier transform of the function η_α.

Using arguments, similar to those for the first case, we obtain

$$\max |f(t)| \geqslant D_3 \lambda, \quad t \in [2\pi k - \tau, -2\pi k + \tau]. \tag{231}$$

It remains to put $C_3(\alpha) = D_3$ in formulas (230) and (231).

This completes the proofs of lemma VII and of theorems XXIII and XXIV.

§9. ε-entropy of spaces of real functionals.

1. In this subsection we consider spaces of functionals satisfying the Hölder condition.

Suppose X is a metric space, $F(x)$ is a real functional on X. The functional $F(x)$ is said to *satisfy the Hölder condition with function* $\omega = \omega(\lambda)$, if

$$|F(x_1) - F(x_1)| \leqslant \omega(\rho(x_1, x_2)), \tag{232}$$

where $\omega(\lambda)$ is a monotone positive function defined for positive λ.

We shall say that $F(x)$ *satisfies the Hölder condition with exponent* β ($o < \beta \leqslant 1$), if $\omega(\lambda) = L\lambda^\beta$.

By $\mathcal{D}_\omega^A(C)$, $A \subset X$, we denote the set of all functionals over X that satisfy the Hölder condition with function ω and are bounded on A by the constant C.

All the functionals will be considered, as always, in the uniform metric

$$\rho(F_1, F_2) = \sup_{x \in A} |F_1(x) - F_2(x)|.$$

We have the following

Theorem XV. *If A is completely bounded and*

$$\log \log(1/\varepsilon) << \mathcal{H}_\varepsilon(A) \approx \mathcal{H}_{\varepsilon k}(A)$$

for any $k > 0$, then

$$\log \mathcal{H}_\varepsilon \mathcal{D}_\omega^A(C)) \approx \mathcal{H}_{\omega^{-1}(\varepsilon)}(A).$$

Let us prove that under the assumptions of the theorem, we have the inequality

$$\mathcal{H}_{\delta_2}(A) \leqslant \log \mathcal{H}_\varepsilon \mathcal{D}_\omega^A(C)) \leqslant \mathcal{H}_{\delta_1}(A) + \log \log s, \tag{233}$$

where $\delta_1 = 1/2\omega^{-1}(\varepsilon/2)$, $\delta_2 = 2\omega^{-1}(2\varepsilon)$, $s = 2[2C/\varepsilon] + 1..$

Indeed, for any $\delta_1 > 0$, there exists a δ_1-covering of A consisting of $N_{\delta_1}(A)$ elements: $U_1, \ldots, U_{N_{\delta_1}}$. Choose a point x_i in each of the elements of the covering U_i. The functional $F(x)$ on the set A will be approximated by the functional:

$$\tilde{F}(x) = \left[\frac{2F(x_i)}{\varepsilon}\right] \frac{\varepsilon}{2} \qquad \text{for } x \in U_i$$

It is not difficult to verify that if we take $\delta_1 = 1/2\omega^{-1}(\varepsilon/2)$, then

$$\rho(F, \tilde{F}) \leqslant \varepsilon.$$

The functional $\tilde{F}$ assumes no more than U_i values on each set s and therefore the number of all functionals is no greater than the number $s^{N_{\delta_1}}$, i.e.,

$$N_\varepsilon(\mathcal{D}_\omega^A(C)) \leqslant \left(2\left[\frac{2C}{\varepsilon}\right] + 1\right)^{N_{\delta_1}(A)}. \tag{234}$$

Suppose, on the other hand, that the points $x_1, \ldots, x_{M_{\delta_2}(A)}$ constitute a maximal $(\delta_2/2)$-distinguishable set in A, consisting of $M_{\delta_2}(A)$ elements. Put $\delta_2 = 2\omega^{-1}(2\varepsilon)$. About each x_i taken as the centre, construct the sphere S_i of radius $\delta_2/4$. The spheres thus constructed are obviously non-intersecting.

Let us construct the set of functionals depending on the family $\alpha_1, \ldots, \alpha_{M_{\delta_2}(A)}$, where α_i assumes the value 0 or 1:

$$F_{\alpha_1, \ldots, \alpha_{M_{\delta_2}(A)}}(x) = \begin{cases} 2\varepsilon\left(1 - \dfrac{\omega(\rho(x, x_i))}{2\varepsilon}\right) & \text{for } \alpha_i = 1, \ x \in S_i, \\ 0 & \text{for all other } x \text{ or } \alpha_i = 0. \end{cases}$$

The number of all functionals is obviously equal to $2^{M_{\delta_2}(A)}$ and by construction they are all 2ε-distinguishable. It is easy to check that they satisfy the Hölder condition with function ω, i.e. $2\varepsilon < C$ implies

$$M_{2\varepsilon}(\mathcal{D}_\omega^A(C)) \geqslant 2^{M_{\delta_2}(A)}. \tag{235}$$

The inequality (233) is obtained from (234) and (235) by taking logarithms twice. Let us apply the theorem just obtained and inequality (233) to the space of functionals satisfying the Hölder condition with exponent β over the spaces that we have studied previously.

1) A is the unit n-dimensional cube D_n. The space $\mathcal{D}_p^A(C)$ coincides with the spaces $F_\beta^{D_n}(C)$ or $F_\beta^n(C)$ in the notations of §5.

The inequality (233) implies that

$$\mathcal{H}_\varepsilon(F_\beta^n(C)) \asymp (1/\varepsilon)^{n/\beta}. \tag{236}$$

Inequality (232) is a particular case of the main theorem which we deduced in §5.

2) A is the space $F_q^n(C_1, \ldots, C_p, L) = F_q^n$ introduced in §5. The result of §5

$$\mathcal{H}_\varepsilon(F_q^n) \asymp \left(\frac{1}{\varepsilon}\right)^{n/q}$$

implies that our theorem is applicable, hence

$$\log \mathcal{H}_\varepsilon(\mathcal{D}_\beta^{F_q^n}) \asymp \left(\frac{1}{\varepsilon}\right)^{n/q\beta}. \tag{237}$$

3) A is one of the spaces of analytical functions $A_G^K(C)$, for which

$$\mathcal{H}_\varepsilon(A_G^K(C)) \sim \tau(K, G) \left(\log \frac{1}{\varepsilon}\right)^{\lambda+1}$$

(see §7). Inequalities (233) imply

$$\log \mathcal{H}_\varepsilon(\mathcal{D}_\beta^{A_G^K(C)}) \sim \frac{\tau(K, G)}{\beta^{\lambda+1}} \left(\log \frac{1}{\varepsilon}\right)^{\lambda+1}.$$

2. In this subsection we give some unessential specifications for inequality (233) for functionals satisfying Lipschitz condition.

We shall say that the functional $F(x)$ *satisfies the Lipschitz condition* if it satisfies the Hölder condition with exponent $\beta = 1$.

Further for simplicity we assume that the Lipschitz constant is $L = 1$. The space of functional satisfying the Lipschitz conditions over the space A will be denoted by $\mathcal{D}_1^A$.

Let us show that for a connected completely bounded set A contained in a centrable[24] space and for any integer s, we have the following inequalities:

$$2^{M_{2\varepsilon}(A)} \leqslant N_\varepsilon(\mathcal{D}_1^A(C)) \leqslant \left(\left[\frac{C(s+1)}{\varepsilon}\right] + 1\right)(s+1)^{N_{s\varepsilon/(s+1)}(A)}. \tag{238}$$

The inequality on the right is obtained in the following way. Consider the $(s\varepsilon/(s+1))$-covering of A consisting of $N_{s\varepsilon/(s+1)}(A) = N$ elements $U_1, \ldots, U_N$. By x_i denote the centre of the set U_i. The fact that the set A is connected makes it possible to join any two sets U_i and U_j by a chain of intersecting sets U_k. Now construct the approximating functional $\tilde{F}$ for the function F. For its value on the set U_1, we take the number

$$\left[\frac{F(x_1)(s+1)}{\varepsilon}\right]\frac{\varepsilon}{s+1},$$

i.e. approximate the value of $F(x_1)$ with precision $\varepsilon/(s+1)$. It is easy to see that the values of the functional F at the centres x_i of the sets U_i adjacent to U_1 satisfy the inequality

$$|F(x_i) - \tilde{F}(x_1)| \leqslant |F(x_i) - F(x_1)| + |F(x_1) - \tilde{F}(x_1)| \leqslant \frac{s\varepsilon}{s+1} + \frac{\varepsilon}{s+1} = \varepsilon. \tag{239}$$

---

[24]For the definition see §1.

Inequality (239) means that for the approximation of $F(x_i)$ with precision $\varepsilon/(s+1)$ at the centres x_i of the sets U_i adjacent to U_1 we shall require knowing the value of $F(x_1)$ one of $s+1$ numbers:

$$\tilde{F}(x_1) \pm (2k-1)\frac{\varepsilon}{s+1} \quad (k = 1, \ldots, \frac{s+1}{2}) \qquad \text{for } s \text{ odd}$$

or

$$\tilde{F}(x_1) \pm 2k\frac{\varepsilon}{s+1} \quad (k = 1, \ldots, \frac{s}{2}) \qquad \text{for } s \text{ even}).$$

Since we assumed that the set A is connected, we can construct the entire functional $\tilde{F}$ in a unique way, passing in turn from one centre to the next, choosing at each step one of $s+1$ values. Thus we obtain

$$\rho(F, \tilde{F}) \leqslant \varepsilon,$$

i.e. an ε-net on the space $\mathcal{D}_1^A(C)$. The inequality on the right-hand side of (238) means that all possible functionals $\tilde{F}$ have been enumerated.

To obtain the inequality in (238) on the left, let us construct a maximal 2ε-distinguishable set $x_1, \ldots, x_M$ consisting of $M_{2\varepsilon}(A) = M$ elements. As in the proof of Theorem XXV, let us construct a set of functionals depending on $\alpha_1, \ldots, \alpha_M$, $\alpha_i = \pm 1$:

$$F_{\alpha_1,\ldots,\alpha_M}(x) = \begin{cases} \varepsilon(1 - \rho(x, x_i)/\varepsilon) & \text{for } \alpha_i = 1, \ \rho(x, x_i) \leqslant \varepsilon, \\ -\varepsilon(1 - \rho(x, x_i)/\varepsilon) & \text{for } \alpha_i = -1, \ \rho(x_i, x) \leqslant \varepsilon, \\ 0 & \text{for all other } x. \end{cases}$$

One proves in a similar way that all these functionals belong to $\mathcal{D}_1^A(C)$ and are 2ε-distinguishable. The number of all these functionals is $2^{M_{2\varepsilon}(A)}$, which proves the inequality (238).

Let us apply the inequalities obtained to calculations in specific spaces.

1) The space $F_1^n(C)$ of functions given on the unit cube and satisfying the Lipschitz condition

$$|f(x) - f(x')| \leqslant \max_{1 \leqslant i \leqslant n} |x_i - x_i'|.$$

In this case the inequalities (238) yield

$$2^{(1/2\varepsilon)^n} \leqslant N_\varepsilon(F_1^n(C)) \leqslant \left(2\left[\frac{C(s+1)}{\varepsilon}\right] + 1\right)(s+1)^{((s+1)/2\varepsilon s)^n} \tag{240}$$

for any s.

Hence for expressions similar to the densities in §4

$$\delta = \lim_{\varepsilon \to 0} \inf \mathcal{H}_\varepsilon(F_1^n(C))\varepsilon^n, \quad \bar{\delta} = \lim_{\varepsilon \to 0} \sup \mathcal{H}_\varepsilon(F_1^n(C))\varepsilon^n$$

we obtain

$$2^{-n} \leqslant \delta = \leqslant \bar{\delta}\left(1 + \frac{1}{s}\right)^n \log s 2^{-n} \leqslant \frac{e \log n}{2^n}.$$

Stronger estimates of $\bar{\delta}, \delta$ and of their asymptotics, as well as the question as to whether they coincide or not, have not been studied and constitute a problem of some interest.

2) The space $\mathcal{D}_1^{F_1^{\Delta}(C)}$ of functionals satisfying the Lipschitz conditions on the space $F_1^{\Delta}(C)$ of functions $f(x)$ on the segment $\Delta = [0,1]$ satisfying the Lipschitz conditions (see §2).

In this case the inequalities (238) yield

$$2^{2^{[1/\varepsilon]-1}} \leqslant N_\varepsilon(\mathcal{D}_1^{F_1^{\Delta}(C)}(C_1)) \leqslant (2\left[C_1(s+1)\varepsilon\right]+1)(s+1)^{2^{[(1+s)/s\varepsilon]+1}}. \tag{241}$$

Putting $s = [1/\varepsilon]$, we obtain

$$2^{1/\varepsilon} \preceq \mathcal{H}_\varepsilon(\mathcal{D}_1^{F_1^{\Delta}(C)}) \preceq \log(1/\varepsilon)2^{1/\varepsilon}. \tag{242}$$

It follows in particular from (242) that

$$\log \mathcal{H}_\varepsilon(\mathcal{D}_1^{F_1^{\Delta}(C)}) \sim 1/\varepsilon; \tag{243}$$

(243) is a strengthening of formula (237) for the space under consideration.

Appendix I

A. G. Vitushkin's Theorem
ON THE IMPOSSIBILITY OF REPRESENTING FUNCTIONS
OF SEVERAL VARIABLES AS SUPERPOSITIONS
OF FUNCTIONS OF A LESSER NUMBER OF VARIABLES

The first draft of the theory developed in this article was written by one of the authors as commentary to A.G.Vitushkin's paper [2] devoted to the representations of functions of several variables by functions of a lesser number of variables.

In this appendix we give an exposition of the problem which was studied by A. G. Vitushkin and its solution, using the results of §5 of the present article.

Suppose the function

$$y = f(x_1, \ldots, x_n)$$

is given by the sequence of formulas S:

$$
\begin{aligned}
y &= \varphi(y_1, \ldots, y_m), \\
y_k &= \varphi_k(y_1^k, \ldots, y_{m_k}^k), \\
y_{k_2}^{k_1} &= \varphi_{k_1 k_2}(y_1^{k_1 k_2}, \ldots, y_{m_{k_1 k_2}}^{k_1 k_2}), \\
&\cdots \\
y_{k_s}^{k_1, \ldots, k_{s-1}} &= \varphi_{k_1, \ldots, k_s}(y_1^{k_1, \ldots, k_s}, \ldots, y_{m_{k_1, \ldots, k_s}}^{k_1, \ldots, k_s}), \\
y_k^{k_1, \ldots, k_s} &= x_{r_k^{k_1, \ldots, k_s}}.
\end{aligned}
\tag{244}
$$

The domain of definition of the function f is the set of all points $(x_1, \ldots, x_n)$ such that, after we compute all the y with $r+1$ indices, in the step-by-step computation according to formulas (244), we always get a point from the domain of definition of the function φ with r indices. The function f is said to be the *superposition of the function φ of order s* constructed from the scheme S. Here we assume that the superposition scheme is defined by fixing the following table of natural numbers T:

$$n$$

$$m$$

$$m_k, \ k = 1, \ldots, m,$$

$$m_{k_1 k_2}, \ k_1 = 1, \ldots, m; \ k_2 = 1, 2, \ldots, m_{k_1},$$

$$\ldots$$

$$m_{k_1, \ldots, k_s}, \ k_1 = 1, 2, \ldots, m; \ldots; \ k_s = 1, 2, \ldots, m_{k_1, \ldots, k_{s-1}},$$

$$r_k^{k_1, \ldots, k_s} = 1, \ldots, n; \ k_1 = 1, 2, \ldots, m; \ k_s = 1, \ldots, m_{k_1, \ldots, k_{s-1}}.$$

We shall agree that all the functions $\varphi_{k_1, \ldots, k_r}$ are defined on unit cubes

$$0 \leqslant y_k^{k_1, \ldots, k_r} \leqslant 1$$

of the corresponding dimensions. Then automatically the domain of definition of f is contained in an n-dimensional cube. However, we shall consider only those superpositions which are defined on the entire cube.

Now to each sequence of indices $k_1, \ldots, k_r$ with $0 \leqslant r \leqslant s$ (when $r = 0$ we consider the "empty" sequence of indices), satisfying the conditions of table T, we assign the class $\Phi_{k_1, \ldots, k_r}$ of functions $\varphi_{k_1, \ldots, k_r}$ in $m_{k_1, \ldots, k_r}$ variables. The class of functions f thus obtained will be denoted by $S(\Phi)$.

Theorem XXVI. *Suppose all the functions $\varphi = \varphi_{k_1, \ldots, k_r}$ of $m = m_{k_1, \ldots, k_r}$ of the classes $\Phi_{k_1, \ldots, k_r}$ satisfy the Lipschitz condition*

$$|\varphi(y_1, \ldots, y_m) - \varphi(y_1', \ldots, y_m')| \leqslant C \sum_{k=1}^{m} |y_k - y_k'|$$

with one and the same constant C; then in the uniform metric

$$\|f\| = \sup |f|$$

(both for the functions f and for the functions φ)

$$\mathcal{H}_\varepsilon(S(\Phi)) \preceq \mathcal{H}_{\varepsilon'}(\Phi_{k_1, \ldots, k_r}),$$

where

$$\varepsilon' = \varepsilon / C^s$$

and sums are taken over all families of indices $k_1, \ldots, k_r$ satisfying the conditions of Table T.

For the proof let us cover each $\Phi_{k_1, \ldots, k_r}$ by

$$N_{k_1, \ldots, k_r} = N_{\varepsilon'} \Phi_{k_1, \ldots, k_r}$$

sets $U_{k_1,\ldots,k_r}$ of diameter $2\varepsilon'$. Choosing for each family of indices $k_1,\ldots,k_r$ one specific set $U_{k_1,\ldots,k_r}$, let us constitute the set U of all functions f which can be obtained by superpositions according to the scheme S from the functions $\varphi_{k_1,\ldots,k_r}$ belonging to the chosen $U_{k_1,\ldots,k_r}$. It is easy to see that U is of diameter $\leqslant 2\varepsilon$. The number of different elements $U_{k_1,\ldots,k_r}$ is obviously equal to

$$N = \prod_{k_1,\ldots,k_r} N_{k_1,\ldots,k_r}.$$

Taking logarithms, we obtain the required estimate.

Now assume (this assumption will further be called condition (A)) that part of the classes $\Phi_{k_1,\ldots,k_r}$ consists of a single function (in each class) satisfying the Lischitz condition with the corresponding constant $Z_{k_1,\ldots,k_r}$ while part are the classes

$$F_{q_{k_1,\ldots,k_r}}^{m_{k_1,\ldots,k_r}}(C_0^{k_1,\ldots,k_r},\ldots,C_p^{k_1,\ldots,k_r},C^{k_1,\ldots,k_r})$$

of all functions of smoothness $q = q_{k_1,\ldots,k_r} = p + \alpha \geqslant 1$ with given constants $C_0,\ldots,C_p,C^{11}$.

It is easy to see that Theorem XXVI implies

$$\mathcal{H}_\varepsilon(S(\Phi)) \preceq (1/\varepsilon)^\rho, \tag{245}$$

where

$$\rho = \sup_{k_1,\ldots,k_r} (m_{k_1,\ldots,k_r}/q_{k_1,\ldots,k_r}).$$

Consider the intersection of $S(\Phi)$ of the class F_q^n of functions f of smoothness q defined on the n-dimensional unit cube. In F_q^n introduce the topology induced by the norm $\|f\|$ (see Remark 3 at the end of §5).

We have the following

Lemma VIII. *Under the conditions* $(A)\,\rho < n/q$, *the intersection* $S(\Phi) \cap F_q^n$ *is nowhere dense in* F_q^n.

The lemma is a direct consequence of the estimate (245) and Remark 3 at the end of §5. Since the space F_q^n with the metric indicated there is a complete space, the complement to the union of a countable number of nowhere dense sets in F_q^n is everywhere dense. This is what allows us to deduce Theorem XXVII stated below from our lemma; this theorem is a certain generalization of the theorem proved in 1954 by A.G. Vitushkin who used basically different methods.

Denote by

$$\Sigma_\rho^n(\psi_1,\ldots,\psi_t)$$

the class of functions f defined in a n-dimensional unit cube and representable in the form of superpositions (according to any scheme of any finite order s) of a finite number of fixed functions

$$\psi_1,\ldots,\psi_t$$

and of a finite number of functions φ for each of which the ratio m_φ/q_φ of the number of variables to the smoothness is $\leqslant \varphi$.

---

[1]See §5.

Theorem XXVII. *If $\rho < n/q$ and the functions $\psi_1, \ldots, \psi_t$ satisfy the Lipschitz conditions (with any constants $Z_1, \ldots, Z_t$), then the set of functions f from F_q^n which do not occur in $\Sigma_\rho^n(\psi_1, \ldots, \psi_t)$ are everywhere dense in F_q^n.*

For the proof one need only represent $\Sigma_\rho^n(\psi_1, \ldots, \psi_t)$ in the form of a finite union of sets which satisfy the assumptions of the lemma. For such sets we can take the sets of functions satisfying conditions (A) for some (fixed for each set) scheme of superposition, fixed families of indices such that $\Phi_{k_1, \ldots, k_r}$ consists of some of the functions $\psi_1, \ldots, \psi_t$ and fixed natural values of constants $C_0^{k_1, \ldots, k_r}, \ldots, C_p^{k_1, \ldots, k_r}, C^{k_1, \ldots, k_r}$. Obviously the number of different sets of this type is countable.

APPENDIX II

RELATIONSHIP WITH THE PROBABILISTIC THEORY OF APPROXIMATE SIGNAL TRANSMISSION

1. Each element x of the set A contained in the metric space R will be viewed as a possible "input signal" of a transmitting or memorizing device. We shall assume that the corresponding "output signal" is a random element η of the space R with the probability distribution

$$\mathcal{P}(\eta \in Q) = \Pi(x, Q),$$

depending on x. By means of the kernel $\Pi(x, Q)$, from any distribution

$$\mathcal{P}(\xi \in B) = \mathcal{P}_\xi(B)$$

on the set A let us construct the distribution

$$\mathcal{P}(\eta \in Q) = \int_A \mathcal{P}_\xi(dx)\Pi(x, Q) = \mathcal{P}_\eta(Q)$$

and the joint distribution

$$\mathcal{P}((\xi, \eta) \in U) = \int \int_U \mathcal{P}_\xi(dx)\Pi(x, dy) = \mathcal{P}_{\xi\eta}(U).$$

Further suppose

$$\mathcal{J}_{\mathcal{P}_\xi}(\xi, \eta) = \int_A \int_R \mathcal{P}_{\xi\eta}(dx, dy) \log \frac{\mathcal{P}_{\xi\eta}(dx, dy)}{\mathcal{P}_\xi(dx)\mathcal{P}_\eta(dy)}$$

is the Shannon amount of information contained in η in the conditional distribution Π with respect to ξ distributed according to the law $\mathcal{P}_\xi$ (see XXVII). In accordance to Shannon, the expression

$$\mathcal{E}(\Pi) = \sup_{\mathcal{P}_\xi} \mathcal{J}_{\mathcal{P}_\xi}^\Pi$$

is the capacity of the device determined by the distribution Π. We shall be concerned with the case when the output signal η with probability 1 is at a distance no greater than ε from the input signal x:

$$\Pi(x, \{y : \rho(x, y) \leqslant \varepsilon\}) = 1. \tag{246}$$

Under the condition (246), we shall say that the kernel Π belongs to the class K_ε.

Theorem XXVIII. *If $\Pi \in K_\varepsilon$, then*

$$\mathcal{E}(\Pi) \geqslant \mathcal{E}_{2\varepsilon}(A).$$

Indeed, if we place the input signal at $M_{2\varepsilon}(A)$ points of maximal (with respect to the number of elements) 2ε-distinguishable sets with probability $M_{2\varepsilon}^{-1}(A)$ at each point, then the value of η will be determined from the value of ξ with error equal to zero, i.e. the amount of information $\mathcal{J}(\xi, \eta)$ will be equal to the entropy of the distribution ξ

$$\log M_{2\varepsilon}(A) = \mathcal{E}_{2\varepsilon}(A).$$

Theorem XXIX. *There exists a kernel $\Pi \in K_\varepsilon$ such that*

$$\mathcal{E}(\Pi) = \mathcal{H}_\varepsilon^R(A).$$

To construct a kernel satisfying Theorem XXIX, it suffices to take in R the set U of $N_\varepsilon^R(A)$ points which is a ε-net for A and assign to each $x \in A$ with probability 1 one of the points $y \in U$ satisfying $\rho(x, y) \leqslant \varepsilon$ as the output signal η. Then the entropy of η, and therefore also $\mathcal{J}(\xi, \eta)$, will be no greater than $\mathcal{H}_\varepsilon^R(A)$ (independently of the choise of $\mathcal{P}_\xi$).

Now consider the expression

$$\mathcal{E}_\varepsilon^R(A) = \inf_{\Pi \in K_\varepsilon} \mathcal{E}(\Pi).$$

This is the guaranteed capacity of the transmitting device with precision of transmission ε (when besides (246) nothing is known about the distribution Π). Theorems XXVIII and XXIX directly imply the estimates

$$\mathcal{E}_{2\varepsilon}(A) \leqslant \mathcal{E}_\varepsilon^R(A) \leqslant \mathcal{H}_\varepsilon^R(A). \tag{247}$$

2. Together with the expression $\mathcal{E}(\Pi)$, the following expression

$$\mathcal{H}_\varepsilon^R(\mathcal{P}_\xi) = \inf_{\Pi \in K_\varepsilon} \mathcal{J}_{\mathcal{P}_\xi}^\Pi,$$

is relevant in questions of signal transmission with precision up to ε; it is known as the "entropy of the distribution $\mathcal{P}_\xi$ under condition (246) on the precision of transmission" or briefly ε-entropy of the distribution $\mathcal{P}_\xi$.

Theorem XXX.

$$\tilde{\mathcal{H}}_\varepsilon^R(A) = \sup_{\mathcal{P}_\xi} \mathcal{H}_\varepsilon^R(\mathcal{P}_\xi) \leqslant \mathcal{E}_\varepsilon^R(A)$$

Theorem XXX follows directly from the well-known general formula

$$\sup_x \inf_y F(x, y) \leqslant \inf_y \sup_x F(x, y).$$

Theorem XXXI. *There exists a* $\mathcal{P}_\xi$ *for which*

$$\mathcal{H}_\varepsilon^R(\mathcal{P}_\xi) \geqslant \mathcal{E}_{2\varepsilon}(A).$$

To prove this, it suffices to consider the distribution $\mathcal{P}_\xi$ which assigns to each of $M_{2\varepsilon}(A)$ points of a maximal 2ε-distinguishable set U the probability $M_{2\varepsilon}^{-1}(A)$. From Theorems XXX and XXXI and formula (247), we get

$$\mathcal{E}_{2\varepsilon}(A) \leqslant \tilde{\mathcal{H}}_\varepsilon^R(A) \leqslant \mathcal{H}_\varepsilon^R(A). \tag{248}$$

Theorems XXX and formulas (247), (248) and Theorem IV finally imply

$$\mathcal{E}_{2\varepsilon}^R(A) \leqslant \tilde{\mathcal{H}}_\varepsilon^R(A) \leqslant \mathcal{E}_\varepsilon^R(A). \tag{249}$$

The inequality on the right of (249) coincides with Theorem XXX. To prove the left inequality, we must note that by (247), Theorems IV, and formula (248), we have

$$\mathcal{E}_{2\varepsilon}^R(A) \leqslant \mathcal{H}_{2\varepsilon}^R(A) \leqslant \mathcal{E}_{2\varepsilon}(A) \leqslant \tilde{\mathcal{H}}_\varepsilon^R(A).$$

December 15, 1958

REFERENCES

[1] K. I. Babenko, *On the entropy of a class of analytical functions*, Nauch. Dokl. Vyssh. Shk. **1,2** (1958).

[2] A. G. Vitushkin, *On Hilbert's thirteenth problem*, Dokl. Akad. Nauk SSSR **95,4** (1954), 701-704.

[3] A. G. Vitushkin, *Absolute entropy of metric spaces*, Dokl. Akad. Nauk SSSR **117,5** (1957), 745-748.

[4] V. D. Yerokhin, *On the asymptotics of ε-entropy of analytical functions*, Dokl. Akad. Nauk SSSR **120,5** (1958), 949-952.

[5] A. N. Kolmogorov, *Estimates of the minimal number of elements of ε nets in various functional classes and their applications to the question of representing functions of several variables as superpositions of functions of a lesser number of variables*, Uspekhi Mat. Nauk **10,1** (1955), 192-194.

[6] A. N. Kolmogorov, *On certain asymptotic characteristics of completely bounded metric spaces*, Dokl. Akad. Nauk SSSR **108,3** (1956), 385-389.

[7] V. M. Tikhomirov, *On ε-entropy of certain classes of analytic functions*, Dokl. Akad. Nauk SSSR **117,2** (1957), 191-194.

[8] W. Hurewicz, G. Wallman, *Dimension theory*, Princeton Univ. Press, 1948.

[9] F. Hausdorff, *Dimention und äusseres Mass*, Math. Ann. **79** (1919), 157-179.

[10] N. S. Bakhvalov, *Concerning the question of the number of arithmetical operations needed to solve the Poisson equation for squares by the method of finite differences*, Dokl. Akad. Nauk SSSR **113,2** (1957), 252-255.

[11] N. S. Bakhvalov, *On setting up equations in finite differences for approximate solutions of the Laplace equation*, Dokl. Akad. Nauk SSSR **114, 6** (1957), 1146-1149.

[12] A. G. Vitushkin, *Certain esimates of the theory of tables*, Dokl. Akad. Nauk SSSR **114,5** (1957), 923-926.

[13] A. N. Kolmogorov, *On the linear dimension of topological vector spaces*, Dokl. Akad. Nauk SSSR **120,2** (1958), 239-241.

[14] V. D. Yerokhin, *On conformal maps of annuli on the main basis of the space of functions analytical in an elementary neighbourhood of an arbitrary continuum*, Dokl. Akad. Nauk SSSR **120,4** (1958), 689-692.

[15] P. S. Alexandrov, *Introduction to the general theory of sets and functions*, Moscow, Leningrad, Gostekhizdat, 1958.

[16] P. S. Urysohn, *Works on topology and other branches of mathematics*, vol. 2, Moscow, Gostekhizdat, 1951.

[17] S. Banach, *Théorie des opération linéaires*, Warszawa, 1932.

[18] S. N. Bernstein, *Probability theory*, Moscow, Leningrad, Gostekhizdat, 1946.

[19] L. F. Toth, *Disposition in the plane, sphere and in a space*, Moscow, Fizmatgiz, 1958.

[20] H. Davenport, *The covering of space by spheres Reneli*, Rend. Circ. Palermo **1,2** (1952), 92-107.

[21] L. Watson, *The covering of space by spheres*, Rend. Circ. Palermo **5,1** (1956), 93-100.

[22] H. F. Blichfeldt, *The minimum value of quadratic forms, and the closest packing of spheres*, Math. Ann. **101** (1929), 605-608.

[23] N. I. Akhiezer, *Lectures in approximation theory*, Moscow, Leningrad, Gostekhizdat, 1947.

[24] G. M. Fikhtengolts, *A course in differential and integral calculus*, vol. 3, Moscow, Gostekhizdat, 1949.

[25] A. I. Markushevich, *Theory of analytical functions*, Moscow, Leningrad, Gostekhizdat, 1950.

[26] V. A. Kotelnikov, *On the transmission possibilities of "ether" and wires in electrical communications*. Inthebook *Materials to the 1st All-Union Congress in questions of technical reconstructions in communications and the development of weak current industry*, Moscow, 1933.

[27] C. Shannon, *A mathematical theory of communication.*, Bell Syst. Techn. J. **27,4** (1948), 623-656.

[28] B. Ya. Levin, *Distribution of roots of entire functions*, Moscow, Gostekhizdat, 1956.

[29] M. A. Lavrentyev, B. V. Shabat, *Methods of the theory of functions of complex variables*, Moscow, Fizmatgiz, 1958.

8
VARIOUS APPROACHES TO
ESTIMATING THE COMPLEXITY OF APPROXIMATE
REPRESENTATION AND CALCULATION OF FUNCTIONS

In order to avoid technical complications, we shall consider only real functions

$$y = f(x)$$

defined in the interval

$$\Delta = \{x \; : \; -1 \leqslant x \leqslant +1\}.$$

All more striking features of the problem we take up can be explained by the example of the function classes W_p, i.e., the class of functions satisfying the inequality

$$\text{vrai max}\,|f^{(p)}(x)| \leqslant 1.$$

A_r is the class of functions which can be analytically extended to the interior of an ellipse with focuses at the points -1 and $+1$, and with sum of semi–axes $r > 1$.

We shall study methods of defining and calculating functions of the classes indicated to within a certain fixed $\varepsilon > 0$.

§ 1. Approximation by algebraic polynomials.

We denote by $n_a(f, \varepsilon)$ the smallest n for which there exists a polynomial

$$P(x) = c_1 + c_2 x + \ldots + c_n x^{n-1}$$

satisfying on Δ the inequality

$$|f - P| \leqslant \varepsilon$$

and for a class of functions F, we put

$$n_a(F, \varepsilon) = \sup_{f \in F} n_a(f, \varepsilon).$$

These are deep results on the asymptotic behaviour of the function $n_a(\varepsilon)$ for the classes W_p and A_r, which are obtained by the classical theory on best approximation:

$$n_a(W_p, \varepsilon) \sim C_p \left(\frac{1}{\varepsilon}\right)^{1/p}, \tag{1}$$

where the C_p are certain constants whose analytic expression we won't present (Nikolski, Bernstein, 1946–47);

$$n_a(A_r, \varepsilon) \sim \frac{1}{\log r} \log \frac{1}{\varepsilon} \qquad (2)$$

(Bernstein, 1913)[1]

§ 2. Approximation by linear forms.

It is natural to ask how succesful the choice of algebraic polynomials is as the means for approximate representation of functions from the classes W_p and A_r (it seems that a problem of this kind was first posed in [3]). Denote by $n(F, \varepsilon)$ the smallest n for which it is possible to find functions

$$\phi_1, \phi_2 \cdots, \phi_n,$$

such that for any function $f \in F$ there is a linear combination

$$L(x) = c_1 \phi_1(x) + c_2 \phi_2(x) + \ldots + c_n \phi_n(x)$$

that satisfies on $\triangle$ the equality

$$|f - L| \leqslant \varepsilon.$$

It is easy to see that

$$n(F, \varepsilon) \leqslant n_a(F, \varepsilon)$$

always holds.

For the class A_r,

$$n(F, \varepsilon) \sim n_a(F, \varepsilon) \sim \frac{1}{\log r} \log \frac{1}{\varepsilon} \qquad (3)$$

(Vitushkin, Yerokhin, 1957–58).

For the class W_p, algebraic polynomials do not arise by this rational method of approximation, since

$$n(F, \varepsilon) \sim C_p' \left(\frac{1}{\varepsilon}\right)^{1/p}, \qquad (4)$$

where

$$\frac{C_p'}{C_p} = \frac{2}{\pi} \qquad (5)$$

(Tikhomirov, 1960).

In other words, when choosing functions ϕ_k for linear approxiamtion, one can limit oneself to forms containing

$$\frac{\pi}{2} = 1,57\ldots$$

times less terms than under approximation by polynomials to within the same ε.

---

[1]For a modern discussion of this problem §§ 1–2, see [1]. Instead of (2), in [1] generalizations obtained later are given. In a more outdated formulation, (2) can be already found in § 46 of the classical paper by Bernstein [2]. In (1), (2) one further has

$$n(\varepsilon) \sim m(\varepsilon), \quad \text{if } n(\varepsilon)/m(\varepsilon) \to 1 \text{ for } \varepsilon \to 0; \; n(eps) \succeq m(\varepsilon), \text{ if } \lim_{\varepsilon \to 0} \inf n)\varepsilon)/m(\varepsilon) > 0;$$

$$n(\varepsilon) \asymp m(\varepsilon), \text{ if at the same time } n(\varepsilon) \succeq m(\varepsilon) \text{ and } m(\varepsilon) \succeq n(\varepsilon).$$

§ 3. Necessary quantity of information.

In order to determine to within ε a function f belonging to the class F, where all functions permit a representation by linear forms

$$L(x) = c_1\phi_1(x) + c_2\phi_2(x) + \ldots + c_n\phi_n(x)$$

with an error not exceeding

$$|f - L| \leqslant \frac{\varepsilon}{2},$$

it suffices to give the coefficients l_k with an error not higher than

$$\varepsilon' = \frac{\varepsilon}{2nM},$$

where M is the upper limit of absolute values of the function ϕ_k.

Approximate values of coefficients can be written in binary system. Thus, specifying a function amounts to giving a definite number of symbols each of which can only take the values 1 or 0. If we limit ourselves to functions subordinate to the additional condition

$$|f(x)| \leqslant 1,$$

then under the approximation of functions by algebraic polynomials of the classes W_p and A_r for $\varepsilon \to 0$, the number of symbols which is needed to define the coefficients, is of order $\log 1/\varepsilon$.

This enables us to obtain the following estimates of the "quantity of information" which is necessary for specifying an arbitrary function f of the class A_r or the class W_p to within ε:

$$I(A_r, \varepsilon) \preceq \left(\log \frac{1}{\varepsilon}\right)^2, \tag{6}$$

$$I(W_p, \varepsilon) \preceq \left(\frac{1}{\varepsilon}\right)^{1/p} \log \frac{1}{\varepsilon}. \tag{7}$$

The first of these estimates cannot be strengthened in the sense of the order of the value $I(A_r, \varepsilon)$, but it can be made more precise:

$$I(A_r, \varepsilon) \sim \frac{1}{2\log r} \left(\log \frac{1}{\varepsilon}\right)^2 \tag{8}$$

(Vitushkin, 1957).

Estimate (7) presents a slightly overstated order of the value of $I(W_p, \varepsilon)$. The true order is the following:

$$I(W_p, \varepsilon) \asymp \left(\frac{1}{\varepsilon}\right)^{1/p} \tag{9}$$

(Kolmogorov, 1955).

For more explicit information on the quantity $I(F, \varepsilon)$, see the paper [4]. It is shown that $I(F, \varepsilon)$ is the integer defined by the inequalities

$$\log_2 N(F, \varepsilon) \leqslant I(F, \varepsilon) < \log_2 N(F, \varepsilon) + 1, \tag{10}$$

where $N(F, \varepsilon)$ is the minimal number of elements of the covering of the set F by the sets of diameter 2ε in the metric

$$\rho(f_1, f_2) = \sup_{x \in \Delta} |f_1(x) - f_2(x)|.$$

§ 4. Computational complexity.

In this section, we limit ourselves, for simplicity, to functions which in addition to the condition

$$|f(x)| \leqslant 1$$

introduced in the previous section, also satisfy the Lipshitz condition

$$|f(x)' - f(x')| \leqslant 1.$$

Let

$$s \geqslant \log \frac{1}{\varepsilon} + 3.$$

It is easy to see that any table that to each binary number

$$x = -1 + x_1, x_2, \ldots, x_s \quad (x_i = 0, 1)$$

associates the binary number

$$y = -1 + y_1, y_2, \ldots, y_s = f_\varepsilon(x), \quad (y_i = 0, 1) \tag{11}$$

satisfying the following condition

$$|f(x) - f_\varepsilon(x)| \leqslant \frac{\varepsilon}{2},$$

can be regarded as a table defining the values $y = f(x)$ for any $x \in \Delta$ to within ε.

The dependence (11) can be interpreted as a discrete vector function

$$Y = \Psi(X)$$

mapping s–dimensional vectors

$$X = (x_1, x_2, \ldots, x_s)$$

to s–dimensional vectors

$$Y = (y_1, y_2, \ldots, y_s),$$

where the variables x_j and y_i can only take the values 1 or 0.

This kind of vector "function of the algebra of logic" can be naturally represneted in the form of a superposition of elementary functions

$$\zeta = f(\xi, \eta),$$

where the auxiliary variables ξ, η, ζ can only take the values 0, 1. It is known that there are 16 such elementary functions in total. The "complexity" of a vector function $Y = \Phi(X)$ of the algebra of logic is the minimal number of terms in a superposition of elementary functions, i.e., in the form that represents the function (see [5]).

The complexity of ε–representation of a function f is the minimal complexity of a function Φ corresponding to tables which define $y = f(x)$ to within ε.

Combining the results of the notes in [5] and [6] with certain methods developed by Shannon and the estimates stated in section 3, Ofman established that the maximal ε-complexities of functions from W_p and A_r admit the estimates

$$K(W_p, \varepsilon) \asymp \left(\frac{1}{\varepsilon}\right) \Big/ \log \frac{1}{\varepsilon} \tag{12}$$

$$\left(\log \frac{1}{\varepsilon}\right)^2 \Big/ \log\log \frac{1}{\varepsilon} \preceq K(A - r, \varepsilon) \preceq \left(\log \frac{1}{\varepsilon}\right)^\alpha, \quad \alpha = \log_2 3 + 1 = 2,58\ldots \tag{13}$$

The incompleteness of estimates (13) is related to the fact that the asymptotics of the number $m(s)$ of elementary operations over pairs of binary numbers which is needed in multiplicaton of two numbers of s length, is, evidently, unknown up to now. The usual school formula of multiplication gives only

$$m(s) \preceq s^2,$$

and the Ofman and Karacuba estimates are as follows:

$$s \preceq m(s) \preceq s^{\log_2 3}. \tag{14}$$

§ 5. **Estimates of complexity of specific functions.**

In the classical theory of approximations by algebraic polynomials, "inverse theorems" are known indicating that no specific function that does not possess sufficient "smoothness", can be well approximated by polynomials of not too high degree on the whole interval Δ.

Everything that has been said in sections 2 and 3 has to do, in an essential way, with classes of functions, and not with specific functions. But under the approach given in section 4, the problem of estimating ε-complexities $K(f, \varepsilon)$ of specific functions is meaningful and very interesting. Unfortunately, to obtain lower bounds in this case is, appearently, very difficult. Violating smoothness does not necessarily make the function complicated. For example, the estimate

$$K(f, \varepsilon) \preceq (\log 1\varepsilon)^2 \tag{15}$$

of the Van der Waerden function with no derivatives in any of the points, is easily obtained.

REFERENCES

[1] V. M. Tikhomirov, *Diameters of sets and functional spaces, and the theory of bet approximations*, Uspekhi Mat. Nauk, **15** (1960), 81–120.
[2] S. Bernstein, *Sur la valeur asymptotique de la meilleure approximation des fonctions analytiques admettant des singularités données*, Bull. Acad. Roy. Belg., ser. 2, **4**, (1913), 76-90.
[3] A. N. Kolmogorov, *Ueber die beste Annäherung von Funktionen einer gegebenen Funktionenklasse*, Ann. Math., **37** (1936), 107-110.
[4] A. N. Kolmogorov, and V. M. Tikhomirov, *ε-entropy and ε-capacity of sets in functional spaces*, Uspekhi Mat. Nauk, ser 2 **14**, (1954), 3-86.
[5] Yu. Ofman, *On algoritmic complexity of discrete functions*, Dokl. Akad. Nauk, SSSR, **145**, 1 (1962), 48-51.
[6] A. Karatsuba, *Multiplication of multi-valued numbers in automata*, Dokl. Akad. Nauk, SSSR, **145**,2 (1962), 293-294.

9

ON TABLES OF RANDOM NUMBERS[*][1]

From the Author

The editors of *Semiotika i Informatika* have felt it appropriate to publish a Russian translation of my article, which reflects a certain stage of my attempts to give meaning to the frequency interpretation of probability due to R. von Mises. The article was written during my stay in India in 1962 and published by the Indian journal *Sankhyā*, a very authoritative periodical in mathematical statistics in English-speaking countries, but little known here in the USSR. The reader familiar with the work of von Mises will perhaps notice the wider definition of the algorithm A used for selecting samples. But the main difference from von Mises is the strictly finite character of the whole approach and the introduction of a quantitative estimate of the stab ility of frequencies. The problem of bringing together estimates (4.3) and (4.4) , despite its elementary character, apparently still awaits its solution. Further stages of my search in this direction are reflected in two notes, published in 1965 and 1969 in the journal *Problemy Peredachy Informatsii* . In *Semiotika i Informatika*, see V.V.Vyugin's survey (1981, vyp. 16).

1. INTRODUCTION

The set-theoretic axioms of probability theory, in whose formulation it was my lot to take part [1], allowed to eliminate most of the difficulties in constracting the mathematical apparatus appropriate for numerous applications of probabilistic methods, and so succesfully, that the problem of finding the causes of the applicability of mathematical probability theory was felt by many researches to be of secondary importance.

I have already expressed the point of view [1, chap.1][2] that the basis of the applicability of the mathematical theory of probability to random events of the real

---

[*]Sankhyā. Indian J. Statist. Ser.A. 1963, vol.25, N 4, 369-376

[1]This article was published in Russian in the periodic collection *Semiotika i Informatika* (Moscow, VINITI,1982, vyp. 18, 3-13) and was prefaced by the short text especially written by A.N.Kolmogorov which appears here under the title "From the Author" . The text of the article itself is not the original (published in *Sankhyā*), but a new translation of the Russian text that appeared in *Semiotika* (*Translator's Note*).

[2]This book first appeared in German: Kolmogoroff A., Grundbegriffe der Wahrscheinlighkeit-srechnung. Berlin, 1933. Russian transation: Kolmogorov A. N., Main concepts of probability theory, Moscow-Leningrad, ONTI, 1936; 2d edition (same title) Moscow, Nauka, 1974. The English edition is: Foundations of the theory of probability, Chelsea, 1950. (*Translator's note*).

world is the *frequency approach to probability* in one form or another, which was so strongly advocated by von Mises. Nevertheless for a long time I believed that:

(1) the frequency approach based on the idea of *a limiting frequency* as the number of trials tends to infinity does not give a foundation for the applicability of the results of probability theory to practical questions, where we deal with a finite number of trials;

(2) the frequency approach in the case of a large but finite number of trials cannot be developed in a rigorous formal purely mathematical way.

Accordingly, I sometimes put forward a frequency approach to probability that included the conscious use of certain not entirely formal considerations about "practical certainty", "approximately constant frequency for large series of trials", without any precise description of what series are to be considered "sufficiently large", etc. (see[1–3]:"*Foundations in the theory of probability*", Chap. 1., or in more detail in the *Great Soviet Encyclopedia* the article "Probability" or the chapter on probability theory in the book "*Mathematics, its content, methods and meaning*").

I still maintain the first of the two positions stated above. Concerning the second, I have come to the conclusion that the notion of random distribution can be introduced in a purely formal way, namely: it can be shown in sufficiently large collections the distribution of a certain property may be such that the frequency of its appearance will be aproximately the same for all sufficiently large samples, if only the *law is sufficiently simple*. The complete development of this approach involves introducing a measure of complexity of algorithms. I intend to discuss this question in another article. In the present one I will only use the fact that *the number of simple algorithms cannot be very large*.

To be definite, let us consider a table

$$T = (t_1, \ldots, t_N)$$

consisting of N zeros and ones: $t_k = 0$ or 1. This table will be called *random*, if for different choices of a sufficiently large subset A of $\{1, \ldots, N\}$ the frequency

$$\pi(A) = \frac{1}{n} \sum_{k \in A} t_k$$

of the appearance of ones in A is approximately the same. One can select, for example, for A:

(a) the set of the first n even numbers;
(b) the set of the first n prime numbers $p_1, \ldots, p_n$, etc.

The ordinary notion of randomness does not lead to a constant value of frequences under selection methods independent of the structure of the table. For example, one can select for A

(c) the set of the first n values $k \geqslant 2$ such that $t_{k-1} = 0$;
(d) the set of the first n values $k > 1$ such that $t_{k-1} = a_1, t_{k-2} = a_2, \ldots, t_{k-s} = a_s$;
(e) the set of the first n even numbers $k = 2i$ such that $t_i = 1$;

(f) the set of numbers chosen according to the rule

$$k_1 = 1$$
$$k_{i+1} = k_i + 1 + t_{k_i} p_i$$

and so on. A precise definition of the notion of "admissable selection algorithm" will be given in section 2.

If for a table of sufficient size N at least one simple randomness test of this type with a sufficiently large sample size n shows a "considerable" discrepancy from the principle of constant frequency, we immediately reject the conjecture that the table is of "purely random" origin.

2. ADMISSIBLE ALGORITHMS OF CHOICE AND (n, ε)-RANDOMNESS

An admissible selection algorithm for a subset

$$A = R(T) \subset \{1, \ldots, N\}$$

corresponding to a table T of size N is determined by functions[3]

$$F_0, \quad G_0, \quad H_0;$$
$$F_1(\xi_1, \tau_1), \quad G_1(\xi_1, \tau_1), \quad H_1(\xi_1, \tau_1);$$
$$F_2(\xi_1, \tau_1; \xi_2, \tau_2), \quad G_2(\xi_1, \tau_1; \xi_2, \tau_2), \quad H_2(\xi_1, \tau_1; \xi_2, \tau_2);$$
$$\cdots\cdots\cdots\cdots\cdots\cdots\cdots\cdots\cdots\cdots\cdots\cdots$$
$$F_{N-1}(\xi_1, \tau_1; \xi_2, \tau_2; \ldots; \xi_{N-1}, \tau_{N-1}),$$
$$G_{N-1}(\xi_1, \tau_1; \xi_2, \tau_2; \ldots; \xi_{N-1}, \tau_{N-1}),$$
$$H_{N-1}(\xi_1, \tau_1; \xi_2, \tau_2; \ldots; \xi_{N-1}, \tau_{N-1}),$$

where the variables τ_k and the functions G_k and H_k take the values 0 and 1, while the variables ξ_k and the functions F_k take values in $\{1, \ldots, N\}$. The functions F_k must satisfy the additional condition

(2.1) $$\qquad\qquad F_k(\xi_1, \tau_1; \ldots; \xi_k, \tau_k) \neq \xi_i.$$

The application of the algorithm consist in generating the sequence

(2.2) $$\qquad\qquad x_1 = F_0,$$
$$x_2 = F_1(x_1, t_{x_1}),$$
$$x_3 = F_2(x_1, t_{x_1}; x_2, t_{x_2}),$$
$$\cdots\cdots\cdots\cdots\cdots\cdots\cdots\cdots$$
$$x_s = F_{s-1}(x_1, t_{x_1}; \ldots; x_{s-1}, t_{x_{s-1}}),$$

---

[3]Functions in the first row are constants (i.e. functions of an empty set of variables).

and determining what elements of the sequence are to be included in A. The sequence terminates as soon as the value[4]

$$(2.3) \qquad H_s(x_1, t_{x_1}; \ldots; x_s, t_{x_s}) = 1$$

appears. In this case the last term of the sequence will be x_s. If

$$H_k(x_1, t_{x_1}; \ldots; x_k, t_{x_k}) = 0$$

for all $k < N$, then the sequence ends with the element x_s for $s = N$, i.e. uses all the elements of $\{1, \ldots, N\}$; indeed, according to condition (2.1), all the elements of the sequence (2.2) are different.

The elements x_k for which

$$G_{k-1}(x_1, t_{x_1}; \ldots; x_{k-1}, t_{x_{k-1}}) = 1,$$

constitute the set A. I believe that this construction correctly reflects the von Mises approach in all its generality; while we have the basic restriction: when we determine whether $x \in \{1, \ldots, N\}$ belongs to A or not, we do not make use of the value of t_x.

Suppose we are given a system

$$\mathcal{R}_N = \{R\}$$

of admissable selection algorithms (the size N of the table is fixed).

Definition. A table T of size N is said to be (n, ε)-*random* with respect to the system $\mathcal{R}_N$, if there exists a constant p, $0 \leqslant p \leqslant 1$, such that for any $A = R(T)$, $R \in \mathcal{R}_N$ having $v \geqslant n$ elements, the frequency

$$(2.4) \qquad \pi(A) = \frac{1}{v} \sum_{k \in A} t_k$$

satisfies the inequality $|\pi(A) - p| \leqslant \varepsilon$.

It is sometimes convenient to speak of (n, ε, p)-randomness if the number p is fixed.

Theorem 1. *If the number of elements of the system $\mathcal{R}_N$ is no greater than*

$$(2.5) \qquad \tau(n, \varepsilon) = \frac{1}{2} e^{2n\varepsilon^2 (1-\varepsilon)},$$

than for any p, $0 \leqslant p \leqslant 1$, there exists a table T of size N which is an (n, ε, p)-random with respect to $\mathcal{R}_N$.

The meaning of the estimate appearing in the theorem will become clearer if we introduce the binary logarithm $\lambda(\mathcal{R}_n) = \log_2 \rho(\mathcal{R}_N)$ of the number of elements ρ of the system $\mathcal{R}_N$. The number $\lambda(\mathcal{R}_N)$ is equal to the amount of information needed

---

[4]In particular, if $H_0 = 1$, the choice never begins and the set A is void.

to indicate a specific element R from $\mathcal{R}_N$. Clearly in the case of a large $\lambda(\mathcal{R}_N)$, the system $\mathcal{R}_N$ must contain algorithms whose specification (and not only whose actual execution) is complicated (requires no less than $\lambda(\mathcal{R}_N)$ binary digits).

In our theorem the existence condition of (n, ε)-random tables with respect to $\mathcal{R}_N$ for an arbitrary π may be written in the form of the following inequality

$$\lambda(\mathcal{R}_N) \leqslant 2(\log_2 e)n\varepsilon^2(1 - \varepsilon) - 1.$$

Such a quantitative formulation of the result contained in the theorem is instructive in itself. If the ratio λ/n is small enough than for any ε chosen a priori, for all N and p, and any system of admissable algorithms satisfying

$$\lambda(\mathcal{R}_N) \leqslant \lambda$$

there exist an (n, ε)-random tables with respect to $\mathcal{R}_N$.

The proof of this theorem will be given in section 3. In section 4 we study the possibility of improving the estimates given in the theorem. But first let us make two remarks.

Remark 1. Since the selection algorithm is determined by the functions F_k, G_k, H_k, is natural to consider that two algorithms coincide if and only if the corresponding functions F_k, G_k, H_k coincide. Already from this point of view the number of possible selection algorithms is finite for a fixed N.

It is possible to take a different point of view, considering two algorithms different if and only if they yield different sets $A = R(T)$ for at least one table T. From this point of view the number of algorithms decreases further and does not exceed

$$(2^N)^{2^N} = 2^{N \cdot 2^N}.$$

The question of finding an exact estimate of the number of admissable algorithms (from the second point of view) is not so simple. It is only easy to estimate the number of algorithms constructing the set A independently of the properties of the table T. The number of such distinct algorithms equals 2^N, to the number of subsets $A \subset \{1, \ldots, N\}$.

Remark 2. Admissable selection algorithms from the set of all possible natural numbers were considered by A. Church [4]. In our definition let us replace the finite table T by an infinite sequence of zeros and ones

$$t_1, t_2, \ldots.$$

We assume that the values of the variables ξ_k and of the functions F_k are natural numbers. However we drop the requirement of stopping the selection process when $s = N$, assuming instead that the (infinite) sequence of functions F, G_k, H_k is computable in the traditional formal sense for definitions of this kind. The starting point of Church's study is the existence (for any p) of a sequence $t_1, t_2, \ldots, t_N, \ldots$ whose density equals p for any infinite[4] set A obtained by means of an admissable algorithm.

---

[4]In this notion of Church, only infinitely continued selection algorithms are of significance, hence the functions G_k and everything related to them must be omitted

3. PROOF OF THEOREM 1

This result concerns finite objects; its formulation does not involve any notions from probability theory. The use of certain results from probability theory in the proof does not contradict its formal character, as long as we limit ourselves to considering distributions "weights" in the set of tables T of size N, attributing the weight

$$P(T) = p^M (1 - p)^{N-M}$$

to any table containing M ones. This method of proof does not affect the logical nature of the theorem itself and does not hamper referring to it in any discussion of the domain of applicability of probability theory.

One can prove the following inequality related to the "Bernouilli scheme":

(3.1) $$\mathbf{P}\left\{\sup_{k \geqslant n}\left|\frac{\mu_k}{k} - p\right| > \varepsilon\right\} < 2e^{-2n\varepsilon^2(1-\varepsilon)}.$$

Here p is the probability of success in each trial in a sequence of independent trials; μ_k is the number of successful trials among the first k trials. We can easily deduce the following consequence of (3.1).

Corollary. Let[5]

$$\mathbf{P}\{\xi_k = 1 \mid k \leqslant \nu, \xi_1, \ldots, \xi_{k-1}\} = p,$$

where $\xi_1, \xi_2, \ldots, \xi_\nu$ is a sequence of a random number of random variables and p is a constant. Then

$$\mathbf{P}\left\{\nu \geqslant n, \left|\frac{\mu_\nu}{\nu} - p\right| > \varepsilon < 2e^{-2n\varepsilon^2(1-\varepsilon)}\right\}$$

Now consider a system $\mathcal{R}_N$ of ρ admissible algorithms.

Consider a randomly constructed table, where $t_x = 1$ with probablity p independently of the values of the other t_m. If we now fix $R \in \mathcal{R}_N$ and denote by

$$\xi_1, \xi_2, \ldots, \xi_\nu$$

those elements of the sequence

$$t_1, t_2, \ldots, t_N$$

whose numbers are contained in $A = R(T)$ (listed in order of their appearance in the process of applying the algorithm), then as can be easily seen the conditions of validity (3.2) will hold. Hence for any given $R \in \mathcal{R}_N$ the probability of the number of elements ν of the set A being no less than n and also of having the inequality $|\pi(A) - p| \geqslant \varepsilon$ is less than $2e^{-2n\varepsilon^2(1-\varepsilon)}$.

If

$$\rho \leqslant \frac{1}{2}e^{2n\varepsilon^2(1-\varepsilon)},$$

---

[5] We consider here the conditional probability $\xi_k = 1$ for $k \leqslant \nu$ and given $\xi_1, \ldots, \xi_{k-1}$.

thàn the sum of the probabilities for which the inequality

$$|\pi(A) - p| \leqslant \varepsilon$$

fails for all the algorithms yielding sets of n or more elements will be less than 1. Hence with positive probability the table T will turn out to be (n, ε, p)-random in the sence of section 2. This implies the existence of (n, ε, p)-random tables with respect to $\mathcal{R}_N$ (without appealing to probabilistic assumptions on the distribution $R(T)$ in the space of tables).

4. POSSIBLE SHARPENING OF THE ESTIMATE GIVEN BY THEOREM 1

Let n, ε, N, p be fixed. Then for each nonnegative ρ we have one of the following statements:

(a) for any system $\mathcal{R}_N$ of ρ admissible selection algorithms there exists a table T of size N which is (n, ε, p)-random with respect to $\mathcal{R}_N$;

(b) there exists a system $\mathcal{R}_N$ of ρ admissable selection algorithms with respect to which no table of size N is (n, ε, p)-random.

It is easy to see that if the situation (a) holds for some ρ, then it also holds for all $\rho' < \rho$. It is also clear that for $\rho = 0$ we have the situation (a). Therefore, the least upper bound

$$\tau(n, \varepsilon, N, p) = \sup_{\rho \in a} \rho$$

of all ρ for which we have (a) exists. For all ρ greater than $\tau(n, \varepsilon, N, p)$ we have (b).

If we put

$$\tau(n, \varepsilon) = \inf_{p, N} \tau(n, \varepsilon, N, p),$$

then the assertion of Theorem 1 from section 2 can be written in the form of the following inequality

(4.1) $$\tau(n, \varepsilon) \geqslant \frac{1}{2} e^{2n\varepsilon^2(1-\varepsilon)}.$$

Taking logarithms

$$l(n, \varepsilon, N, p) = \log_2 \tau(n, \varepsilon, N, p), \quad l(n, \varepsilon) = \log_2 \tau(n, \varepsilon),$$

we can write (2.6) in the form

(4.2) $$l(n, \varepsilon) \geqslant 2n\varepsilon^2(1 - \varepsilon) - 1.$$

Actually, the main interest is in an asymptotically exact estimate of $l(n, \varepsilon)$ for small ε and large n and $l(n, \varepsilon)$. As $\varepsilon \to 0$, $n\varepsilon^2 \to \infty$, (4.2) implies

(4.3) $$l(n, \varepsilon) \geqslant 2n\varepsilon^2 + o(n\varepsilon^2).$$

On the other hand, as we shall see, as $\varepsilon \to 0$, $n\varepsilon \to \infty$, we have the relation

(4.4) $$l(n, \varepsilon) \leqslant 4n\varepsilon + o(n\varepsilon).$$

Unfortunately, I have not succeeded in filling the gap between the powers of ε in (4.3) and (4.4).

The estimate (4.4) is a simple consequence of the following theorem, whose statement, unfortunately, is rather complicated and only becomes clear after the proof has been given.

Theorem 2. *If* $k < (1 - 2\varepsilon)/4\varepsilon$, $n \leqslant (k-1)m$, $N \geqslant km$, *then*

$$\tau(n, \varepsilon, N, 1/2) = k \cdot 2^m.$$

To prove the theorem, it suffices, under the conditions

$$k < \frac{(1 - 2\varepsilon)}{4\varepsilon} \quad n = (k-1)m, \quad N = km$$

to construct a system $\mathcal{R}_N$ of $\rho = k \cdot 2^m + 1$ admissable algorithms for which there exists $(n, \varepsilon, 1/2)$-random table T

Let us partition the set $\{1, \ldots, N\}$ into k sets Δ_i, $i = 1, \ldots, k$, containing m elements each. Each Δ_i has 2^m subsets. Construct the set A_{is}, $i = 1, 2, \ldots, k$; $s = 1, 2, \ldots, 2^m$ by taking the union of all Δ_j for $j \neq i$ and the $s - th$ subset of Δ_i. The system $\mathcal{R}_N$ consist of:

$k \cdot 2^m$ algorithms selecting the sets A_{is},

one algorithm, selecting the set $A = \{1, \ldots, N\}$.

We shall prove that no table is $(n, \varepsilon, 1/2)$-random with respect to $\mathcal{R}_N$.

Assume that the table T is $(n, \varepsilon, 1/2)$-random with respect to $\mathcal{R}_N$. It must contain at least $(1/2 - \varepsilon)N$ zeros and $(1/2 - \varepsilon)N$ ones. Therefore we can choose i and j so that Δ_i will contain $\alpha \geqslant (1/2 - \varepsilon)m$ zeros and Δ_j will contain $\beta \geqslant (1/2 - \varepsilon)m$ ones.

Denote by γ the smallest of the numbers α and β; than $\gamma \geqslant (1/2 - \varepsilon)m$. There exist an algorithm $R' \in \mathcal{R}_N$ ($R'' \in \mathcal{R}_N$), selecting the set A' (A'') containing all the elements of $\{1, \ldots, N\}$ except γ elements of Δ_i (Δ_j) corresponding to the zeroes (ones) of the table T. It is easy to see that the corresponding frequencies are

$$\pi(A') = \frac{M}{N - \gamma}, \quad \pi(A'') = M - \gamma N - \gamma,$$

where M is the total number of ones in the table T. Let us estimate the difference between these frequencies:

$$\pi(A') - \pi(A'') = \frac{\gamma}{N - \gamma} \geqslant \frac{(1/2 - \varepsilon)m}{km} > 2\varepsilon.$$

This estimate contradicts the system of inequalities

$$\left| \pi(A') - \frac{1}{2} \right| \leqslant \varepsilon, \quad \left| \pi(A') - \frac{1}{2} \right| \leqslant \varepsilon$$

that follow from the assumption of $(n, \varepsilon, 1/2)$-randomness of the able T. This contradiction proves the theorem.

REFERENCES

1. Kolmogorov A. N., *Foundations of the theory of probability*, Chelsea, 1950.
2. Kolmogorov A. N., *Probability*, Great Soviet Encyclopeadia, 2nd edition, vol. 7, 1951, pp. 508–510.
3. Kolmogorov A. N., *Probability theory*, Mathematics, its Content, Methods, and Meaning, vol. 2, Academy of Sciences publishers,, Moscow, USSR, 1956, pp. 252–284; English translation published by MIT Press.
4. Church A., *On the concept of a random sequence*, Bull. Amer. Math. Soc. **46** no. 2 (1940), 130–135.

THREE APPROACHES TO THE DEFINITION OF
THE NOTION OF AMOUNT OF INFORMATION *

§1. The combinatorial approach.

Suppose the variable x can assume values belonging to a finite set X which consists of N elements. We then say that the "entropy" of the variable x is equal to

$$H(x) = \log_2 N.$$

By indicating a specific value

$$x = a$$

of the variable x, we "remove" this entropy, communicating the "information"

$$I = \log_2 N.$$

If the variables $x_1, x_2, \ldots, x_k$ range independently over sets which consist respectively of $N_1, N_2, \ldots, N_k$ elements, then

$$H(x_1, x_2, \ldots, x_k) = H(x_1) + H(x_2) + \cdots + H(x_k). \tag{1}$$

To transmit the amount of information I, we must use I' binary digits, where

$$I' = \begin{cases} I & \text{when } I \text{ is an integer,} \\ [I] + 1 & \text{when } I \text{ is a fraction} \end{cases}$$

For example, the number of distinct "words" consisting of k zeros, k ones and one two equals

$$2^k(k + 1).$$

Therefore the amount of information in a communication of this type is equal to

$$I = k + \log_2(k + 1),$$

i.e. in order to "code" such a word in a pure binary system we require[1]

$$I' \approx k + \log_2 k$$

---

*Problemy Peredachi Informatsii, 1965, vol. 1, No.1, pages 3-11.

[1]Everywhere below $f \approx g$ means that the difference $f - g$ is bounded while $f \sim g$ means that the ratio $f : g$ tends to one.

zeros and ones.

In an exposition of information theory, one does not usually dwell for long on such a combinatorial approach to things. But I feel it is essential to stress its logical independence of any probabilistic assumptions. Suppose for example we are interested in the problem of coding messages written in an alphabet consisting of s letters and it is known that the frequencies

$$p_r = s_r/s \tag{2}$$

of appearance of separate letters in messages of length n satisfy the inequality

$$\chi = -\sum_{r=1}^{s} p_r \log_2 p_r \leqslant h. \tag{3}$$

It is easy to calculate that for large n the logarithm in base two of the number of messages satisfying the requirement (3) has the asymptotic estimate

$$H = \log_2 N \sim nh.$$

Therefore when transmitting such messages it suffices to use approximately nh binary digits.

The universal method of coding which allows the transmission of any sufficiently long message in an alphabet in s letters using not much more than nh binary digits is not necessarily very complicated; in particular it does not necessarily begin with the determination of the frequencies p_r in the entire message. To understand this, it suffices to notice that in partitioning the message S into m segments $S_1, S_2, \ldots, S_m$, we obtain the inequality

$$\chi \geqslant n^{-1}[n_1\chi_1 + n_2\chi_2 + \cdots + n_m\chi_m]. \tag{4}$$

Actually I do not want to enter here into the details of this special problem. It was only important for me to show that the mathematical questions arising from the purely combinatorial approach to the measurement of the amount of information are not limited to trivia.

The purely combinatorial approach is quite natural when we are concerned with the notion of "entropy of speech" having in mind to estimate the "flexibility" of language, i.e., to find a number assessing the possibility of "branching" in continuing speech for a fixed dictionary and given fixed rules of constructing sentences. For the binary logarithm of the number N of Russian printed texts, consisting of words included in S. I. Ozhegov's Dictionary of the Russian Language and only satisfying the requirements of "grammatical correctness" of length n expressed in the "number of signs" (including spaces between words), M. Ratner and N. Svetlova obtained the estimate

$$h = (\log_2 N)/n = 1,9 \pm 0,1.$$

This is considerably larger than the estimate from above for the "entropy of literary texts" obtained by means of various methods of "guessing the continuation".

This discrepancy is quite natural, since literary texts must not always satisfy the requirement of "grammatical correctness".

It is more difficult to calculate the combinatorial entropy of texts satisfying certain meaningful restrictions. It would for example be of interest to estimate the entropy of Russian texts that can be regarded as sufficiently exact translations of a given foreign text from the point of view of their meaning. Only the presence of such a "residual entropy" makes the translation of poetry possible, since the "resources of entropy" used to follow the chosen meter and the character of the rhymes may be computed with a certain precision. One can show that the classical rhymed eight-syllable iamb with certain natural restrictions on the frequency of "transfers" and so on requires the free use of lexical material characterized by "residual entropy" of order 0.4 (for the method of measuring the length of the texts in terms of the "number of symbols, including spaces between words" described above). If we take into consideration, on the other hand, the fact that stylistic limitations of a given genre probably decrease the estimate of the "total" entropy from 1.9 to no more than $1.1 - 1.2$, it follows that the situation becomes noteworthy both in the case of translations as in the case of original poetic creation.

I hope the pragmatic reader will excuse me for this example. As a justification, let me note that the wider problem of estimating the amount of information used in the creative activity of humanity is of great importance.

Now let us see to what extent the purely combinatorial approach allows to estimate the amount of information contained in the variable x with respect to a related variable y. The relationship between variables x and y ranging over the respective sets X and Y is that not all pairs x, y belonging to the Cartesian product $X \times Y$ are "possible". From the set of possible pairs U, one can determine, for any $a \in X$, the set Y_a of those y for which

$$(a, y) \in U.$$

It is natural to define the conditional entropy by the relation

$$H(y|a) = \log_2 N(Y_a) \tag{5}$$

(where $N(Y_x)$ is the number of elements in the set Y_x) and the information in x with respect to y by the formula

$$I(x : y) = H(y) - H(y|x). \tag{6}$$

x	y			
	1	2	3	4
1	+	+	+	+
2	+	−	+	−
3	−	+	−	−

For example, in the case shown in the table, we have

$$I(x = 1 : y) = 0, \quad I(x = 2 : y) = 1, \quad I(x = 3 : y) = 2.$$

It is clear that $H(y|x)$ and $I(x : y)$ are functions of x (while y enters in them as a "bound variable").

It is easy to introduce as a purely combinatorial concept the notion of "amount of information necessary to indicate an object x with given requirements as to the precision" (see in this connection the large amount of literature concerning "ε-entropy" of sets in metric spaces).

Obviously

$$H(x|x) = 0, \quad I(x : x) = H(x). \tag{7}$$

§2. The probabilistic approach.

The possibilities of further development of information theory on the basis of definitions (5) and (6) remained in the background because the interpretation of the variables x and y as "random variables" possessing certain joint probability distributions yields a considerably richer system of notions and relationships. In parallel with the expressions introduced in §1, here we have

$$H_W(x) = -\sum_x p(x) \log_2 p(x), \tag{8}$$

$$H_W(y|x) = -\sum_y p(y|x) \log_2 p(y|x), \tag{9}$$

$$I_W(x : y) = H_W(y) - H_W(y|x). \tag{10}$$

As before, $H_W(y|x)$ and $I_W(x : y)$ are functions of x. We have the inequalities

$$H_W(x) \leqslant H(x), \quad H_W(y|x) \leqslant H(y|x), \tag{11}$$

which become equalities when the corresponding distributions (on X and on Y_x) are uniform. The expressions $I_W(x : y)$ and $I(x : y)$ are not directly related by an inequality of specific sign. As in §1, we have

$$H_W(x|x) = 0, \quad I_W(x|x) = H_W(x). \tag{12}$$

But the difference consists in that it is possible here to consider the expectations

$$\mathbf{E}H_W(y|x), \quad \mathbf{E}I_W(x : y),$$

while the expression

$$I_W(x, y) = \mathbf{E}I_W(x : y) = \mathbf{E}I_W(y : x) \tag{13}$$

characterizes the "strength of the relationship" between x and y in a symmetric way.

One should, however, note the paradox which appears in the probabilistic approach: the expression $I(x : y)$ in the combinatorial approach is always non-negative, as is natural in the naive representation on the "amount of information", while the expression $I_W(x : y)$ can be negative. The true measure of the "amount of information", can only be obtained by averaging the expression $I_W(x : y)$.

The probabilistic approach is natural in the theory of communication along transmission channels of "mass" information consisting of a large number of disconnected or weakly connected messages satisfying certain specific random laws. In questions of this type, the confusion of probabilities with frequencies within the limits of a single sufficiently long time interval, rigorously justified under the hypothesis of sufficiently fast (mixing) that arises in the applications, is practically harmless. In practice we can say, for example, that the question of the "entropy" of a flow of congratulatory telegrams and the "capacity" of a transmission channel required for their transmission without delay and errors may be viewed as correctly set both in its probabilistic interpretation and under the ordinary replacement of probabilities by empirical frequencies. And if some feeling of dissatisfaction remains here, it is only related to the interaction between the mathematical theory of probability and real life "random phenomena" in general.

But what is the real life significance, for example, of saying what the "amount of information" contained in the text of "War and Peace" is? Is it possible to include this novel in a reasonable way into the set of "all possible novels" and further to postulate the existence of a certain probability distribution in this set? Or should we consider separate scenes from "War and Peace" as forming a random sequence with "stochastic relations" disappearing quickly enough within the distance of several pages?

Basically there is no more clarity in the fashionable expression "the amount of genetic information" necessary, say, for the reproduction of the cuckoo bird's species. Again, within the framework of the accepted probabilistic approach, two versions are possible. In the first version we consider the set of all possible species with a probability distribution on this set taken from somewhere or other[2]. In the second version, the characteristic properties of the species are viewed as a family of weakly related random variables. We can argue in favour of the second version by using considerations based on the actual mechanism of mutations. But these considerations are illusions, if we suppose that the result of natural selection gives rise to a system of characteristic properties of the species which are in accord with each other.

§3. The algorithmic approach.

Basically the most meaningful approach has to do with the amount of information in "something" (x) about "something else" (y). It is not by accident that it is precisely this approach which was generalized in probability theory to the case of continuous variables, when entropy is infinite, but where in a wide range of cases the expression

$$I_W(x, y) = \int \int P_{xy}(dx\,dy) \log_2 \frac{P_{xy}(dx\,dy)}{P_x(dx)P_y(dy)}$$

is finite. Real objects which are to be studied are very (unboundedly?) complex, but the relationship between two really existing objects are fully assessed in a simpler

---

[2] The consideration of the set of all species existing or having once existed on earth, even under a purely combinatorial calculation, would give absolutely unacceptedly small estimates from above (something of the order of 100 bytes).

schematic description. If a geographical map gives us important information about a section of the Earth's surface, the microstructure of paper and paint appearing on the paper has no relationship to the microstructure of the section of the Earth's surface shown on this paper.

Practically, we are usually interested in the amount of information in an individual object x relatively to an individual object y. Of course, it is already known in advance that such an individual estimate of the amount of information may have a reasonable meaning only in the case of sufficiently large amounts of information. There is no significance in asking what the amount of information in the sequence

$$0110$$

about the sequence

$$1100.$$

actually is. But if we take out specific table random numbers of the ordinary size used in statistical practice and write out, for each of its digits, the number of units of its square according to the scheme

$$0123456789$$
$$0149656941$$

then the new table will contain approximately

$$(\log_2 10 - \frac{8}{10})n$$

information about the first one (n is the number of digits in the tables).

In connection with the above, the definition of the expression

$$I_A(x : y)$$

given further will contain a certain amount of indeterminacy. Different equivalent versions of this definition will lead to values which are equivalent only in the sense that $I_{A_1} \approx I_{A_2}$, i.e.

$$|I_{A_1} - I_{A_2}| \leqslant C_{A_1 A_2},$$

where the constant $C_{A_1 A_2}$ depends on the definitions of the universal methods of programming A_1 and A_2 underlying the two versions of our definition.

We shall consider a "numbered set of objects", i.e. a countable set

$$X = \{x\},$$

each element of which has been assigned a "number" $n(x)$, i.e., a finite sequence of zeros and ones beginning with a one. Denote by $l(x)$ the length of the sequence $n(x)$; we shall assume that:

1) the correspondence between X and the set D of binary sequences of the type described above is one-to-one;

2) $D \subset X$, the function $n(x)$ on D is recursive [1] and for $x \in D$ we have

$$l(n(x)) \leqslant l(x) + C,$$

where C is a certain constant;

3) the ordered pair (x, y) is contained in X whenever both x and y are, the number of this pair is a recursive function of the numbers of x and y and

$$l(x, y) \leqslant C_x + l(y),$$

where C_x depends only on x.

Not all these requirements are essential but they facilitate our exposition. The final result of the construction is invariant with respect to the passage to a new numbering $n'(x)$ possessing the same properties, if it can be expressed recursively in terms of the old numeration, and with respect to the inclusion of the system X in a wider system X' (under the assumption that the numbers n' in the extended system for elements of the first system can be expressed recursively in terms of the original numbers n). Under these transformations, the new "complexities" and amounts of information remain equivalent to the original ones in the sense of $\approx$.

The "relative complexity" of the object y for a given x will be defined as the minimal length $l(p)$ of the "program" p for obtaining y from x. As stated, this definition depends on the "method of programming". The method of programming is none other than a function

$$\varphi(p, x) = y$$

which assigns the object x to each program p and object y.

In accordance to the universally accepted approach in modern mathematical logic, one should assume that the function φ is partially recursive. For such a function we put

$$K_\varphi(y|x) = \begin{cases} \min_{\varphi(p,x)=y} l(p), \\ \infty, \text{ if there is no } p \text{ such that } \varphi(p, x) = y. \end{cases}$$

Here the function

$$v = \varphi(u)$$

depending on $u \in X$, with values $v \in X$, is called *partially recursive* if it induces a partially recursive function of the corresponding numbers

$$n(v) = \Psi[n(u)].$$

In order to understand the definition, it is important to note that partially recursive functions in general are not everywhere defined. There does not exist a regular process for finding out whether the application of the programme p to the object x will actually yield a result or will not. Hence the function $K_\varphi(y|x)$ is not necessarily effectively computable (general recursive) even in the case when it is necessarily finite for any x and y.

Main theorem. *There exists a partially recursive function $A(p, x)$ such that for any other partially recursive function $\varphi(p, x)$ we have the inequality*

$$K_A(y|x) \leqslant K_\varphi(y|x) + C_\varphi,$$

where the constant C_φ does not depend on x and y.

The proof is based on the existence of a universal partially recursive function

$$\Phi(n, y)$$

possessing the following property: by fixing an appropriate number n, we can obtain, according to the formula

$$\varphi(u) = \Phi(n, y),$$

any other partially recursive function. The function $A(p, x)$ that we require is defined by the formula[3]

$$A((n, q), x) = \Phi(n, (q, x)).$$

Indeed, if

$$y = \varphi(p, x) = \Phi(n, (p, x)),$$

then

$$A((n, p), x) = y, \quad l(n, p) \leqslant l(p) + C_n.$$

The function $A(p, x)$ satisfying the requirements of the main theorem will be called (as well as the method of programming which it defines) *asymptotically optimal*. Obviously for such functions for any x and y the "complexity" $K_A(y|x)$ is finite. For two such functions A_1 and A_2, we have

$$|K_{A_1}(y|x) - K_{A_2}(y|x)| \leqslant C_{A_1 A_2},$$

where $C_{A_1 A_2}$ does not depend on x and y, i.e. $K_{A_1}(y|x) \approx K_{A_2}(y|x)$.

Finally,

$$K_A(y) = K_A(y|1)$$

may be viewed as simply being the "complexity of the object y" and we can define the amount of information in x with respect to y by the formula

$$I_A(x : y) = K_A(y) - K_A(y|x).$$

It is easy to prove[4] that this expression is always essentially positive

$$I_A(x : y) \widetilde{>} 0,$$

---

[3]$\Phi(n, y)$ is defined only in the case $n \in D$, while $A(p, x)$ is defined only in the case when p is of the form (n, q), $n \in D$.

[4]Choosing for the "comparison function" the function $\varphi(p, x) = A(p, 1)$, we obtain $K_A(y|x) \leqslant K_\varphi(y|x) + C_\varphi = K_A(y) + c_\varphi$.

which should be understood in the sense that $I_A(x : y)$ is no less than a certain negative constant C depending only on the modalities of the chosen method of programming. As we have already mentioned, the whole theory is based on applications to large amounts of information by comparison to which $|C|$ will be negligibly small.

Finally

$$K_A(x|x) \approx 0, \ I_A(x : x) \approx K_A(x).$$

Of course it is possible to avoid the indeterminacy related to the constants C_φ, etc. by fixing once and for all the domain of definition of the object X, their numeration and the function A, but it seems doubtful that this can be done in a non-arbitrary way. One should, however, think that various "reasonable" possibilities of choice appearing here will lead to estimates of "complexities" which differ by hundreds rather than by tens of thousands of bits. Therefore such expressions as the "complexity" of the text of the novel "War and Peace" must be viewed as being practically uniquely determined. Experiments on guessing the continuations of literary texts allow to estimate from above the conditional complexity for a given amount of "apriori information" (on language, style, content of the text), which is possessed by the guesser. In experiments carried out at the chair of probability theory of Moscow State University, such estimates from above varied between 0.9 and 1.4. Estimates of the order $0.9 - 1.1$ obtained by N. G. Rychkova, led less successful guessers to talk about her telepathic relationship with the authors of the texts.

I think that for the "amount of genetic information" the approach described above also gives a definition of the notion itself which is correct in principle, no matter how hard the factual estimate of this amount is.

§4. Concluding remarks.

The approach developed in §3 possesses one basic defect: it does not take into account the "difficulties" of transforming the program p and the object x into the object y. By introducing the appropriate definitions, one can prove a purely mathematical statement which can be correctly interpreted as an indication of the existence of cases when the object possessing a very simple program, i.e. possessing a very small complexity $K(x)$, may be recovered by means of short programs only as the result of computations of practically unrealizable length. Elsewhere I shall propose the study of the dependence of the necessary complexity of the program

$$K^t(x)$$

on the admissible difficulty t of its transformation into the object x. The complexity $K(x)$ which can be defined in §3 will then appear as the minimum of $K^t(x)$ when the restrictions on the value of t are removed.

The applications of the approach developed in §3 to a new foundation of probability theory also remains outside this note. Roughly speaking, here I mean the following. If a finite set M consisting of a very large number of elements N possesses a definition by means of a program of length negligibly small by comparison to $\log_2 N$, then almost all elements of M are of complexity $K(x)$ close to $\log_2 N$. The elements $x \in M$ of this complexity are to be viewed as "random" elements of the set M. A version of this idea, not yet in final form, may be found in the article [2].

REFERENCES

[1] V. A. Uspensky, *Lectures on computable functions*, Moscow, Fizmatgiz, 1960.
[2] A. N. Kolmogorov, *On tables of random numbers*, Sankhya. A **25**,4 (1963), 369-376.

ON THE REALIZATION OF NETWORKS
IN THREE-DIMENSIONAL SPACE*[1]

(Jointly with Ya. M. Barzdin)

By a (d, n)-network we shall mean a oriented graph with n numbered vertices $\alpha_1, \alpha_2, \ldots, \alpha_n$ and dn marked edges such that precisely d edges are incident to each vertex and one of them is marked by the weight x_1, another by the weight x_2, etc., and finally the last one by the weight x_d.

Examples of such networks are logical networks and neuron networks. It is precisely for this reason that the question of constructing such networks in ordinary three-dimensional space under the condition that the vertices are balls while the edges are tubes of a certain positive diameter is of importance.

In the sequel, without essential loss of generality (in view of the fact that our estimates will be carried out only up to order) we shall limit ourselves to considering only $(2, n)$-networks, which will simply be called *networks*.

We shall say that the network $\mathfrak{A}$ is *realized* in three-dimensional space[2] if:

1) to each vertex $\alpha \in \mathfrak{A}$ a certain point $\varphi\alpha$ of the given space is assigned (for distinct α and β the points $\varphi\alpha$ and $\varphi\beta$ are different); this point will be called a φ-*point*;

2) to each edge $p = (\alpha, \beta)$ originating at α and ending at β we assign a certain continuous curve K_p joining the points $\varphi\alpha$ and $\varphi\beta$; this curve will be called a *conductor*;

3) the distance between any two distinct φ points is no less than 1;

4) the distance between any conductors corresponding to non-incident edges[3] is no less than 1;

5) the distance between any two conductors corresponding to edges of the form (α, β) and (γ, α) or (β, α) and (γ, α), $\beta \neq \gamma$, respectively is no less than 1 outside a neighbourhood of radius 1 of the point $\varphi\alpha$.

---

*Problemy Kibernetiki, 1967, N19, p. 261-268.

[1]At the beginning of the sixties A. N. Kolmogorov proved Theorem 1 for nets with bounded branching and also the following approximation to Theorem 2: there exists a net with $n > 1$ elements any realization of which has diameter greater than $C\sqrt{n}/\log n$, where C is a certain constant not depending on n. The final versions of the theorems given here belong to Ya. M. Barzdin. *Author's note.*

[2]We have in mind ordinary three-dimensional Euclidean space.

[3]Two edges (α, β) and (γ, δ) are assumed incident iff $(\alpha = \gamma) \vee (\alpha = \delta) \vee (\beta = \gamma) \vee (\beta = \delta)$.

Suppose R is a certain realization of the network $\mathfrak{A}$ in three-dimensional space. We shall say that the solid W *contains* the realization R if all the φ-points and conductors appearing in this realization are contained in the solid W and are located at a distance no less than 1 from its surface. By the volume of the realization R we shall mean the minimal volume of a solid containing a given realization R. By the *volume* $V(\mathfrak{A})$ of the network $\mathfrak{A}$ we will mean the minimal volume of its realizations in three-dimensional space.

We shall say that almost all networks possess a certain property E if

$$E(n)/S(n) \to 1 \text{ as } n \to \infty,$$

where $S(n)$ is the number of all pairwise distinct[4] networks in n vertices and $E(n)$ is the number of all pairwise distinct networks in n vertices which possess the property E.

Theorem 1 (Estimate from above). *For all networks $\mathfrak{A}$ with n vertices we have*

$$V(\mathfrak{A}) \leqslant C_1 n \sqrt{n},$$

where C_1 is a certain constant not depending on n.

Theorem 2 (Estimate from below). *For almost all networks $\mathfrak{A}$ with n vertices we have*

$$V(\mathfrak{A}) \geqslant C_2 n \sqrt{n},$$

where C_2 is a certain positive constant > 0 not depending on n.

Theorem 1 follows from the stronger statement:

Theorem 1*. *Any network with n vertices may be realized in a sphere of radius $C\sqrt{n}$, where C is a constant not depending on n.*

Proof of Theorem 1*. We first prove Theorem 1* for networks such that each vertex has no more than two edges originating from it. Such networks will be called networks with bounded branching.

Thus, suppose that $\mathfrak{A}$ is a certain network with n vertices and bounded branching. Let us show that it can be realized in the parallelepipedon $ABCDA'B'C'D'$ shown on Figure 1 (without loss of generality we shall assume that $\sqrt{n}$ is an integer). This will mean that $\mathfrak{A}$ can be realized in a sphere of radius $C\sqrt{n}$, where $C = \sqrt{21}$.

Suppose U is a set on points of the lower base $ABCD$ of the form $(4i + 2, 2j, 0)$, $i = 0, 1, \ldots, \sqrt{n} - 1$; $j = 0, 1, \ldots, \sqrt{n} - 1$.

Assign to each vertex α of the network $\mathfrak{A}$ its own points $\varphi\alpha$ from U and to each arc $p = (\alpha, \beta)$ the polygonal line $K_p(\zeta) = (\varphi\alpha)abcdefgh(\varphi\beta)$ (Figure 1) depending on the natural parameter ζ in the following way: suppose the coordinates of the points $\varphi\alpha$ and $\varphi\beta$ are respectively $(4i + 2, 2j, 0)$ and $(4i' + 2, 2j', 0)$, then:

the coordinates of the point h are $(4i' + 2 + \tau, 2j', 0)$, where $\tau = 1$, if the arc p has the weight x_1 and $\tau = -1$, if the arc p has weight x_2,

---

[4]Two networks $\mathfrak{A}$ and $\mathfrak{B}$ are assumed identical iff any two vertices with the same numbers both in $\mathfrak{A}$ and $\mathfrak{B}$ are joined by similarly oriented edges having the same weights.

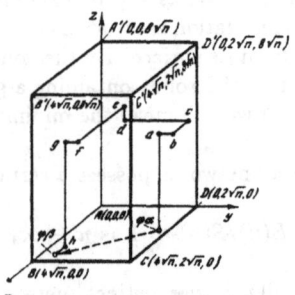

FIGURE 1

the coordinates of the points a and g are respectively $(4i + 2, 2j, 2\zeta - 1)$ and $(4i' + 2 + \tau, 2j', 2\zeta)$,

the coordinates of the points b and f are respectively $(4i + 2, 2j + 1, 2\zeta - 1)$ and $(4i' + 2 + \tau, 2j' + 1, 2\zeta)$,

the coordinates of the points c are $(4i, 2j + 1, 2\zeta - 1)$,

the coordinates of the points d and e are respectively $(4i, 2j' + 1, 2\zeta - 1)$ and $(4i, 2j' + 1, 2\zeta)$.

The arcs $p = (\alpha, \beta)$ and $p' = (\alpha', \beta')$ will be called related, if either the points $\varphi\alpha$ and $\varphi\alpha'$ have identical abscissas, or the points $\varphi\beta$ and $\varphi\beta'$ have identical ordinates, or both. It follows from the boundedness of branching that the number of arcs from the network $\mathfrak{A}$ related to the arc p is no greater than $4\sqrt{n}$.

The definition of the polygonal line $K_p(\zeta)$ has the following important consequences.

1) If the arcs $p = (\alpha, \beta)$ and $p' = (\alpha', \beta')$ are not related, then the corresponding polygonal lines $K_p(\zeta)$ and $K_{p'}(\zeta')$ for any values of the parameters ζ and ζ' are located at a distance $\geqslant 1$ from each other (excluding neighbourhoods of radius 1 of the point $\varphi\alpha$ ($\varphi\beta$) in the case when $\alpha = \beta'$ ($\beta = \alpha'$)).

2) If the arcs $p = (\alpha, \beta)$ and $p' = (\alpha', \beta')$ are related, then the corresponding polygonal lines $K_p(\zeta)$ and $K_{p'}(\zeta')$ possess the property that whenever $\zeta \neq \zeta'$ these lines are located at a distance $\geqslant 1$ from each other (except in the neighbourhood of radius 1 of the point $\varphi\alpha$ ($\varphi\beta$) in the case when $\alpha = \beta'$ ($\beta = \alpha'$ or $\beta = \beta'$)).

Hence to prove the fact that the network $\mathfrak{A}$ can be realized in the parallelepipedon $ABCD\,A'B'C'D'$, it suffices to establish that for the arcs of the network $\mathfrak{A}$ we can choose a natural number as the value of the parameters ζ so that:

a) these values for distinct related arcs are different,

b) the corresponding polygonal lines are all inside the parallelepipedon $ABCDA'B'C'D'$, i.e. $1 \leqslant \zeta \leqslant 4\sqrt{n}$.

Such a choice of the values of parameter ζ is always possible in view of the fact that for any arc p the number of arcs related to it is no greater than $4\sqrt{n}$, i.e. is no greater than the number of admissible distinct values of the parameter ζ.

Thus we have shown that Theorem 1* holds for networks with bounded branching. The validity of Theorem 1* for arbitrary networks follows from the fact that any network $\mathfrak{A}$ with n vertices may be reconstructed into a network $\mathcal{U}'$ with bounded branching by adding no more than $2n$ new vertices. For example, arcs of the form shown on Figure 2 may be replaced by a tree of the form shown on Figure 3 (the γ_i are new vertices). As a result we obtain a network $\mathfrak{A}'$ with bounded branching and $n' \leqslant 3n$ vertices. This network may be realized in a sphere of radius $C\sqrt{n'}$. In order to obtain the realization of the original network $\mathfrak{A}'$ in the sphere from the realization of the network $\mathfrak{A}$, we must interpret the points $\varphi\gamma_i$ as branching points on polygonal lines joining $\varphi\alpha$ to $\varphi\beta_1,...,\varphi\beta_k$.

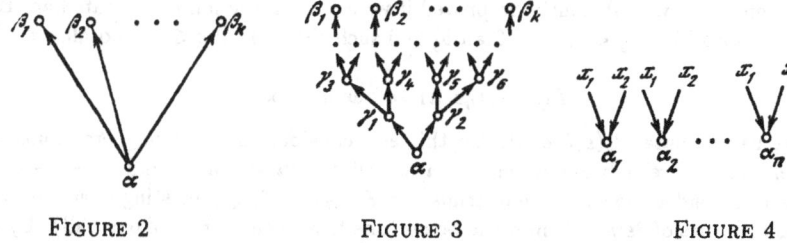

FIGURE 2   FIGURE 3   FIGURE 4

Suppose ω is a certain partition of the vertices of the network $\mathfrak{A}$ into three ordered parts which are denoted respectively by $\omega_1, \omega_2, \omega_3$. The partition ω of the network $\mathfrak{A}$ with n vertices will be called a (ε, δ)-partition, if ω_1 contains $[\varepsilon n]$ vertices, ω_2 contains $[\delta n]$ vertices and ω_3 contains $n - [\varepsilon n] - [\delta n]$ vertices. By the degree of connectivity $S(\mathfrak{A}, \omega)$ of the network $\mathfrak{A}$ with respect to the partition ω we shall mean the maximal number of non-incident arcs joining the vertices of ω_3 to the vertices of ω_1.

Lemma 1. *There exist $0 < \varepsilon_0 < 1$, $a_0 > 0$ and $b_0 > 0$ such that for all $\varepsilon \leqslant \varepsilon_0$ and $a \leqslant a_0$ the degree of connectivity of almost all networks $\mathfrak{A}$ with n vertices with respect to any $(\varepsilon, a\varepsilon)$-partition is no less than $b_0 \varepsilon n$.*

Let us prove this lemma. By the degree of incidence $Z(\mathfrak{A}, \omega)$ of the network $\mathfrak{A}$ with respect to the partition ω, we shall mean the number of vertices belonging to ω_3 from which there are arcs going to vertices belonging to ω_1. Since no more than two arcs can enter any one vertex, it follows that $S(\mathfrak{A}, \omega) \geqslant (1/2)Z(\mathfrak{A}, \omega)$. Thus our lemma will be proved if we prove a similar statement for the degree of incidence (let us denote this statement by A).

To this end the process of constructing networks with n vertices $\alpha_1, \alpha_2, \ldots, \alpha_n$ will be described in the following way. First we are given n vertices $\alpha_1, \alpha_2, \ldots, \alpha_n$ such that precisely two arcs with free second extremities enter into each one of them (Figure 4). Then the free extremity of each arc is randomly joined to one of the vertices[5] from $(\alpha_1, \ldots, \alpha_n)$. As the result, we obtain a certain network with vertices $\alpha_1, \alpha_2, \ldots, \alpha_n$.

Suppose at first a certain (ε, δ)-partition ω of the set $(\alpha_1, \ldots, \alpha_n)$ was fixed. Denote by $P_\omega(\varepsilon, \delta, c, n)$ the probability of obtaining, as the result of the procedure

---

[5]It is assumed that the probability of choosing any vertex is the same, i.e. equal to $1/n$.

described above, a network which has a degree of incidence with respect to ω satisfying $Z(\mathfrak{A}, \omega) < cn$. Note that for fixed ε and δ the probability $P_\omega(\varepsilon, \delta, c, n)$ for all (ε, δ)-partitions ω of the set $(\alpha_1, \ldots, \alpha_n)$ is the same and that the number of all possible (ε, δ) partitions of the set $(\alpha_1, \ldots, \alpha_n$ is equal to[6] $\binom{n}{\varepsilon n}\binom{n-\varepsilon n}{\delta n}$. Denote by $P(\varepsilon, \delta, c, n)$ the probability of obtaining, as the result of the procedure described above, a network which for at least one (ε, δ)-partition has a degree of incidence $< cn$. From the above it follows that

$$P(\varepsilon, \delta, c, n) < \binom{n}{\varepsilon n}\binom{n-\varepsilon n}{\delta n} P_\omega(\varepsilon, \delta, c, n). \tag{1}$$

Statement A will obviously be proved if we establish the following statement B: There exist $0 < \varepsilon_0' < 1$, $a_0' > 0$ an $b_0' > 0$ such that for all $\varepsilon \leqslant \varepsilon_0'$ and all $a \leqslant a_0'$ we have

$$P(\varepsilon, a\varepsilon, b_0'\varepsilon, n) \to 0 \text{ as } n \to \infty.$$

Let us estimate $P_\omega(\varepsilon, \delta, c, n)$. To this end consider the following probabilistic model. There are n boxes of which εn are white, δn are black and $(1 - \varepsilon - \delta)n$ are red. Consider the sequence of trials $\Omega = \xi_1, \xi_2, \ldots, \xi_{2\varepsilon n}$ consisting in successive random throws of $2\varepsilon n$ balls into the boxes (one trial is the throw of one ball). By a successful result of a trial, we mean the event in which the ball falls into an empty red box. Denote by $P'(\varepsilon, \delta, c, n)$ the probability that the number of successful outcomes in a series of trials Ω is $< cn$. Obviously

$$P_\omega(\varepsilon, \delta, c, n) = P'(\varepsilon, \delta, c, n) \tag{2}$$

(in the role of the balls we are considering free extremities of the arcs entering into the vertex ω_1, in the role of the white boxes the vertices of ω_1, in the role of the black boxes the vertices ω_2, in the role of the red boxes the vertices of ω_3).

Let us estimate $P'(\varepsilon, \delta, c, n)$. The probability of the successful outcome of the trial ξ_i in the sequence Ω, although it does depend on the outcomes of the previous trials, always satisfies $> 1 - 3\varepsilon - \delta$ (the smallest probability of a successful outcome may be obtained by the trial $\xi_{2\varepsilon n}$ under the condition that the outcomes of all the previous trials were successful; in this case as can be easily seen, this probability is

$$n^{-1}[(1 - \varepsilon - \delta)n - (2\varepsilon n - 1)] > 1 - 3\varepsilon - \delta).$$

Therefore

$$P'(\varepsilon, \delta, c, n) < P''(\varepsilon, \delta, c, n), \tag{3}$$

where $P''(\varepsilon, \delta, c, n)$ is the probability that in $2\varepsilon n$ Bernoulli trials, each of which has probability of successful outcome equal to $1 - 3\varepsilon - \delta$, the number of successful outcomes is $\leqslant cn$.

Let us estimate $P''(\varepsilon, \delta, c, n)$. It is known that

$$P''(\varepsilon, \delta, c, n) = \sum_{i=0}^{cn} \binom{2\varepsilon n}{i}(1 - 3\varepsilon - \delta)^i(3\varepsilon + \delta)^{2\varepsilon n - i}.$$

---

[6] Here and in the sequel for simplicity we shall assume that $\varepsilon n, \delta n$ and cn are integers.

Further estimates will be carried out under the following assumptions:

a) $0 < \varepsilon \leqslant 1/2$ and $0 < \delta < 1 - \varepsilon$;

b) $0 < cn \leqslant$ expectation of $2\varepsilon n(1 - 3\varepsilon - \delta)$, i.e. $0 < c \leqslant 2\varepsilon(1 - 3\varepsilon - \delta)$.

In this case

$$P''(\varepsilon, \delta, c, n) < cn \binom{2\varepsilon n}{cn}(1 - 3\varepsilon - \delta)^{cn}(3\varepsilon + \delta)^{2\varepsilon n - cn} < n\binom{2\varepsilon n}{cn}(3\varepsilon + \delta)^{2\varepsilon n - cn}.$$

Hence it follows from (1)-(3) that under the assumptions (a) and (b) we have

$$P(\varepsilon, \delta, c, n) < \binom{n}{\varepsilon n}\binom{n - \varepsilon n}{\delta n}n\binom{2\varepsilon n}{cn}(3\varepsilon + \delta)^{2\varepsilon n - cn}.$$

Now let us introduce parameters a and b such that $\delta = a\varepsilon$ and $c = b\varepsilon$. Then, instead of the assumptions (a) and (b), we obtain the following inequalities:

(a') $0 < \varepsilon \leqslant 1/2$ and $a > 0$, $1 - \varepsilon - a\varepsilon > 0$;

(b') $0 < b \leqslant 2 - 6\varepsilon - 2a\varepsilon$. Moreover, the Stirling formula implies that for sufficiently large k and r we have*)

$$\binom{k}{r} = \frac{k!}{r!(k - r)!}$$
$$< \exp\{k \ln k + 1/2 \ln k - r \ln r - 1/2 \ln r - (k - r)\ln(k - r) - 1/2\ln(k - r)\}.$$

Hence for sufficiently large n:

$$P(\varepsilon, a\varepsilon, b\varepsilon, n) < \binom{n}{\varepsilon n}\binom{n - \varepsilon n}{a\varepsilon n}n\binom{2\varepsilon n}{b\varepsilon n}(2\varepsilon + a\varepsilon)^{2\varepsilon n - b\varepsilon n} <$$
$$< \exp\{[n \ln n + 1/2 \ln n - \varepsilon n \ln \varepsilon n - 1/2 \ln \varepsilon n - (n - \varepsilon n)\ln(n - \varepsilon n) -$$
$$- 1/2\ln(n - \varepsilon n)] + [(n - \varepsilon n)\ln(n - \varepsilon n) + 1/2\ln(n - \varepsilon n) -$$
$$- a\varepsilon n \ln a\varepsilon n - 1/2 \ln a\varepsilon n - (n - \varepsilon n - a\varepsilon n)\ln(n - \varepsilon n - a\varepsilon n) -$$
$$- 1/2\ln(n - \varepsilon n - a\varepsilon n)] + \ln n + [2\varepsilon n \ln 2\varepsilon n + 1/2 \ln 2\varepsilon n - b\varepsilon n \ln b\varepsilon n -$$
$$- 1/2 \ln b\varepsilon n - (2\varepsilon n - b\varepsilon n)\ln(2\varepsilon n - b\varepsilon n) - 1/2\ln(2\varepsilon n - b\varepsilon n)] +$$
$$+ (2\varepsilon n - b\varepsilon n)\ln(3\varepsilon + a\varepsilon)\} = \exp\{n[\varepsilon(1 - a - b)\ln \varepsilon -$$
$$- \varepsilon a \ln a + \varepsilon \ln 4 - \varepsilon b \ln b - \varepsilon(2 - b)\ln(2 - b) + \varepsilon(2 - b)\ln(3 + a) -$$
$$- (1 - \varepsilon - \varepsilon a)\ln(1 - \varepsilon - \varepsilon a)] + [-1/2 \ln n - 1/2 \ln \varepsilon - 1/2 \ln a -$$
$$- 1/2 \ln \varepsilon - 1/2 \ln(1 - \varepsilon - \varepsilon a) + 1/2 \ln 2 + 1/2 \ln \varepsilon - 1/2 \ln b -$$
$$- 1/2 \ln \varepsilon - 1/2 \ln \varepsilon - 1/2 \ln(2 - b)]\} <$$

(since $\varepsilon, a/b$ do not depend on n)

$$< \exp\{n[\varepsilon(1 - a - b)\ln \varepsilon - \varepsilon a \ln a + \varepsilon \ln 4 - \varepsilon b \ln b - \varepsilon(2 - b)\ln(2 - b) +$$
$$+ \varepsilon(2 - b)\ln(3 + a) - (1 - \varepsilon - \varepsilon a)\ln(1 - \varepsilon - \varepsilon a)]\} <$$

---

*)ln means natural logarithm (*Translator's note*).

(by assumptions (a') and (b'))

$$< \exp\{n[\varepsilon[-(1 - a - b)|\ln\varepsilon| - a\ln a + \ln 4 - b\ln b - (2 - b)\ln(2 - b) +$$
$$+ (2 - b)\ln(3 + a)] + |\ln(1 - \varepsilon(1 - a))|]\}.$$

Suppose $a = a_0' > 0$ and $b = b_0' > 0$ are fixed in an arbitrary way but so that $1 - a_0' - b_0' > 0$. Now let us take into consideration the facts that $|\ln(1 - \varepsilon(1 + a))| \to \varepsilon(1 + a)$ and $|\ln\varepsilon| \to \infty$ as $\varepsilon \to 0$. Now it is easy to see that we can find an ε_0' satisfying the inequalities (a') and (b') so that the expression

$$[\varepsilon_0'[- (1 - a_0' - b_0')|\ln\varepsilon_0'| - a_0'\ln a_0' + \ln 4 - b_0'\ln b_0' - (2 - b_0')\ln(2 - b_0') +$$
$$+ (2 - b_0')\ln(3 + a_0')] + |\ln(1 - \varepsilon_0'(1 + a_0'))|]$$

will be less than a certain negative constant. This will mean that for the given a_0', b_0' and ε_0' we have $P(\varepsilon_0', a_0', \varepsilon_0'b_0'\varepsilon_0', n) \to 0$ as $n \to \infty$.

To obtain a complete proof of statement B, it only remains to verify that $P(\varepsilon, a\varepsilon, b_0'\varepsilon, n) \to 0$ as $n \to \infty$ for all $\varepsilon \leqslant \varepsilon_0'$ and $a \leqslant a_0'$. The validity of this follows from the fact that if the expression

$$[\varepsilon[- (1 - a - b)|\ln\varepsilon| - a\ln a + \ln 4 - b\ln b - (2 - b)\ln(2 - b) +$$
$$+ (2 - b)\ln(3 + a)] + |\ln(1 - \varepsilon(1 + a))|]$$

is less than a certain negative constant for $\varepsilon = \varepsilon_0'$, $a_0 = a_0'$ and $b = b_0'$, then it will possess the following property also when $\varepsilon \leqslant \varepsilon_0'$ and $b = b_0'$.

The lemma is proved.

Now let us take into consideration the following two circumstances:

1) according to the definition of realization, the distance between any two conductors corresponding to non-incident arcs is $\geqslant 1$;

2) according to Lemma 1, almost all networks with n vertices are such that any $\varepsilon_0 n$ vertices are joined to the remaining vertices by at least $b_0\varepsilon_0 n$ non-incident arcs (the constants ε_0 and b_0 come from Lemma 1).

This implies the validity of the following lemma.

Lemma 2. *Almost all networks $\mathfrak{A}$ with n vertices possess the following property: for any realization of the network $\mathfrak{A}$ in three-dimensional space, there are less than $\varepsilon_0 n$ points in any parallelepipedon of surface area $L \leqslant 1/2b_0\varepsilon_0 n$ (the constants ε_0 and b_0 from Lemma 1).*

Proof of Theorem 2. Suppose μ and s are certain numbers (their values will be chosen later). Suppose $\mathfrak{A}$ is a certain network with n vertices for which we have the assertions of Lemmas 1 and 2 (almost all networks are such). Suppose further that the network $\mathfrak{A}$ is realized in three-dimensional space and let T be a certain parallelepipedon whose edges are of length $\geqslant s\mu\sqrt{n}$ and which contains all the φ points and conductors arising for the given realization of the network $\mathfrak{A}$ (Figure 5).

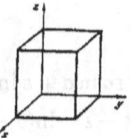

FIG.5

Then:

a) By means of the planes $z = i\mu\sqrt{n}\,(i = 0, 1, 2, \ldots)$ let us divide the parallelepipedon T into parallel slices of thickness $\mu\sqrt{n}$; without loss of generality we can assume that the last slice is also of thickness $\mu\sqrt{n}$. Let us number the slices in the direction of the Oz axis, using the numbers 1, 2, 3... (thus the lowest slice will be supplied with the number 1, the next one with the number 2, and so on). Then using this partition, let us construct the partition of the parallelepipedon T into the parts $A_1, A_2, \ldots, A_s$; in A_i we choose only those slices that have numbers of the form $ks + i$, where k is an integer. Then from all these parts let us choose the one which contains the smallest number of φ-points. Suppose it is the part A_z. It is easy to see that A_z contains no more than $n/s\varphi$-points. Slices belonging to A_z will be called distinguished.

b) By means of the planes $y = i\mu\sqrt{n}\,(i = 0, 1, 2, \ldots)$ let us partition the parallelepipedon T into parallel slices, then construct a partition $B_1, B_2, \ldots, B_s$ similar to the one in a) and let us call the slices belonging to the part B_y containing the least number of φ-points distinguished.

c) By means of the planes

$$x = i\mu\sqrt{n}\,(i = 0, 1, 2, \ldots)$$

let us partition the parallelepipedon T into parallel slices and construct a partition $C_1, C_2, \ldots, C_s$ as in a) and call distinguished those slices that belong to the parts C_x containing the least number of φ-points.

It follows from the above that the distinguished slices of the parallelepipedon T contain no more than $3n/s$ φ-points.

Now suppose the numbers μ and s are such that the cubes bounded by distinguished slices have a surface area no greater than $1/2b_0\varepsilon_0 n$ (here we mean each cube separately), i.e. suppose μ and s satisfy the inequality

$$6)s - 1)^2\mu^2 n \leqslant 1/2b_0\varepsilon_0 n \tag{4}$$

(the constants ε_0 and b_0 come from Lemma 1).

Then according to Lemma 2, each of the above-mentioned cubes contains less than $\varepsilon_0 n$ points. This in turn implies that if the number of φ points contained in non-distinguished slices is no less than $\varepsilon_0 n$, i.e.,

$$n - 3n/s \geqslant \varepsilon_0 n, \tag{5}$$

then one can construct a set G of these cubes which contains no less than $1/2\varepsilon_0 n$ and no more than $\varepsilon_0 n$ points[7].

Suppose further that the number s is such that the distinguished slices contain no more than $a_0(\varepsilon_0/2)n\varphi$-points, i.e. let s satisfy the inequality

$$3n/s \leqslant a_0(\varepsilon/2)n. \tag{6}$$

Consider the following (ε, δ)-partition of the network $\mathfrak{A}$:

---

[7]Here and in the sequel when we say that "the set G contains φ-points", we mean that elements of the set G (i.e. cubes) contain φ-points.

a) ω_1 is the set of vertices of the network $\mathfrak{A}$ such that in the realization of the network $\mathfrak{A}$ under consideration these points are assigned points from G; the number of elements εn of the set ω_1 is equal to the number of φ-points contained in G, i.e. $\varepsilon_0/2 \leqslant \varepsilon \leqslant \varepsilon_0$;

b) ω_2 is the set of all vertices of the network $\mathfrak{A}$ which in the realization of a network $\mathfrak{A}$ under consideration are assigned points from distinguished slices; thus the number of elements δ_n of the set ω_2 is equal to the number of φ-points contained in the distinguished slices, i.e., $\delta \leqslant a_0 \varepsilon_0/2$.

It follows from Lemma 1 that the network $\mathfrak{A}$ for such an (ε, δ)-partition has a degree of connectivity $\geqslant b_0 \varepsilon n \geqslant b_0(\varepsilon/2)n$. This means that the φ-points contained in the set of cubes G and φ-points contained in the other cubes bounded by distinguished slices are joined to each other by at least $b_0(\varepsilon_0/2)n$ non-incident conductors. The set of these conductors will be denoted by E. According to the definition of realization, the distance between conductors from E must be $\geqslant 1$. Since the thickness of the distinguished slices is equal to $\mu\sqrt{n}$, it follows that each of the conductors from the set E joins points located at a distance $\geqslant \mu\sqrt{n}$. Hence we obtain that any solid W' which contains all the conductors from E and in such a way that they are at a distance $\geqslant 1$ from its surface, has a volume greater than

$$b_0(\varepsilon_0/2)n\pi(1/2)^2\mu\sqrt{n} = (\pi/8)b_0\varepsilon_0\mu n\sqrt{n}.$$

Now let us recall that these estimates were obtained under the condition that s and μ satisfy the inequalities (4), (5) and (6). It is easy to verify that there exists s and $\mu > 0$ not depending on n that satisfy these inequalities. This means that the volume of the network satisfies

$$V(\mathfrak{A}) > (\pi/8)b_0\varepsilon_0\mu n\sqrt{n} = C_2 n\sqrt{n},$$

where C_2 is a certain constant > 0 not depending on n.

Theorem 2 is proved.

TO THE LOGICAL FOUNDATIONS
OF THE THEORY OF INFORMATION
AND PROBABILITY THEORY*

1. We shall mainly be concerned with the basic notions of information theory. Our starting point will be the notion of conditional entropy of the object x for a given object y, $H(x|y)$, which can be interpreted as the amount of information necessary to determine the object x in the situation when the object y is already known. Denoting by $\emptyset$ the "necessarily known object", we obtain the unconditional entropy

$$H(x|\emptyset) = H(x).$$

The information contained in the object y about the object x can be formally defined by means of substraction

$$I(x|y) = H(x) - H(x|y).$$

Naturally we then have

$$I(x|x) = H(x).$$

Ordinarily one uses the probabilistic definition of entropy which, however, is not related to individual objects but to "random" ones, i.e. essentially to the probability distribution in certain classes of objects. In order to stress this difference, we shall denote random objects by Greek letters. Limiting ourselves to the case of discrete distributions, let us recall the standard definition

$$H(\xi|\eta = -\sum_{x,y} p\{\xi = x, \eta = y\} \log_2 p(\xi = x|\eta = y)... \qquad (1)$$

It is well known that "probabilistic" statements can be actually interpreted in terms of statistical statements. In other words, our definition (1) may be used in practice only when applied to very large statistical families of pairs of objects

$$(x_1, y_1), (x_2, y_2), \ldots, (x_n, y_n), \ldots$$

Not all the applications of information theory can be reasonably described within the framework of this interpretation and its basic notions. I think that the necessity

---

*Problemy Peredachi Informatsii, 1969, vol. 5, No.3, p. 3-7.

of giving a definite meaning to the expressions $H(x|y)$ and $I(x|y)$ in the case of individual objects x and y (not regarded as realizations of random trials with specific distribution laws) was understood a long time ago by many researchers working in information theory.

As far as I know, the first publication that appeared containing an outline of the resconstruction of information theory satisfying the requirements described above, is Solomonoff's article [1] published in 1964. I developed similar ideas, before I learned about Solomonoff's work, in 1963-1964, and published my first note on this topic [2] in the beginning of 1965. The young Swedish mathematician Martin-Löf began developing this new approach with energy in the years 1964-65 in Moscow. The lectures [3] which he delivered in Erlangen in 1966 are, it seems, the best introduction to the entire circle of ideas of my articles.

The underlying idea of the definition is very simple: the entropy $H(x|y)$ is the minimal length of a "program" P written as a sequence of zeros and ones which allows to construct the object x when the object y is in our possession

$$H(x|y) = \min_{A(P,y)=x} l(P). \qquad (2)$$

The exact implementation of this idea is based on the general theory of "computable" (partially recursive) functions, i.e. on the general theory of algorithms in its most general form. We shall return later to the explanation of the notation $A(P, y) = x$.

Although Martin-Löf and I felt that the importance of the new approach is doubtless, its development was slowed down by the fact that the simplest formulas which, in the probabilistic approach, are obtained as the result of very simple algebraic transformations of the expression (1), could not be recovered in the new approach. Among the formulas which could not be recovered is the following one

$$I(x|y) = I(y|x). \qquad (3)$$

Actually, if we think a little its real content is in fact quite non-trivial and should in fact seem doubtful in its unconditional applicability as are the formulas of "decomposition" of information

$$H(x, y) = H(x) + H(y|x). \qquad (4)$$

Formulas (3) and (4) are so habitual that we do not at once notice that in the new approach they are simply wrong, and may be recovered only in the form of approximate relations

$$|I(x|y) - I(y|x)| = O(\log_2 H(x, y)), \qquad (3')$$
$$H(x, y) = H(x) + H(y|x) + O(\log_2 H(x, y)). \qquad (4')$$

2. Consider a long sequence

$$x = (x_1, x_2, \ldots, x_l), \quad l = l(x),$$

consisting of zeros and ones. It is easy to understand that there exist sequences for which the entropy is no less than their length

$$H(x) \geqslant l(x).$$

Such sequences cannot be determined by a program which is shorter than their length. In order to obtain them, they must be written out. For these sequences, there are just no simpler laws of construction which determine them. It is natural to recall that the absence of a laws of construction from the point of view of common sense is a feature of randomness. We practically start from the assumption that it is on this basis that "tables of random numbers" appearing in textbooks on mathematical statistics and probability theory are constructed.

In order to move a little further, let us consider how we imagine a sequence of zeros and ones appearing as the result of independent trials with probability p of obtaining a one at each trial. If l is large, then the number of ones is approximately equal to lp, i.e. the frequency $k/l \approx p$ gives an understanding of a certain regularity which appears in the sequence x. In view of this regularity, the conditional entropy x may be estimated by the inequality

$$H(x|l,k) \leqslant \log_2 \binom{k}{l} + O(1)$$

(the meaning of the additional term $O(1)$ in our formula will be discussed further). If the entropy $H(x|l,k)$ is close to this upper bound, then there does not exist an essentially more economic method of determining x other than to indicate l,k and the number of the sequence x among all the $\binom{l}{k}$ sequences with given l and k. This is the way we imagine "Bernoulli sequences" in which different signs are "independent" and arise with a certain "probability" p.

Thus we see that a certain analogue of the notion of a "Bernoulli" sequence may be stated in the language of algorithmic information theory as described above.

Bernoulli sequences in this new sense are especially studied in the paper by Martin-Löf [4]. Naturally, the notion of finite Bernoulli sequence must be made relative. For finite sequences, there is no clear bound between "regular" and "random". According to Martin-Löf, a sequence is m-Bernoulli if

$$H(x|l(x), k(x)) \geqslant \log_2 \binom{l(x)}{k(x)} - m.$$

A strict differentiation between "Bernoulli" and "non-Bernoulli" sequences is possible only after we pass to the limit for infinite sequences of zeros and ones

$$x = (x_1, x_2, \ldots, x_n, \ldots).$$

For such a sequence we shall denote by x^l the initial segment of length l

$$x^l = (x_1, \ldots, x_l).$$

It would seem desirable to define x as being "Bernoulli" by requiring the existence of an m such that all x^l are m-Bernoulli, i.e. if we always have

$$H(x^l|l, k_l) \geqslant \log_2 \binom{l}{k_l} - m.$$

But as Martin-Löf showed, no such infinite sequences of zeros and ones actually exist. The reasons are easy to explain. It follows from traditional probability theory that in actual random sequences one meets with parts consisting of only ones or only zeros of arbitrary large length. Clearly the description of such parts of the infinite sequence may be simplified considerably as compared to the standard one.

The most natural definition of infinite Bernoulli sequence is the following: x is said to be m-Bernoulli if there exists an m such that all the x^l are initial segments of finite m-Bernoulli sequences. Martin-Löf gives another, possibly somewhat more restrictive, definition. The more restrictive approach (with which Martin-Löf actually began) to $(1/2)$-Bernoulli sequences was independently developed by Chaitin [5].

Possibly some readers have already noticed that, in considering infinite sequences, we are actually studying a problem which was already set by Mises in his approach to what he called "collectives". It is known that the Mises approach was formalized in the language of the theory of computable functions by Church [6]. There exists an extension of the Mises approach (which generalizes the notion of admissible system of choice) given in my paper [7]. A strictly formal exposition of the theory generalized in this way may be found in the articles by Loveland [8,9].

However, the class of Bernoulli sequences in the sense of Church and Loveland is too wide. Their segments may still be fairly "regular". There exist sequences that are Bernoulli in the sense of both of these definition with segments which have only logarithmically increasing entropy:

$$h(x^l) = O(\log_2 l).$$

Conversely, Bernoulli sequences in the sense of Martin-Löf (and according to the definition given above) have segments whose complexity is almost maximal in the sense of the inequality

$$H(x^l) \geqslant l - O((\log_2 l)^{1+\varepsilon}), \text{ where } \varepsilon > 0 \text{ is arbitrary.}$$

Sequences that are Bernoulli according to Martin-Löf possess all the constructive properties that are proved in modern probability theory (for any probability p) " with probability one". Nothing similar can be said about sequences which are Bernoulli according to Church or Loveland.

3. The preceding brief exposition should justify the following general theses:

1) the main notions of information theory must and can be founded without the use of probability theory and so that the notion of "entropy" and "amount of information" turn out to be applicable to individual objects;

2) the notions of information theory thus introduced may lie at the foundations of a new concept of randomness corresponding to the natural idea that randomness is the absence of regularity.

Let us make an important remark.

The exposition in the first part of the article (information) was somewhat simplified. Only in the second part of the article we stressed the necessary relativization of the difference between "random" and "non-random" in the application to finite objects. The situation in the foundations of information theory is similar. Essentially it is applicable to the consideration of large arrays of information when the initial information contained in the method of constructing the theory itself is "disappearingly small". Our main formula (2) assumes a "universal method of programming" A. Such methods exist in the sense that there are programming methods A which possess the property $H_A(x|y) \leqslant H_{A'}(x|y) + C_{A'}$ and allow to program "anything" by a program of length no greater than that obtained by any other programming method by no more than a constant depending only on this second programming method and not on x and y. It is in the fact that we noticed this rather simple circumstance that I see the achievement of Solomonoff and myself here.

Hence all the assumptions of algorithmic information theory in their general formulation are true only up to terms of the form $O(1)$. Applying them to formulas (3) and (4), the only unexpected thing is the appearance of terms of logarithmic order.

It is only important to understand that when we turn to probability theory, we are using a considerable rougher relativization. The actual interpretation of probabilistic results is always statistical and the estimates of errors obtained on the application of probabilistic results to finite objects is considerably less delicate than that in the exposition of information theory developed by us here.

REFERENCES

[1] R. Solomonoff, *A formal theory of inductive inference.1*, Inform. and Contr. **7**,1 (1964), 1-22.

[2] A. N. Kolmogorov, *Three approaches to the definition of notion of "amount of information"*, Problemy Peredachy Informatsii **1**,1 (1965), 3-11.

[3] P. Martin-Löf, *Algorithms and random sequences*, Erlangen, Univ. Press, 1966.

[4] P. Martin-Löf, *The definition of random sequances*, Inform. and Contr. **9**,6 (1966), 602-619.

[5] G. Chaitin, *On the length of programs for computing finite binary sequences*, J. Assoc. Comput. Mach. **13**,4 (1966), 547-569.

[6] A. Church, *On the concept of a random sequence*, Bull.Amer.Math. Soc. **46** (1940), 130-135.

[7] A. N. Kolmogorov, *On tables of random numbers*, Sankhya. A **25**,4 (1963), 369-376.

[8] D. A. Loveland, *A new interpretation of the von Mises concept of random sequence*, Ztschr. math. Log. und Grundl. Math. **12** (1966), 279-294.

[9] D. Loveland, *The Kleene hierarchy classification of recursively random sequences*, Trans. Amer. Math. Soc. **125**, 3 (1966), 497-510.

13

THE COMBINATORIAL FOUNDATIONS
OF INFORMATION THEORY
AND THE PROBABILITY CALCULUS*[1]

§1. On the increasing role of finite mathematics.

I should like to begin with some considerations which go beyond the framework of the main topic of my report. Mathematics formalized according to Hilbert is nothing other than the theory of operations on schemes of special form consisting of a finite number of symbols disposed in a certain order and related by various connections. For example, according to N. Bourbaki's approach, the entire theory of sets studies only expressions consisting of the signs

$$\tau, \vee, \neg, =, \in, \supset$$

and "letters" which are related by "links" [diagram 1] as for example in the expression called

the "empty set". Remaining on the finite point of view, it would be logical to use some standard notation for an unbounded sequence of "letters", for example

$$\Pi_0, \Pi_1, \Pi_{10}, \Pi_{11}, \Pi_{100}, \ldots$$

It is curious that in the presence of "links"   , mathematical expressions formalized according to Bourbaki are not words stretched out into a line, as are for example, words in the theory of normal algorithms of A.A. Markov, but are essentially one-dimensional complexes with vertices supplied with special symbols.

But for the real intuitively understood contents of mathematics, such an approach to the subject as reconstruction of one-dimensional complexes according to special rules is really quite irrelevant. N. Bourbaki notes that in his approach the

---

*Uspekhi Mat. Nauk, 1983, vol. 38, vyp. 4, p. 27-36.

[1] The text being published was prepared in 1970 in connection with my report at the International Mathematical Congress in Nice.

expression whose significance is "the integer 1" contains several hundred thousand symbols, but nevertheless the notion "the integer 1" does not therefore become unaccessible to our intuitive understanding.

Pure mathematics was successful developed mainly as the science of the infinite. And the founder of the formalized and entirely finite approach to mathematics, Hilbert himself, undertook his titanic task only in order to ensure that mathematics would have the right to remain in the "Cantorian paradise" of set theory. Apparently this state of affairs is deeply rooted in our consciousness, which operates with great ease with an intuitively clear understanding of unbounded sequences, limiting processes, continuous and even smooth manifolds, etc.

Until the most recent time, the dominating trend in the mathematical sciences was the modelling of real phenomena by means of mathematical models constructed on the basis of infinite and continuous mathematics. For example, studying the process of molecular heat conduction, we imagine a continuous medium in which temperature satisfies the equation

$$\frac{\partial u}{\partial t} = K\left(\frac{\partial^2 u}{\partial x^2} + \frac{\partial^2 u}{\partial y^2} + \frac{\partial^2 u}{\partial z^2}\right). \tag{1}$$

Mathematicians are used to considering the corresponding difference scheme

$$\Delta_t u = K(\Delta_{xx} u + \Delta_{yy} u + \Delta_{zz} u) \tag{2}$$

only as appearing in the approximate solution of the "exact" equation (1), but the actual process of heat conductivity is no more similar to its continuous model expressed by equation (1) than to the discrete model expressed directly by equation (2).

It is quite probable that with the development of modern computing techniques, it will become understood that in many cases it is reasonable to study real phenomena without making use of the intermediate step of their stylization in the form of infinite and continuous mathematics, passing directly to discrete models. This is especially true for the study of complexly organized systems capable of processing information. In the most developed systems of this sort, the attraction to discrete methods of work is due at this time to sufficiently well explained causes. A paradox requiring an explanation in this context is the fact that the brain of a mathematician works essentially according to discrete principles and nevertheless a mathematician has a better intuitive grasp of, say, the properties of geodesic lines on smooth surfaces than that of combinatorial schemes that would approximately replace these properties.

Using his brain as given by God, the mathematician can avoid being interested in the combinatorial basis of his work. But the artificial intellect of the computer must be created by human beings who must necessarily enter into the field of combinatorial mathematics in order to do this. At present it is still too early to give a final assessment of what this will mean for the general architecture of mathematics in the future.

§2. Information theory.

The storage and processing of information in discrete form is the basic one. It is on this form that we base the actual measure of the "amount of information" expressed in bits, the number of binary digits. In view of what we said above, the discrete part of information theory plays the role of the leading organizing part in the development of combinatorial finite mathematics. From the general considerations developed above, it is not clear why information theory must be so essentially based on probability theory as it would seem from most textbooks. My goal is to show that this dependence on a probability theory created in advance is not in fact necessary. I shall limit myself, however, to only two simple examples.

The actual content of the formula for entropy[2]

$$H = -\sum p_i \log p_i \tag{1}$$

is the following. If we carry out a large number n of independent trials with probability distribution

$$(p_1, p_2, \ldots, p_s)$$

with s results of each trial, then for a fixed result of the entire series of n trials we shall need approximately nH binary digits. But this result remains valid under considerably weaker and purely combinatorial assumptions. To write down the result of our trials, it suffices to indicate that each of the results appears respectively

$$m_1, m_2, \ldots, m_s$$

times and then indicate the number of the one outcome (of the

$$C(m_1, m_2, \ldots, m_s) = \frac{n!}{m_1! m_2! \ldots m_s!}$$

possible results) that actually took place. To do, this we shall require no more than

$$s \log n + \log C(m_1, m_2, \ldots, m_s)$$

binary digits, while for large n this will be approximately

$$n\left(-\sum \frac{m_i}{n} \log \frac{m_i}{n}\right) \sim nH. \tag{2}$$

In view of the law of large numbers in the case of independent trials with the probability distribution indicated above, we have $m_i/n \sim p_i$. But our assumptions in deriving formula (2) are considerably weaker.

The second example, which deals with the entropy of Markov chains, is quite similar. Here also the assumption that one needs to write down information on the realization of Markov process is strongly redundant.

§3. Definition of "complexity".

If some object has a "simple" structure, then for its description it suffices to have a small amount of information; but if it is "complex", then its description must contain a lot of information. For certain reasons (see §7 below) it is convenient for us to call the quantity that we are now introducing "complexity".

---

[2]The logarithms are always binary.

The standard method for presenting information are binary sequences beginning with 1

$$1, 10, 11, 100, 101, 110, 111, 1000, 1001, \ldots$$

which are binary presentations of natural numbers. We shall denote by $l(n)$ the length of the sequence for n.

Suppose we are dealing with a certain domain of objects D in which we already have a certain standard numeration of objects x by their numbers $n(x)$. However, the indication of the number $n(x)$ will not always be the most economical method for distinguishing the object x. For example, the binary notation of the number

$$9^{9^{9^{9^9}}}$$

is extremely long, but we have defined it in a very simple way. We must carry out a comparative study of various methods of presenting objects from D. It suffices to limit ourselves to methods of presentation in which to each binary notation of the number p we assign a certain number

$$n = S(p).$$

This method of presenting objects from D becomes none other than a function S of a natural argument assuming natural values. A little further we shall consider the case when this function is computable. Such methods of presentation will be called "effective". But at this time let us remain in a most general situation. For each object from D it is natural to consider the number p of least length $l(p)$ which presents this object. This least length will be the "complexity" of the object x under the "method of presentation S":

$$K_S(x) = \min\{l(p), \ S(p) = n(x)\}.$$

In the language of computater science, we can call p the "program" and S the "method of programming". Then we can say that p is the minimal length of the program using which one may obtain the object x by the method of programming S.

If we have several different methods for presenting elements of D

$$S_1, S_2, \ldots, S_r,$$

it is easy to construct a new method S which will give us the object $x \in D$ with complexity $K_S(x)$ and which will be approximately larger by $\log r$ than the minimum of complexities

$$K_{S_1}(x), K_{S_2}(x), \ldots, K_{S_r}(x).$$

The construction of such a method is very simple. We must reserve a sufficient number of initial signs of the sequence p for fixing the method S_i that we shall follow, using as the program the remaining p signs.

Let us say that the method S "engulfs the method S' with precision up to l" if we always have

$$K_S(x) \leqslant K_{S'}(x) + l.$$

Above we showed a construction method for S which engulfs with precision up to l any of the methods $S_1, S_2, \ldots, S_r$, where we have approximately $l \sim \log r$.

The methods S_1 and S_2 are called "l-equivalent" if each of them l-engulfs the other. This entire construction would not be very productive if the hierarchy of methods concerning the engulfing of one of them by the other would be too complicated. Only recently it was noticed that for certain sufficiently natural assumptions this is not so. Further I follow my own paper [1], but more or less similar ideas can be found in the papers [3-5]; in the paper [3], however, they appear in a somewhat veiled form.

Theorem. *Among the computable functions $S(p)$ there exist optimal ones, i.e., functions such that for any other computable functions $S'(p)$ we have*

$$K_S(x) \leqslant K_{S'}(x) + l(S, S').$$

It is clear that all the optimal methods of presenting objects from D are equivalent

$$|K_{S_1}(x) - K_{S_2}(x)| \leqslant l(S_1, S_2).$$

Thus from the asymptotical point of view, the complexity $K(x)$ of an element x for bounded effective methods of presentation does not depend on random properties of the chosen optimal method. Of course the purely practical interest of this result depends on how important the discrepancies in the complexities are for different sufficiently viable (but at the same time sufficiently convenient) natural methods of programming.

§4. Regularity and randomness.

The view that "randomness" consists of the absence of "regularity" is quite traditional. But apparently only now has it become possible to give a basis to exact statements on the conditions of applicability to real phenomena of results from mathematical probability theory directly based on this simple consideration.

Any results of observations may be made into a protocol, in the form of a finite, although sometimes very long, notation. Hence when we speak of the absence in the results of observations of any regularity, we have in mind the absence of a *sufficiently simple* regularity. For example, the sequence of 1,000 digits

$$1274031274031274031\ldots,$$

which reccurs with period a of six digits will certainly be called a "regular", not a "random" sequence. A sequence of the first thousand decimal digits of the fractional part of the number π

$$1415\ldots,$$

as is known possesses many properties of "random sequences". Knowing its rule of formation, we also refuse to admit it as being "random". But if we are given a polynomial of 999-th degree whose values for $x = 1, 2, 3, \ldots, 1000$ give a sequence of integers $p(x)$ ranging between 0 and 9 obtained as the result of specific random trials such as those obtained by playing a roulette, the presence of such a polynomial will not forbid us from considering such a sequence "random". If in one way or another we have come to the conclusion that the sequence of results of certain trials cannot be fully described at an appropriate level of complexity, then we say that this sequence is only partially regular and partially "random". But this is not the type of "randomness" that is needed to make the application of probability theory possible. Applying probability theory, we do not limit ourselves to negating regularity, we use the hypotheses of randomness of observed phenomena to make definite positive conclusions from them.

We shall soon see that practical conclusions from probability theory may be founded as consequences of hypotheses about the limiting complexity of the phenomena under the given limitations.

§5. Stability of frequencies.

Following von Mises, it is often claimed that the basis of the applications of probability theory is the hypothesis of stability of frequencies. In a form closer to practice, this notion is also accepted in my well-known book on the foundations of probability theory (1933). Suppose the outcomes of a sequence of a large number N of trials is written in the form of a sequence of zeros and ones

$$11010010111001011...$$

It is said that the appearance of one is random with probability p, if the ratio of ones to the total number of digits is

$$M/N \sim p \tag{1}$$

and this frequency cannot be essentially changed by choosing any subsequence that is not too short, the selection of the subsequence being carried out by a simple rule such that that the choice of any specific element from the main sequence does not involve the value of this element[3].

But it turns out that this requirement may be replaced by another one, which can be stated in a much simpler way. The complexity of the sequence of zeros and ones satisfying condition (1) cannot be considerably more than

$$nH(p) = n(-p \log p - (1 - p) \log(1 - p)).$$

It can be proved that the von Mises stability of frequencies is automatically satisfied if the complexity of our sequence is sufficiently close to the upper bound indicated above.

I do not have the possibility here to give a quantitative formulation of this result[4] and consider the more complicated problems of probability theory from the same point of view. But the principle is general: for example, assuming that a sequence of zeros and ones constitutes a Markov chain with the following matrix of transition probabilities

$$\begin{pmatrix} p_{00} & p_{01} \\ p_{10} & p_{11} \end{pmatrix}$$

---

[3]For a more accurate finite formulation of this principle of von Mises, see the paper [17].

[4]See the paper [17], although it does not contain the definition of complexity which appeared in [1].

we essentially define an approximate value of the frequency of ones after ones, ones after zeros, zeros after ones and zeros after zeros. We can compute the maximal complexity of such a sequence of length n if the complexity of a specific sequence with given frequencies of transition is close to this maximal one, then it automatically satisfies all the predictions of the probability theory of Markov chains.

§6. Infinite random sequences.

The programme described above has not yet been carried out, although I have no doubts about the possibility of its implementation. It is precisely its implementation that must be carried out (in a more precise manner than constructions of the von Mises type) and relate the mathematical theory of probability to its applications. Here I have in mind that it is not necessary to replace the existing construction of the mathematical theory of probability based on the general theory of measure. I do not feel that one can consider the research work that I shall now review as the necessary foundations of probability theory, but in itself it is certainly very interesting.

For a mathematician, an attractive problem is to define what infinite sequences of zeros and ones should be called "random". I shall limit myself to the simplest case of sequences of zeros and ones of frequency equal to $1/2$. We should like to require that the sequence

$$x = (x_1, x_2, \ldots, x_n, \ldots),$$

be such that all its finite segments

$$x^n = (x_1, x_2, \ldots, x_n)$$

have complexity

$$K(x^n) \geqslant n - C,$$

where C is a certain constant (differing for different x). However, Martin-Löf proved the following theorem.

Martin-Löf's first theorem. *If $f(n)$ is a computable function such that*

$$\sum 2^{-f(n)} = \infty, \tag{1}$$

then for any binary sequence

$$x = (x_1, x_2, \ldots, x_n, \ldots)$$

there exists an infinite number of values n for which

$$K(x^n) < n - f(n).$$

The assumptions of the theorem are satisfied, for example, by the function $f(n) = l(n)$, but if the series

$$\sum 2^{-f(n)} \tag{2}$$

"constructively converges"[5] then almost all (in the sense of diadic measure) sequences x possess the property

$$K(x^n) \geqslant n - f(n) \tag{3}$$

beginning from a certain n. It would not be logical to take property (3) for the definition of a random sequence. The definition given by Martin-Löf is deeper. I do not have the possibility to present it here in detail. Random binary sequences according to Margin-Löf possess all "effectively verifiable"[6] properties which, from the point of view of ordinary modern probability theory, are satisfied with "probability one" in the case of independent trials such that $x_n = 1$ with probability $1/2$. For such random sequences Martin-Löf proves the following theorem.

Martin-Löf's second theorem. *For random sequences of zeros and ones we have inequality (3) beginning with some n if only the function f is such that the series (2) constructively converges.*

I have mentioned these delicate but fairly special results due to Martin -Löf to show that here a new field of very interesting mathematical studies has appeared (in this connection see other papers by Martin-Löf and Shnorr, for example [9]).

§7. Conditional complexity and amount of information.

The complexity in the presentation of some objects may be decreased if some other object is already given. This fact reflects the following definition of conditional complexity of the object x for a given object y:

$$K_S(x|y) = \min_{S(n(y),p)=n(x)} l(p).$$

Here the method of conditional definition S is a function of two arguments: the number of the object y and the number p of the program computing the number $n(x)$ when y is given. Concerning conditional complexity we can repeat everything that was mentiond in §3.

If the conditional probability $K(x|y)$ is much less than the unconditional probability $K(x)$, then it is natural to understand this state of affairs as implying that the object y contains some "information about the object x". The difference

$$I_S(x|y) = K_S(x) - K_S(x|y)$$

is naturally taken to be a quantitative measure of information about x contained in y.

Let us take the number zero as the second argument of the function $S(n,p)$ and put

$$S(n,0) = n$$

(the zeroth program produces n from n). Then

$$K_S(x|x) = 0, \quad I_S(x|x) = K_S(x).$$

---

[5] For the details see [8].
[6] Loc cit.

Thus the complexity $K_S(x)$ may well be called the information contained in an object about itself.

Our definition of the amount of information from the point of view of applications has the advantage of dealing with individual objects rather than objects considered as parts of a set of objects with a given probability distribution on this set. The probabilistic definition may convincingly be applied to information contained, for example, in a series of season's greetings telegrams, but would not be very clear how to apply, say, to an estimate of the amount of information contained in a novel or in a translation of a novel into another language as compared to the original. I feel that this new definition is capable of introducing at least some clarity in principle into similar applications of the theory.

The following question arises: does the new definition allow us to prove the main statements of information theory known at this stage? It is clear in advance that they will hold only up to additive constants corresponding to the indefiniteness which already arises in §3. We cannot expect, for example, to have the exact relation

$$I(x|y) = I(y|x), \tag{1}$$

but first it seemed that the difference between the left and the right-hand sides here must be bounded. Actually, Kolmogorov and Levin established only a weaker inequality of the form

$$|I(x|y) - I(y|x)| = O(\log K(x, y)) \tag{2}$$

(see [16]). It was shown there that this difference can actually be precisely of this order.

But in the applications that allow the probabilistic approach, relation (2) successfully replaces (1). Indeed, the exact relation (1) of the probabilistic theory of information allows to make real conclusions only in the case of a large number of pairs (x_i, y_i), i.e. essentially about information

$$(x_1, x_2, \ldots, x_r)$$

with respect to

$$(y_1, y_2, \ldots, y_r)$$

and vice versa. This kind of conclusion can also be obtained from (2), where in this case the expression in the right-hand side turns out to be negligibly small.

§8. Barzdin's theory.

A new series of interesting notions, going beyond the limits of probability theory and of the applied theory of information, have also appeared. To give an example, let me give the exposition of a theorem due to Barzdin. It has to do with infinite binary sequences

$$x = (x_1, x_2, \ldots, x_n, \ldots),$$

in which the set of numbers n such that $x_n = 1$ is recursively enumerable. If the complementary set were also recursively enumerable, then the function $f(n) = x_n$ would be computable and the conditional complexity $K(x^n|n)$ would be bounded. But in the general case, when the set of ones is recursively enumerable, $K(x^n|n)$ can increase unboundedly.

Barzdin's theorem [15]. *For any binary sequence with a recursively enumerable set of ones M, we have*

$$K(x^n|n) \leqslant \log n + C_M;$$

there exist sequences for which for any n we have

$$K(x^n|n) \geqslant \log n.$$

I feel that this result is of great importance from the point of view of the foundations of mathematics. To be definite, let us consider the following problem. Number all the Diophantine equations by natural numbers. It was recently proved by Matyasevich that there exists no general algorithm for deciding whether the equation D_n has integer solutions or not. But one might put the question of the existence of an algorithm which would answer the question of the existence or non-existence of solutions for the first n Diophantine equation by using some additional information $\mathcal{I}_n$ for various orders of growth of the amount of this information when n increases. Barzdin's theorem shows that this growth can be very slow:

$$\log n + C.$$

§9. Conclusion.

This report was of necessity very incomplete. A detailed bibliography of works in this direction may be found in [16]. Let me repeat certain conclusions.

1. Information theory must come before probability theory and not be based on the latter. The foundation of information theory, by the very essence of this discipline, must be of finite combinatorial character.

2. The applications of probability theory may obtain a unified foundation. We always mean conclusions of some hypotheses on the impossibility of decreasing the complexity of the objects studied by using certain fixed means. Naturally, this approach to the question does not hamper the development of probability theory as a part of mathematics included in the general theory of measure.

3. The notions of information theory in their application to infinite sequences make possible some very interesting research that, although it is not necessary from the point of view of the foundations of probability theory, may have a certain significance in the study of the algorithmic aspect of mathematics as a whole.

M.V. Lomonosov Moscow State University
August 20, 1982

References

[1] A. N. Kolmogorov, *Three approaches to the definition of the notion of "amount of information"*, Problemy Peredachy Informatsii(in Russian) **1,1**, (1965), 3–11.

[2] A. N. Kolmogorov, *Logical basis for information theory and probability theory*, IEEE Trans. Inform. Theory **IT-14** (1968), 662–664.

[3] R. L. Solomonoff, *A formal theory of inductive inference*, Inform. and Contr. **7,1** (1964), 1–22.

[4] G. Chaitin, *On the length of programs for computing finite binary sequences*, J. Assoc. Comput. Mach. **13,4** (1966), 547–569.

[5] D. Loveland, *A new interpretation of the von Mises concept of random sequence*, Ztschr. math. Log. und Grundl. Math. **12** (1966), 279–294.

[6] A. Church, *On the concept of a random sequence*, Bull. Amer. Math. Soc. **46** (1940), 130–135.

[7] P. Martin-Löf, *The definition of random sequences*, Inform. and Contr. **9,6** (1966), 602–619.

[8] P. Martin-Löf, *Algorithms and random sequences*, Erlangen, Univ. Press, 1966.
[9] C. P. Schnorr, *Ein Bemerkung zum Bergiff der zufalligen Folge*, Ztschr. Wahrscheinlichkeits-theor. verw. Geb. **14** (1969,), 27–35.
[10] B. A. Trakhtembrot, *Complexity of algorithms and computations*, Novosibirsk, Nauka, 1967.
[11] A. N. Kolmogorov, V. A. Uspensky, *To the definition of algorithm*, Uspekhi Mat. Nauk **13,4** (1958), 3–28.
[12] Ya. M. Barzdin, *Universality problems in the theory of growing automata*, Dokl. Akad. Nauk SSSR **157,3** (1964), 542–545.
[13] Yu. P. Ofman, *A universal automaton*, Trudy Mosk. Mat. Obshchestvo 14 (1965), 186–199.
[14] Yu. P. Ofman, *Modelling self-constructing systems on the universal automaton*, Problemy Peredachy Informatsii **2,1** (1966), 68–73.
[15] Ya. M. Barzdin, *Complexity of programs, determining if a natural number no greater than n belongs to a given recursively ennumerable set*, Dokl. Akad. Nauk SSSR **182, 6** (1968), 1249–1252.
[16] A. R. Zvonkin, L. A. Levin, *Complexity of objects and the foundation of the notions of information and randomness by means of the theory of algorithms*, Uspekhi Mat. Nauk **25,6** (1970), 85–127.
[17] A. N. Kolmogorov, *On tables of random numbers*, Sankhya. A **25,4** (1963), 369–376.

COMMENTS AND ADDENDA

ON WORKS IN INFORMATION THEORY AND SOME OF ITS APPLICATIONS

A. N. KOLMOGOROV

The appearance of information theory as an independent science is related to the ideas of cybernetics. In its more general form, cybernetics studies control processes, and a rational system of control necessitates the definition of information, its obtention, storage, transmission along channels of communication and its processing. It is therefore natural that the cycle of my works on information theory was largely influenced by the publications of Norbert Wiener and Claude Shannon (1948) at the end of the 50s and the 60s[1].

A considerable part of these works can be conveniently reviewed in connection with the presently existing three approaches to the introductin of the main notions of information theory:

I — purely combinatorial approach,

II — probabilistic approach,

III — algorithmic approach.

The mathematical apparatus of information theory also has applications which lie beyond the limits of these three interpretations. Such are the works where this apparatus is applied to the theory of dynamical systems. Among my papers the one appearing here under No.5 should be ranked among those (see section IV below).

I. THE PURELY COMBINATORIAL APPROACH.
WORKS N 7-10, 12, 13

In the combinatorial approach, the amount of information transmitted by indicating a definite element in a set of N elements is taken to be equal to the binary logarithm of N (R. Hartley, 1928). For example, there are

$$C(m_1, \ldots, m_s) = \frac{n!}{m_1! \ldots m_s!}$$

---

[1]Norbert Wiener, with great generosity, has accepted my precedence in the spectral theory of stationary processes and their extrapolation. As to the general ideas of cybernetics, I myself, in numerous publications, have only been involved in the propaganda of these ideas. The aim of the present edition is only to publish those papers which constitute my original contribution to the mathematical theory of information.

different words in the alphabet consisting of s elements containing m_i entries of the i-th letter of our alphabet $(m_1 + \cdots + m_s = n)$. Therefore the amount of information that interests us is equal to

$$H = \log_2 C(m_1, \ldots, m_s).$$

When $n, m_1, \ldots, m_s$ tend to infinity, we have the asymptotic formula

$$H \sim n \left(\sum_{i=1}^{s} \frac{m_i}{n} \log_2 \frac{m_i}{n} \right). \tag{1}$$

The reader probably has already noticed the similarity of this formula with the formula from probabilistic information theory

$$H = n \left(\sum_{i=1}^{s} p_i \log_2 p_i \right). \tag{2}$$

If the word was obtained from the well-known scheme with independent trials, then the asymptotic formula (1) is an obvious corollary to formula (2) and to the law of large numbers, but the range of applications of formula (1) is much wider (see, for example, papers on the transmission of information via non-stationary channels). Generally, I feel it is an important problem to free oneself, whenever possible, of unnecessary probabilistic assumptions. I have stressed the independent value of the purely combinatorial approach to problems of information theory many times in my lectures.

The purely combinatorial approach to the notion of entropy is the basis of my work and the work of my collaborators on ε-entropy and ε-capacity of compact classes of functions. Here the ε-entropy $H_\varepsilon(K)$ is the amount of information necessary to single out some individual function from the given class of functions, while the ε-capacity $C_\varepsilon(K)$ is the amount of information which may be encoded by elements of K under the condition that elements of K located from each other at a distance no less than ε can certainly be distinguished (concerning my own work, papers Nos 7 and 8 in this volume, see the comments by V. A. Tikhomirov). At the end of my paper No.8 certain of my results on the complexity of approximate computations of functions from various functional classes are presented without proof. In the work of E. A. Asarin, complete proofs of the theorem appearing in my Stockholm report (No.8 of the present edition) and some of their strengthenings and generalizations are given. It is interesting to note the unexpected result that the estimate of the complexity of computations of analytical functions can be considerably decreased by using fast multiplication algorithms.

In the paper No.9, on the level of elementary computations in the framework of the purely combinatorial approach to the notion of information, the possibility of giving a formal definition of the notion of randomness is analyzed. Naturally, it was not possible to give any kind of absolute definition of randomness, but only to define randomness with respect to a bounded class of rules for choosing subsequences and with respect to the measure of ε-stability of frequencies (Remark 2 in §2 of this paper goes beyond the limits of the strictly finite approach and is discussed in the commentary by A. Kh. Shen to my papers Nos. 10, 12, 13).

II. THE PROBABILISTIC APPROACH.
WORKS NO. 2-4

A very complete commentary to these papers is given by R. L. Dobrushin.

III. THE ALGORITHMIC APPROACH.
WORKS NO. 10, 12, 13

The algorithmic approach is founded on the application of the theory of algorithms to the definition of the notion of entropy or complexity of a finite object and of the notion of information in one finite object about another. The intuitive difference between "simple" and "complicated" objects has apparently been perceived a long time ago. On the way to its formalization, an obvious difficulty arises: something that can be described simply in one language may not have a simple description in another and it is not clear what method of description should be chosen. The main discovery, made by myself and simultaneously by R. Solomonoff, is that by means of the theory of algorithms it is possible to limit this arbitrary choice, defining complexity in an almost invariant way (the replacement of one method of description by another only leads to the addition of a bounded summand)[2]. Among my papers the most complete exposition of these ideas is contained in the paper No.13 (see also the commentary by A. Kh. Shen to the papers Nos. 10, 12, 13).

IV

The paper No.5 contains certain appliations of the mathematical apparatus of information theory to the theory of dynamical systems. The notion of amount of information introduced in that paper does not get any kind of real interpretation. It is only used as a mathematical mechanism in the theory of dynamical systems (see the commentary to this paper by Ya. G. Sinai).

---

[2] As far as I know, the first published paper containing a reconstruction of information theory on an algorithmic basis is the article by R. Solomonoff published in 1964. I came to a similar approach in 1963-1964 before I learned about Solomonoff's work and published my first note (No.10 in this volume) about this at the beginning of 1965.

INFORMATION THEORY
(PAPERS NO. 2-4)

(R. L. DOBRUSHIN)

There is one trait that A. N. Kolmogorov's cycle of papers on information theory has in common with any really important research work. After a certain amount of time goes by, their main ideas are understood by experts as truisms, while the situation was hardly the same in the period when these papers were being created. One should remember the atmosphere of the 50s when A. N. Kolmogorov became interested in questions of information theory. This was a time when, under the pressure of the approaching advent of computers, uncritical distrust of the ideas of cybernetics in our country were suddenly replaced by just as uncritical glorification of the subject and, as the result, many superficial and speculative ideas came to the forefront. This noisy eulogy in its turn generated a natural negative reaction.

The theory of information was then regarded as one of the branches of cybernetics. Actually it had been created by the extremely profound work (see [1]) of the American scientist Shannon, inspired by the concrete requirements of communications technology, before Wiener coined the term "cybernetics". The theory of information immediately solicited great interest both among communications engineers and among representatives of many other sciences. However, this boom was not really so justified. What seemed interesting at first glance was that an actual mathematical description of the amount of information is possible and yet there was not enough real understanding that such a general and varied object as information does possess a unified method of numerical measurement, while the ideas of Shannon were founded only in their application to the quite important, but nevertheless limited, situation when one considers optimal methods of coding and uncoding information in order to transmit it along communication channels or storing it. As Shannon himself wrote at the time (see [2]): "As a consequence it has perhaps ballooned to an importance beyond its actual accomplishments. ... In short, information theory is currently partaking of a somewhat heady draught of general popularity." However, this fashionable interest did not include mathematicians. The first papers by Shannon were written on the level of physical rigor, which is in contradiction with the learning possibilities of average mathematicians. As A. N. Kolmogorov wrote in his preface to the Russian translations of Shannon's articles published in 1963, "The value of Shannon's work for pure mathematics was not sufficiently appreciated at the time. I remember that even at the International Congress of Mathematicians in Amsterdam (1954), my American colleagues, experts in probability theory, felt that my interest to Shannon's work was somewhat exaggerated because this was mainly technology and not mathematics. Today such opinions hardly need any refutation. However, the rigorous mathematical "foundation" of Shannon's ideas were left by their author to his successors in all the remaining difficult cases. However, his mathematical intuition is remarkably precise." The situation was further complicated by the fact that the only edition of the Russian translation of Shannon's article [3] available in the fifties contained arbitrary cuts, and not only in connection with the generally accepted idea on the impossibility of applying mathematics to the humanities in principle, but simply because certain of the mathematically most interesting parts of the article seemed too abstract to the editors of the translation (see the remark at the beginning of the article [4]).

What was then necessary was work at a mathematical level of rigor which would establish the foundations of a mathematical theory of information. This problem was successfully solved in the work of Khinchin [5,6], Gelfand, Kolmogorov and Yaglom [7]. Another necessity was the popularization of the ideas of information theory, together with an authoritive explanation of the limits of its applicability. This problem was solved by the well-known report by A. N. Kolmogorov (8) at the Session of the Academy of Sciences of the USSR devoted to the automation of production. The first pioneering work by Khinchin was devoted to the proof of the main theorems of information theory for the discrete case. In the subsequent papers by Gelfand, Kolmogorov and Yaglom [4,7], the more general case of continuous systems, which was only touched upon slightly in the original work of Shannon, was considered. General properties of the amount of information, explicit formulas for informational characteristics in the Gaussian case, formulations of coding theorems for transmitting messages with given precision, obtained in those papers, remain the invariant part of all modern papers in information theory. And, what may be even more important, these papers by Moscow scientists have begun a tradition of presenting results of information theory on the mathematical level of rigor, which, since then, is invariably present in the work of those experts in information theory who consider themselves mathematicians as well as of those who think of themselves as engineers. This level of mathematical rigor also appears in the later, and also very meaningful, work of Shannon (see [9]).

It often happens that after a kernel of ideas has been mathematically described in terms that come from another science, it turns out that unexpected relationships of these ideas with other branches of mathematics appear. This happened with information theory. A. N. Kolmogorov discovered some very important and fruitful possibilities for applying the notion of entropy to the problem of the isomorphism of dynamical systems (see [10]) and to the entropy characterization of functional spaces (see [11]). The thesis that A. N. Kolmogorov suggested was the following: besides the already accepted probabilistic approach to the definition of the amount of information, other approaches in many situations are possible and more natural: the combinatorial and the algebraic one led to the creation of a new branch of science – the algorithmic theory of information. These studies are characterized in detail in the other commentaries.

Now let us pass to a more detailed discussion of further research related to the ideas of A. N. Kolmogorov about the probabilistic theory of information. The list of subsequent papers in which the notions of information theory first appearing in the work of Kolmogorov and his coauthors are used is too long to be quoted here. Therefore we shall limit ourselves to just a few works by Moscow scientists directly inspired by A. N. Kolmogorov. A detailed exposition of the results of the report [4] is contained in the article by Gelfand and Yaglom [12]. In the monograph by Pinsker [13], the asymptotic studies of information characteristics of random processes begun by A. N. Kolmogorov was continued. Dobrushin [14] developed A. N. Kolmogorov's idea about the general formulation of the fundamental theorem of information theory for arbitrary sequences of transmitting devices and messages with increasing information. These studies were extended by Pinsker [15]. Let us also note the work of the Chinese scientist Khu Go Din, who was visiting at the time at Moscow University [16,17] and the work of the talented mathematician Yerokhin [18] carried out before his untimely death. In the work of Fitingov [19,20], done under the direct influence of A. N. Kolmogorov, a new approach to the problem of optimal coding of messages in a situation when the statistics of the messages is unknown was developed. In A. N. Kolmogorov's own work [4], he notes the following open question: is the natural explicit formula for the mean amount of information valid for an arbitrary pair of stationarily related Gaussian processes? In the paper [21] it was discovered that this may not be so if both processes are singular. Studies of this difficult problem are still under way in recent years, thus Solev (see [22]) found some very general conditions for the validity of this formula for processes with continuous time.

REFERENCES

[1] Shannon C. E., *A mathematical theory of communication*, Pt I, II - Bell Syst. Techn. J. **27,4** (1948), 623–656.

[2] Shannon C. E., *The bandwagon*, IRE Trans. Inform. Theory **IT-2,1** (1956), 3.

[3] Shannon C. E., *The mathematical theory of communication*.

[4] I. M. Gelfand, A. N. Kolmogorov, A. M. Yaglom, *The amount of information and entropy for continuous distributions*, Trudy III All-Union Math. Congress Moscow Academy of Sciences of the USSR Publications, vol. 3, 1958, pp. 300–320.

[5] A. Ya. Khinchin, *The notion of entropy in probability theory*, Uspekhi Mat. Nauk **8,3** (1953), 3–20.

[6] A. Ya. Khinchin, *On the basic theorems of information theory*, Uspekhi Mat. Nauk **11,1** (1956), 17–75.

[7] I. M. Gelfand, A. N. Kolmogorov, A. N. Yaglom, *To the general definition of the amount of information*, Dokl. Akad. Nauk SSSR **111,4** (1956), 745–748.

[8] A. N. Kolmogorov, *The theory of information transmission*, Plenary Session, Moscow, Editions of the Academy of Sciences of the USSR (1957), 66–69, Session of the Academy of Sciences of the USSR on scientific problems of the automatization of production.

[9] Claude Shannon, *Work in the theory of information and cybernetics*, Moscow, Foreign Literature Editions, 1963.

[10] A. N. Kolmogorov, *A new metric invariant of transitive dynamical systems and automorphisms of Lebesgue spaces*, Dokl. Akad. Nauk SSSR **119,5** (1958), 861–864.

[11] A. N. Kolmogorov, *On the entropy in unit of time as a metric invariant of automorphism*, Doklady AN SSSR **124,4** (1959), 754–755.

[12] S. I. Gelfand, A. M. Yaglom, *On the computation of the amount of information of a random function contained under such functinon*, Uspekhi Mat. Nauk **12,1** (1957), 3–52.

[13] M. S. Pinsker, *Information and informational stability of random variables and processes*, Moscow, Academy of Sciences of the USSR Editions, 1960.

[14] R. L. Dobrushin, *The general formulation of the main theorem of Shannon in information theory*, Uspekhi Mat. Nauk **14,6** (1959), 3–104.

[15] M. S. Pinsker, *Sources of messages.*, Problemy Peredachy Informatsii 4 (1963), 5–20.

[16] Khu Go Din, *Three inverse theorems to Shannon's theorem in information theory*, Acta Math. Sinica (in Chinese) 11,3 (1961), 260–294.

[17] Khu Go Din, *On informational stability of a sequence of channels*, Teoriya Veroyatn. i Prilozhen. 7,3 (1962), 271–282.

[18] V. D. Yerokhin, *ε-entropy of a discrete random object*, Teoriya Veroyatn. i Prilozhen. 3,1 (1958), 103–107.

[19] B. M. Fitingov, *Optimal coding for an unknown or changing statistics of messages*, Problemy Peredachy Informatsii 2,2 (1966), 3–11.

[20] B. M. Fitingov, *The compression of discrete information*, Problemy Peredachy Informatsii 3, 3 (1967), 26–36.

[21] Grood I. J., Doog K. C., *A paradox concerning rate of information*, Inform. and Contr. 1,2 (1958), 113-126; Inform. and Contr. 2 (1959), 195–197.

[22] V. N. Solev, *On the mean amount of information in unit time contained in one Gaussian stationary process with respect to another*, Zap. Nauch. Seminarov LOMI 29 (1972), 18-26.

ALGORITHMIC INFORMATION THEORY
(TO PAPERS NO. 10, 12, 13)

(A. KH. SHEN)

In this commentary we list the main results obtained by A. N. Kolmogorov and his pupils and followers in the domain of algorithmic information theory.

Recall that $K(x)$ denotes the simple Kolmogorov entropy (=complexity) of the object x. We prefer to use the term "entropy" as in [1] rather than "complexity" as in [2] in order to avoid confusion with "the complexity of computation" – timewise, volumewise, etc. By $K(y|x)$ we denote the conditional Kolmogorov entropy of the object y for the known object x. (When we speak of "objects" we mean, unless otherwise specified, natural numbers or some other constructive objects which are computably coded by natural numbers, for example binary words). As shown in [2], the notion of conditional entropy may be regarded as the basis for the definition of the notion of "amount of information".

Let us see how this "algorithmic approach " is related to the notions of entropy and information within the framework of the two other approaches developed in [2]: the combinatorial and the probabilistic one. (Let us note at once that the combinatorial approach was later developed in the work of Goppa [3,4]).

The combinatorial approach. It is easy to show that the amount of objects whose (simple Kolmogorov) entropy is no greater than n is approximately equal to 2^n (up to a bounded set separated away from zero). On the other hand, if a simply constituted set consists of M objects, then the entropy of its elements is not much greater than $\log_2 M$. One of the versions of a more exact formulation of this statement is the following: suppose $A_0, A_1, \ldots$ is a compuable sequence of recursively enumerable sets (computability is understood in the sense that given i one can indicate the program of an algorithm whose domain of definition is A_i), and the number of elements in A_i is no greater than 2^i. Then there exists a C such that for any i and for any $x \in A_i$ we have the inequality $K(x) \leqslant i + C$. (The statement easily follows from the results of [5]). Roughly speaking, in order to establish the inequality $K(x) \leqslant i$ it is sufficient to include x in a (simply constituted) set of no more than 2^i elements.

The probabilistic approach. The meaning of the results on the relationship of the algorithmic approach with the probabilistic one may be briefly stated as follows: under the algorithmic approach, one takes into consideration not only all the probabilistic rules, but others as well (if they exist); if the other rules do not exist, then the algorithmic approach leads to the same results as the probabilistic one.

Suppose x is a sequence of zeros and ones of length n, where the frequency of ones is equal to p, while the frequency of the zeros is equal to $q = 1 - p$. Suppose H is the corresponding Shannon entropy, i.e.

$$H = -p \log_2 p - q \log_2 q.$$

Then

$$K(x) \leqslant nH + C \log_2 n.$$

(Here K is the simple Kolmogorov entropy, C is a constant depending only on the choice of p, q and on the method of description and independent of n (see [5], Theorem 5.1). Thus the "specific Kolmogorov entropy" $K(x)/n$ cannot be much greater than H, (but can be much lesser than H if the sequence x contains certain regularities which differ from probabilistic ones). The following result shows that if a sequence only has probabilistic regularities, then its specific Kolmogorov entropy is close to the Shannon one. Suppose $p, q \in [0,1]$, $p + q = 1$. Consider, on the set of infinite sequences of zeros and ones, the Bernoulli measure with probability of appearance of zero and one equal to p and q respectively. Then for almost all (with respect to this measure) sequences $x = x_0 x_1 \ldots$, we have the relation

$$\lim(K(x_0 x_1 \ldots x_{n-1})/n) = H,$$

where $H = -p \log_2 p - q \log_2 q$. (This statement easily follows from [5, formula (5.18)].)

Together with simple and conditional Kolmogorov entropy, several authors have introduced other algorithmic versions of the entropy of a finite object, e.g. the *entropy of deciding* (see [5]), the *monotone* and *prefix entropy* (see [6]).

Various forms of entropy turned out to be related to so-called apriori probability (see [5,6]) and to the notion of complexity of a computable function as the logarithm of its number in its "optimal enumeration" as introduced by Schnorr (see [7-9]).

In 1966 Martin-Löf [10] proposed a definition of a random infinite sequence of zeros and ones with respect to a given computable measure P on the space Ω of all infinite sequences of zeros and ones. Let us state this definition.

Suppose P is a measure on the space Ω of infinite sequences of zeros and ones possessing the property of *computability*, i.e. such that the measure of the set

$$\Gamma_x = \{\omega \in \Omega \,|\, x \text{ is the beginning of } \omega\}$$

for any word x in the alphabet $\{0,1\}$ is a computable real number, which can be effectively found from x. Let us define the notion of constructive zero set with respect to the measure P – a notion which is the effectivization of the usual notion of zero measure. Let us say that the set $A \subset \Omega$ is a *constructive zero set* with respect to P if for any rational $\varepsilon > 0$ one can effectively indicate a recursively enumerable set of finite words $\{x_0, x_1, \ldots\}$ for which we have

$$(a) \ A \subset \bigcup \Gamma_{x_i}, \quad (b) \ \sum P(\Gamma_{x_i}) \leqslant \varepsilon.$$

(We have in mind the existence of an algorithm that, given the number ε, indicates the corresponding recursively enumerble set in the standard numeration.) It can be proved (Martin-Löf theorem) that for computable measures there exists a maximal (with respect to inclusion) constructive zero (with respect to the given measure) set. In other words, the union of all constructive zero (with respect to the computable measure P) sets again turns out to be a constructive zero set with respect to P. The proof of this fact can be found in [5] and [6].

According to Martin-Löf, a sequence is called *random with respect to the computable measure* P if it is not contained in a maximal constructive zero set or (which is the same) is not contained in any constructive zero set. Thus for any property A of a sequence of zeros and ones the conditions:

(a) all sequences that are random with respect to the measure P possess the property A;

(b) the set of sequences that do not possess the property A are constructively of zero measure with respect to P;

are equivalent.

The ordinary rules of probability theory (such as the law of large numbers or the iterated logarithm rule) lead to sets of zero measure which are also of constructive zero measure. This gives us ground for saying that sequences that are random according to Martin-Löf possess all the constructive properties which in contemporary probability theory can be proved "with probability one" [1].

Let us note that for the first time the idea of classifying infinite sequences of zeros and ones into "random" and "non-random" ones was proposed by von Mises ([11, 12]). Mises's approach was later specified and modified by various authors ([13,14]); one of these modifications was proposed in [15, 2, remark 2] (later it was proposed again in [16,17]). For the history of the "frequency approach to randomness" which arose from Mises' ideas, see a more detailed exposition in [9, 18]; actually this approach is more of historical significance.

As indicated in [1, page 5], it is natural to try to obtain a randomness criterion for sequences in terms of entropy of its initial segments. (Kolmogorov writes about the difficulties which appear in this approach in [1] in connection with the definition of the notion of Bernoulli sequence). These difficulties were overcome in 1973 independently by Levin and Schnorr (see [19, 8]), who indicated such a criterion in terms of monotone entropy. This criterion is the following: a sequence ω is random with respect to a computable measure P if and only if there exists a constant C such that the monotone entropy $KM(x)$ of any of its initial segments x is no less than $-\log_2 P(\Gamma_x) - C$. It may be proved that $KM(x)$ cannot be much larger than $-\log_2 P(\Gamma_x)$: there exists a C such that for any word x we have the inequality

$$KM(x) \leqslant -\log_2 P(\Gamma_x) + C;$$

thus a sequence ω is random if it satisfies

$$KM(x) = -\log_2 P(\Gamma_x) + O(1)$$

for all x which are beginnings of ω. For the proof of these facts, see [6]. In the case of a uniform Bernoulli measure, when $P(\Gamma_x) = 2^{-l(x)}$, where $l(x)$ is the length of

x, a sequence turns out to be random if the monotone entropy of its initial segment of length n is equal to $n + O(1)$.

In conclusion, let us indicate where one can find the proofs of certain results stated in this commentary to the articles. The proofs of formulas (3) and (4) from [1] are published in [5]. On p. 6 in [1] it is stated that there exist Bernoulli sequences in the sense of Church which have initial segments of complexity that grows logarithmically and also the stronger assertion that there exist Bernoulli sequences in the sense of Loveland with initial segments of complexity that also grows logarithmically. The proof of the first statement can be found in [4], that of the second one is not published. Let us note that in [9] and [18] sequences that are Bernoulli in the sense of Church are called random in the sense of Mises-Church, while Bernoulli sequences in the sense of Loveland are called random in the sense of Mises-Kolmogorov-Loveland.

REFERENCES

[1] A. N. Kolmogorov, *To the logical foundations of information theory and probability theory*, Problemy Peredachy Informatsii **5,3** (1969), 3–7.

[2] A. N. Kolmogorov, *Three approaches to the definition of the notion of "amount of information"*, Problemy Peredachy Informatsii **1,1** (1965), 3–11.

[3] V. D. Goppa, *Non-probabilistic mutual information without memory*, Problemy Upravl. i Teorii Inform. **4,2** (1975), 97–102.

[4] V. D. Goppa, *Information of words (initial approximation - information without memory)*, Problemy Peredachy Informatsii **14,3** (1978), 3–17.

[5] A. K. Zvonkin, A.L. Levin, *Complexity of finite objects and the foundations of the notion of information and randomness by means of algorithm theory*, Uspekhi Mat. Nauk **25,6** (1970), 85–127.

[6] V. V. Vyugin, *Algorithmic entropy (complexity) of finite objects and its application to the definition of randomness and amount of information*, Semiotica i Informatica, vol. 16, Moscow, VINITI, 1981, pp. 14–43.

[7] C. P. Schnorr, *Optimal Godel numberings*, In: Information processing 71: Proc. IFIP. Congr. 71. Lubljana **1** (1971), 56–58.

[8] Schnorr C. P., *Optimal enumerations and optimal Godel numberings*, Math. Syst. Theory **8,2** (1975), 182–191.

[9] V. A. Uspensky, A.L. Semyonov, *Theory of algorithms: its main discoveries and applications*, Novosibirsk: Computing Center of the Siberian Section of the Academy of Sciences of the USSR,part 1 (1982), 99–342, Algorithms in modern mathematics and its aplications. Materials of the International Symposium in Urgench (Uzbek SSR).

[10] P. Martin-Löf, *The definition of random sequence*, Inform. and Contr. **9,6** (1966), 602–619.

[11] R. Mises von., *Grundlagen der Wahrscheinlichkeitsrechnung*, Math. Ztschr. **Bd. 5** (1919,), S. 52–99.

[12] R. Mises von., *Wahrscheinlichkeit, Statistik und Wahrheit*, Wien, J. Springer, 1928.

[13] A. Church, *On the concept of a random sequence*, Bull. Amer. Math. Soc. **46,2** (1940), 130–135.

[14] R. P. Daley, *Minimal-program complexity of pseudorecursive and pseudorandom sequences*, Math. Syst. Theory **9,1** (1975), 83–94.

[15] A. N. Kolmogorov, *On tables of random numbers*, Sankhya. A **25,4** (1963), 369–376.

[16] D. Loveland, *A new interpretation of the von Mises concept of random sequence*, Ztschr. math. Log. und Grundl. Math. **Bd. 12,4** (1966), 279–284.

[17] D. Loveland, *The Kleene hierarchy classification of recursively random sequences*, Trans. Amer. Math. Soc. **125,3** (1966), 497–510.

[18] A. Shen, *The frequency approach to the definition of the notion of random sequence*, Moscow VINITI 18 (1982), 14–42, Semiotica i Informatica.

[19] L. A. Levin, *On the notion of random sequence*, Dokl. Akad. Nauk SSSR **212,3** (1973), 548–550.

$\mathcal{E}$-ENTROPY AND $\mathcal{E}$-CAPACITY
(TO ARTICLES NO.7, 8)

(V. M. TIKHOMIROV)

The topic of the given commentary is directly related to the papers [1-5]. However, it is natural to consider a somewhat wider cycle in which we shall include [6-10]. The backbone bringing together the entire cycle of articles [1-10] is the notion of entropy.

In the first section we consider different definitions of entropy (which appeared in the 50s and 60s under the influence of the work of C. Shannon and A. N. Kolmogorov) in order to make it easier for the reader to find his way in the subject matter of the entire cycle.

The articles [1-3] were of preliminary character, and their contents is in fact swallowed up by the review article [4] contained in the present book (see No.7). The report [5] only indicated a programme of further studies which, according to A. N. Kolmogorov's strategy, was intended to unite the notions of approximation theory and computational mathematics.

The cycle of paprs [1-10] have been reviewed in the two articles [11, 12] written in connection with A. N. Kolmogorov's 60th and 80th anniversaries. The papers [6-8, 10] are commented by R. L. Dobrushin and Ya. G. Sinai. The paper [5] has practically not been developed so far. Therefore in sections 2-10 of the present commentary, we limit ourselves to discussing the paper [4], indicating the main direction of the development of its subject matter.

1. On the notion of entropy. The term "entropy" was introduced (in thermodynamics) by the German scientist R. Clausius in 1865. Later the notion of entropy played a fundamental role in statistical physics. L. Boltzman described the essence of the notion of entropy as the "measure of indeterminacy" of the state of a gas.

Creating the foundations of information theory, Shannon [13] used the term "entropy" to characterize the indeterminacy of a source of messages. Essentially, Shannon's definition of entropy has to do with a random object $\xi = (X, \Sigma, P)$, where X includes a finite number of elementary events

$$X = \bigcup_{i=1}^{N} A_i, \ A_i \bigcap A_j =, \ i \neq j, \ P(A_i) = p_i.$$

The entropy of this random object (according to Shannon) is the expression

$$H(\xi) = \sum_{i=1}^{N} p_i \log_2 \frac{1}{p_i}.$$

Of course this notion can be trivially generalized to the case of a countable number of elementary events. In the case when the number of elementary events is equal to N and all of them are equally probable, we obtain $H(\xi) = \log_2 N$. Hence the number $\log_2 N$ is often called the entropy of a "non-random object" consisting of N elements.

It should be noted that before Shannon attempts were made to introduce expressions of a similar content (R. Fischer in statistics, R. Hartley in the theory of information). But it is precisely Shannon who succeeded in obtaining mathematical results related to the notion that he had introduced and these results constitute the foundations of a new science – the theory of information.

When passing to continuous messages, a difficulty arises. All the natural analogues of Shannon's entropy turn out to be equal to infinity. A. N. Kolmogorov indicated several times (see for example [7]) that the main notion which can be generalized to absolutely arbitrary continuous messages and signals is not the notion of entropy but the notion of *amount of information* $I(\xi, \eta)$ in the random object ξ with respect to the object η. In the simplest cases the magnitude $I(\xi, \eta)$ was defined by Shannon in [13, Appendix 7] and in the general case by A. N. Kolmogorov [6, 7, 4, Appendix II]. In the case of a discrete random object desccribed above, we have $H(\xi) = I(\xi, \xi)$.

Starting from the notion of amount of information, Kolmogorov [7] gives the definition of *ε-entropy of a random object*. Essentially it consists in the following. Suppose a source of information yields a magnitude ξ and we are required *to code the information obtained with precision ε*. This by definition means that from the signal ξ we must choose another signal ξ' so that the joint distributions $P_{\xi\xi'}$ belong to certain class of distributions W_ε determined by the parameter ε. Then the ε-entropy of the source is naturally defined as

$$H_\varepsilon(\xi) = \inf\{I(\xi, \xi') | P(\xi, \xi') \in W_\varepsilon\}.$$

A. N. Kolmogorov's approach to the quantitative characteristics of indeterminacy of continuous objects is very general. Various particular cases have appeared in many papers. Let us present one of them. Suppose $\xi = ((X, \rho), \Sigma, P)$ is a random object, where X is on one hand a probabilistic space and on the other a metric one. Let us split X into N disjoint measurable parts $X = \bigcup_{i=1}^{N} A_i$, $A_i \cap A_j =, i \neq j, P(A_i) = p_i$. The expression

$$H_\varepsilon^1(\xi) = \inf \sum_{i=1}^{N} P(A_i) \log_2 1/P(A_i),$$

where the lower bound is taken over all disjoint measurable partitions into sets of diameter $\leqslant \varepsilon$ is the ε-entropy in the sense of the definition above. The expression $H_\varepsilon^1(\xi)$ was studied, for example, in the paper [14].

In the paper [3], Kolmogorov gives a definition (similar to the one just given above) of the *ε-entropy* $\mathcal{H}_\varepsilon(C)$ *of a non-random object* C when the latter is a subset of a metric space X. That is the name that he gave to the binary logarithm of the expression $N_\varepsilon(C)$, the minimal number of elements of the covering of the set C by sets of diameter $\leqslant 2\varepsilon$.

Together with this ε-entropy (which was later called *absolute*) A. N. Kolmogorov gives another definition, the definition of *relative ε-entropy* $\mathcal{H}_\varepsilon(C, X)$ as the binary logarithm $N_\varepsilon(C, X)$ of the minimal number of elements of an ε-net for X. This expression was studied in detail in [4] and then in many other papers.

Appendix 7 of Shannon's paper [13] mentioned above contained the construction of "the (ε, δ)-entropy" $H_{\varepsilon\delta}(X)$ of a metric space with measure $((X, \rho), \Sigma, P)$, equal to the binary logarithm of the minimal number of elements of a ε-net in a set whose complement is of measure $\leqslant \delta$. This notion was developed in the article [15] and others.

For the class of functions on the entire real line, Kolmogorov introduced the notion of *ε-entropy for unit length* (see 8 in [4]). A similar notion of (ε, δ)-*entropy for unit length* was essentially introduced previously by Shannon (in the same appendix §8 in [13]). These notions will be discussed further in Section 8 below.

In the paper [8] Kolmogorov takes another step in the direction of extending the notion of entropy. He introduces the notion of entropy of a dynamical system. A dynamical system (X, S) is simply a space with measure (X, Σ, μ) which is supplied with a transformation $S : X \rightarrow X$ preserving the measure. To every splitting

$$A = \{A_1, \ldots, A_N\}, \ X = \bigcup_{i=1}^{N} A_i, \ A_i \cap A_j = \emptyset, \ i \neq j$$

we can relate the entropy of this splitting

$$H_1(A) = \sum_{i=1}^{N} \mu(A_i) \log_2 \mu^{-1}(A_i).$$

Denote by $\mu_{i_1,\ldots,i_r}$ the number

$$\mu(A_{i_1} \cap SA_{i_2} \cap \cdots \cap S^{r-1}A_{i_r})$$

and put

$$H_r(A) = \sum_{\{i_1,\ldots,i_r\}} \mu_{i_1,\ldots,i_r} \log \mu_{i_1,\ldots,i_r}^{-1}.$$

According to the well-known MacMillen theorem from information theory, the following limit exists:

$$\bar{H}(A) = \lim_{r\to\infty} H_r(A)/r.$$

The *entropy of the dynamical system* (X, S) *in unit time* is the number

$$\bar{H} = \sup_A \bar{H}(A).$$

'Concerning the remarkable role of the notion introduced above in the development of the theory of dynamical systems, see the commentary by Ya. G. Sinai.

Later one more step was taken. Suppose X is a topological space and S is its homeomorphism. To each splitting $A = \{A_1, \ldots, A_N\}$ of the space X we can relate its "non-random entropy" $H_1(A) = \log_2 N$. Denote by $A_{i_1 \ldots i_r}$ the set $A_{i_1} \cap S A_{i_2} \cap \cdots \cap S^{r-1} A_{i_r}$ and by $H_r(A)$ the binary logarithm of the number of such non-empty sets. It can again be proved that the limit

$$\bar{H}(A) = \lim_{r \to \infty} H_r(A)/r$$

exists. The *topological entropy of the pair* (X, S) *in unit time* is by definition the number

$$\bar{H} = \sup_A \bar{H}(A).$$

Apparently, there is a genetic relationship between all the notion introduced (including perhaps, entropy in statistical physics) and it is possible to expect the appearance of a paper where they will be studied simultaneously.

One of the main goals in introducing these notions was to distinguish different representations in the class of those objects where entropy is defined. The entropy of a source determines the necessary rate of transmission (see [13]), i.e. distinguishes sources by this speed, the entropy of a dynamical system gives a method for constructing invariant dynamical systems (see [8] and the commentary by Ya. G. Sinai), ε-entropy of functional classes allows to distinguish linear topological spaces (see [9] and Section 7 further in this commentary), etc.

In the remaining part of the commentary we discuss the entropy characteristics of expressions of the type of $\mathcal{H}_\varepsilon(C)$ and $\mathcal{H}_\varepsilon(C, X)$.

2. ε-entropy and cross-sectional measures. The introduction to the paper [4] mentions the reasons which led A. N. Kolmogorov to formulate the "general program of study of ε-entropy and ε-capacity, interesting from the point of view of the theory of functions on compact sets in functional spaces." There in the role of original stimuli, he mention the ideas contained in the work of Hausdorff, Pontriagin-Shnirelman, Vitushkin and Shannon.

However, it should be mentioned that the notion of entropy is stated there along the traditional scheme used in topology and in the theory of approximations to define asymptotical characteristics called cross-sectional measures. Let us explain this.

Suppose X is a normed space. C is a subset of X, $\mathfrak{A} = \{A\}$ is a family of approximating sets for the sets A. Let us put

$$\mathcal{E}(C, \mathfrak{A}, X) = \inf_{A \in \mathfrak{A}} \sup_{x \in C} \inf_{y \in A} \|x - y\|.$$

If in the role of $\mathfrak{A}$ we take $\mathcal{L}_N$, the family of all linear subspaces of dimension $m \leqslant N$, the magnitude $\mathcal{E}(C, \mathcal{L}_N, X)$ is called the *Kolmogorov N cross-section C*. This number (usually denoted by $d_N(C, X)$) was introduced by Kolmogorov as early as 1936. But if for $\mathfrak{A}$ we take Σ_N, the family of all N-point sets, then we obtain a value which is denoted by $\varepsilon_N(C, X)$. The binary logarithm of the function inverse to $\varepsilon_N(C, X)$ is none other than the (relative) ε-entropy $\mathcal{H}_\varepsilon(C, X)$.

Suppose further that Z is a certain set, $\varphi : C \to Z$ a map ("coding" elements of C by elements of Z) and $\Phi = \{\varphi\}$ is a certain family of "codings". Let us put

$$\mathcal{K}(C, \Phi) = \inf_{\varphi \in \Phi} \sup_{x \in C} \{\operatorname{diam} \varphi^{-1}(\varphi(x))\}.$$

If in the role of Φ we take the family $\mathcal{U}_N$ of all continuous maps of C and all complexes of dimension $\leqslant N$, then the value $\mathcal{K}(C, \mathcal{U}_N)$ turns out to coincide with the so-called Urysohn N-cross-section, introduced in 1922 (this historically is the first magnitude called a cross-sectional measure).

If in the role of Φ we take the family Φ_N of all maps into the set $\{1, 2, \ldots, N\}$, we obtain a magnitude which is denoted by $\varepsilon^N(C)$. The binary logarithm of the inverse function to $\varepsilon^N(C)$ is none other than the absolute ε-entropy $\mathcal{H}_\varepsilon(C)$. In more detail this is described in [16]. The relationship of cross-sectional measures and entropy was studied by Brudny, Timan, Tikhomirov, Mityagin and others.

3. On exact solutions of the ε-entropy and ε-capacity problems. The question of finding the exact value of ε-entropy or ε-capacity is a very difficult one. Many old problems of geometry, known as difficult, can be interpreted in terms of ε-entropy. As an example, let us state Borsuk's problem (see [17]), which we shall express in terms of the magnitude N_ε.

Suppose X is n-dimensional Euclidean space and $C \subset X$ is a certain set of diameter one. Borsuk's question is the following: what is the minimal number of parts of smaller diameter into which the set C can be partitioned? In other words, we must find the expression

$$\beta(X) = \max\{N_{1-0}(C) | \operatorname{diam} C \leqslant 1\}.$$

Borsuk's conjecture was that $\beta(X) \leqslant n + 1$. In its general formulation it has not been proved at present. Besides the example given in Section 2 of the article [4], the author of this commentary knows no meaningful examples of infinite-dimensional compact sets for which the problem of ε-entropy or ε-capacity is solved exactly. Let us note in passing that an existence theorem for best ε-nets was obtained by Garkavi.

4. On functional dimensions (defined in §3 of the article [4]) of the solutions of differential equations. See Kazaryan's thesis [18], where references to previous papers are presented (Zilezni, Kazaryan, Komura, Tanaka and others).

5. On finite-dimensional problems. Problems equivalent to the asymptotics of ε-entropy and ε-capacity in finite-dimensional spaces have interested geometers from very far back. In the language of geometry, these are problems of best pavings and coverings. Important progress in the estimate of the expression θ_n (§2 in [4]) were obtained in relatively recent times (about this see, for example, Roger's monograph [19]).

Some advances have also been made in the estimate of the expression τ_n from below (§2 from [4]). About this see [20, 21].

6. The entropy of functions of finite smoothness. This was studied in the work of Babenko, Bakhvalov, Birman-Solomyak, Borzov, Brudny, Kotlyar, Golovkin, Dveirin, Kollya, Oleinik, Smolyak, Tikhomirov and others. A result of fundamental importance about ε-entropy of functions of finite smoothness was obtained by Birman and Solomyak for Sobolev classes $\dot{W}_p^r$ given on the unit cube I^n in R^n and considered in the space $L_q(I^n)$: they proved in [22] that if $\dot{W}_p^r$ can be compactly embedded in $L_q(I^n)$, then

$$\mathcal{H}_\varepsilon(\dot{W}_p^r, L_q(I^n)) \asymp \varepsilon^{-n/r}$$

(the metrics p and q do not participate in the answer). It is easy to show that actually the following limit

$$\lim_{\varepsilon \to 0} \varepsilon^{n/r} \mathcal{H}_\varepsilon(\dot{W}_p^r, L_q(I^n)) = \kappa(r, p, q, n)$$

exists.

In §2 of the paper [4] the magnitude $\kappa(1, \infty, \infty, 1)$ was computed (for Lipshitz functions). As far as I know there is no other case when the problem of the strong asymptotics of $\mathcal{H}_\varepsilon$ is solved. (Actually Tikhomirov [15] proved that

$$\kappa(r, p, p, 1) \sim 2r \log e;$$

the case $p = 2$ was considered earlier by Arnold, see §6 in [4]).

7. ε-entropy of classes of analytic functions. Suppose A_G^K is the class of function $x(\cdot)$ on a compact set K possessing an analytic continuation $\tilde{x}(\cdot)$ to the domain $G \supset K$ such that $|\tilde{x}(z)| \leqslant 1 \ \forall z \in G$. The problem of ε-entropy and Kolmogorov cross-sections of the class A_G^K stimulated many studies (Alper, Babenko, Yerokhin, Levin-Tikhomirov, Zakharyuta-Skiba, Nguen Than Van, Fischer-Michelli, Ganelius etc.). The most general result is due to Widom [23], who proved that

$$\lim_{n \to \infty} (d_n(A_G^K, C(K)))^{1/n} = \exp(-1/c(K, G)),$$

where $c(K, G)$ is the Green capacity of the compact set K with respect to G under very small restrictions concerning G and K. This formula implies the following relation for ε-entropy

$$\lim_{\varepsilon \to 0} \frac{\mathcal{H}_\varepsilon(A_G^K, C(K))}{\log^2(1/\varepsilon)} = \exp\left(-\frac{1}{c(K, G)}\right).$$

Other classes of analytical functions were studied, as well as classes of functions in whose definition both smoothness and analycity are involved (Babenko, Tikhomirov, Taikov, Dveirin, Parfyonov, Oleinik, Farkov, Alper, Boyanov, Hubner, Stegbukhner, etc.).

8. Kotelnikov's theorem. In §8 of the article [4] the ε-entropy in unit time of the class of functions with bounded spectra is found. The notion of spectra appeared in physics. It plays an important role in technology and in particular, in the theory of communications. In mathematics functions with bounded spectrum are studied in harmonic and complex analysis. If the function $x(\cdot)$ has a spectrum (i.e., the support of the Fourier transform) on $[-\sigma, \sigma]$, then we have the formula

$$x(t) = \sum_{k \in Z} x\left(\frac{k\pi}{\sigma}\right) \frac{\sin \sigma(t - k\pi/\sigma)}{\sigma(t - k\pi/\sigma)},$$

which may be interpreted as follows: *a function with bounded spectrum is defined by its values computed at equal intervals of length π/σ.* The meaning of this statement from the point of view of information theory (*"the dimension needed in unit time is equal to the doubled width of frequencies"*) was discovered by Kotelnikov in 1933 and in the Soviet literature is known as the "Kotelnikov theorem". Shannon also tried to give meaning to these things. The exact meaning of this kind of result may be interpreted in different ways. One of the interpretations is contained in [4, §8].

On the other hand, Kolmogorov himself studied the meaning of Kotelnikov's statements in the case of *random processes*. It is important to note that a random function with bounded spectrum is always *singular* and its observation is not related to the stationary input of new information. In [7] Kolmogorov gives a formula for the ε-entropy of any Gaussian process with spectral density well-approximated by a density given by a characteristic function. This yields approximate formulas expressing the ideas of Kotelnikov-Shannon.

It should be noted that in Appendix 7 to [13] Shannon indicated a method for obtaining a formula similar to the one that was proved in [4, §8], also as applied to random processes. This method is related to the notion of ε-entropy per unit time which was mentioned above in section 1.

Shannon's formula turned out to be correct [24]. This result was strengthened by Pinsker and Sofman, who showed that ε-entropy (according to Shannon) per unit time $\bar{H}_{\varepsilon\delta}(\xi)$ does not depend on δ and is equal to the ε-entropy of a process ξ according to Kolmogorov $H_\varepsilon(\xi)$ (see Section 1). But then we get

$$\lim_{\varepsilon \to 0} \frac{\bar{H}_{\varepsilon\delta}(\xi)}{\log(1/\varepsilon)} = \operatorname{mes}\{\lambda | f_\xi(\lambda) > 0\} = \operatorname{supp} f_\xi,$$

where f_ξ is the spectral function of the process ξ. But this is precisely what was stated by Shannon as a conjecture in [13].

9. ε-entropy of spaces of functionals. This is the topic of several papers. Let us indicate, in particular, the one by Timan [25]. But it should be noted that many fundamental questions about ε-entropy of functionals of finite smoothness and analytical functionals are still awaiting their solutions.

10. Certain applications of the notion of ε-entropy. In terms of ε-entropy one can give a criterion of the kernel propety of linear topological spaces X (Mityagin): an F-space X is a kernel space if and only if for any compact set K and any symmetric neighbourhood U of zero we have

$$\varlimsup_{\varepsilon \to 0} \log H_\varepsilon(K, X_U)/\log(1/\varepsilon) = 0.$$

(Here X_U is the space X with seminorm generated by U.)

Using the notion of ε-entropy, Kolmogorov introduced a topological invariant – the approximation dimension (somewhat earlier a similar invariant was introduced by Pelchinski); using it, he proved the fact that spaces of analytical functions of different dimension are not isomorphic.

REFERENCES

[1] A. N. Kolmogorov, *Estimates of the minimal number of elements of ε nets in various functional classes and their application to the question of representing functions of several variables as superpositions of functions of a lesser number of variables*, Uspekhi Mat. Nauk **10,1** (1955), 192–194.

[2] A. N. Kolmogorov, *On the representation of continuous functions of several variables as superpositions of continuous functions of a lesser number of variables*, Dokl. Akad. Nauk SSSR **108,2** (1956), 179–182.

[3] A. N. Kolmogorov, *On certain asymptotic characteristics of completely bounded metric spaces*, Dokl. Akad. Nauk SSSR **108,3** (1956), 385–388.

[4] A. N. Kolmogorov, V. M. Tikhomirov, *ε-entropy and ε-capacity of sets in function spaces*, Uspekhi Mat. Nauk **14,2** (1959), 2–86.

[5] A. N. Kolmogorov, *Various approaches to the estimate of the complexity of approximations and computations of functions*, In: Proceedings of the Itern. Congr. Math. Stockholm (1963).

[6] I. M. Gelfand. A. N. Kolmogorov, A. N. Yaglom, *The amount of information and entropy for continuous distributions*, Moscow, Academy of Sciences Publishes 3 (1958), 300–320, Trudy III of the All-Union Mathematical Congress.

[7] A. N. Kolmogorov, *The theory of transmission of information.*, Plenary Session , Moscow, Academy of Sciences Publishers (1957), 66–99, The Session of the Academy of Sciences of the USSR on Scientific Problems of Industrial Automation.

[8] A. N. Kolmogorov, *A new metric invariant of transitive dynamical systems and automorphisms of Lebesgue spaces*, Dokl. Akad. Nauk SSSR 119 (1958), 861–864.

[9] A. N. Kolmogorov, *On the linear dimension of topological vector spaces*, Dokl. Akad. Nauk SSSR 120 (1958), 239–241.

[10] A. N. Kolmogorov, *On the entropy in unit time as a metric invariant of automorphisms*, Dokl. Akad. Nauk SSSR 124 (1959), 754–755.

[11] V. N. Tikhomirov, *The work of A.N. Kolmogorov on ε-entropy of functional classes and superpositions of functions*, Uspekhi Mat. Nauk **18,5** (1963), 55–92.

[12] V. N. Tikhomirov, *Cross-sectional measures and entropy*, Uspekhi Mat. Nauk **38,4** (1983), 91–99.

[13] Shannon C. E., *A mathematical theory of communication*, Pt I, II. - Bell Syst. Techn. J. **27,4** (1948), 623–656.

[14] Renyi A., *On the dimension and entropy of probability distributions*, Acta math. Acad. Sci. Hung. **10,1-2** (1959), 193–215.

[15] Posner E. C., Rodemish E. R., Rumsey H.G., *ε-entropy of stochastic processes*, Ann. Math. Statist. **38,4** (1967), 1000–1020.

[16] V. M. Tikhomirov, *Certain questions of approximation theory*, Moscow, Moscow University Press, 1976.

[17] V. G. Boltyansky, I. Ts. Gokhberg, *The partition of figures into smaller parts*, Moscow, Nauka, 1971.

[18] G. G. Kazaryan, *An inequality between derivatives and differential operators and certain of its application to partial differential equations*, Doctoral Dissertation (1980), Yerevan.

[19] C. A. Rogers, *Packing and covering*, Cambridge Univ. Press, 1964.

[20] V. M. Sidelnikov, *On the best paving of balls on the surface of a n-dimensional Euclidean sphere and the number of vectors of a binary code with given code distance*, Dokl. Akad. Nauk SSSR 213, 5 (1973), 1029–1032.

[21] V. M. Levenshtein, *On the maximal density of fillings of n-dimensional space by different balls*, Mat. Zametki 18, 2 (1975), 301–311.

[22] M. Sh. Birman, M. Z. Solomyak, *Piece-wise polynomial approximations of functions of classes W_p^α*, Matem. Sbornik **73,3** (1977), 331–335.

[23] H. Widom, *Rational approximation and n-dimensional diameter*, J. Approxim. Theory **5,1** (1972), 343–361.

[24] L. B. Sofman, *Shannon's problem*, Dokl. Akad. Nauk SSSR **215,6** (1974), 1313–1316.

[25] A. F. Timan, *On the order of growth of ε-entropy of spaces of real continuous functionals defined on a connected set*, Uspekhi Mat. Nauk **19,1** (1964), 173–177; Funktsional Anal. i

Prilozhen. **5,3** (1971), 104–105; Dokl. Akad. Nauk SSSR **218,3** (1974), 505–507.

TABLES OF RANDOM NUMBERS
(TO THE PAPER NO.9)

(A. Sh. Shen)

1. From the point of view of classical probability theory, the occurence of a given sequence of zeros and ones as the result of n tosses of a symmetric coin has probability 2^{-n}. Nevertheless, the appearance of certain of these sequences (for example, the one consisting of only zeros or the sequence 010101...) seems rather strange to us and leads us to suspect that there is something wrong, while the appearance of others does not lead to such a reaction. In other words, some sequences seem less "random"than others, so one may try to classify finite sequences of zeros and ones according to their "degree of randomness". This classification is the main topic of the article that we are commenting upon here.

When we consider finite sequences, it is possible to speak only of a greater or lesser degree of randomness, whereas for infinite sequences it is natural to try to draw a clear boundary between what is random and non-random, distinguishing in the set of all sequences a certain subset whose elements are called random sequences. The necessity of such a distinction was first indicated by von Mises [8,9]. He also indicated a specific scheme for defining random sequences.

2. Let us describe this scheme, limiting ourselves for simplicity to the case of sequences of zeros and ones. Suppose we are given a certain sequence $x_0 x_1 \ldots$. According to Mises, for the randomness of this sequence it is necessary first of all that there exist mean frequencies of ones, i.e. the limit

$$\lim n^{-1}(x_0 + \cdots + x_{n-1})$$

must exist. (This is a necessary but not a sufficient condition; it is , for example, satisfied by the sequence 01010101... which can hardly be considered random!). Further the mean frequency of ones must be preserved if we pass from the sequence to any of its infinite subsequences obtained by means of any admissible rule of choice. Mises does not give an exact definition of admissible rules, limiting himself to certain examples of admissible rules and to the general indication that "it should be possible to decide whether each flip of the coin in the chosen partial sequence is independent of the result of this flipping, i.e. by means of a certain rule which is related only to the number of the given flip and is established before its result is know" [9, p. 32]. (In the quotation above Mises calls "flipping" what we have been calling terms of our sequence). The restriction thus indicated forbids considering admissible the rule saying "choose all the terms which are equal to one". However,

240

it does not forbid choosing the terms before which you have a one (this rule is considered on p. 36 of Mises's book quoted above [9]) or all the terms whose numbers are prime numbers (loc. cit, p. 32).

3. In 1940 Church [11] proposed a formal definition of randomness, which makes the strategy explained by Mises more specific. Let us give this definition, beginning with an exact definition of admissible rules of choice.

Suppose we have in front of us an infinite (in one direction) series of cards whose top sides are identical, while the bottom sides are supplied with zero or one – the terms of our sequence. We turn over all the cards one after another, beginning from the first, and choose some of them. The numbers written on the lower side of the cards that we have chosen constitute a certain subsequence of the given sequence. Before turning over each of the cards, we declare whether we shall be including it among the chosen ones. While doing this, we can use the information that we have about the numbers written on the bottom sides of the cards that have already been turned over. Thus in this scheme of choice, an admissible rule of choice is given by any set R of finite sequences of zeros and ones; applying this rule to the sequence $x_0 x_1 \ldots$ means choosing a subsequence consisting of those terms x_{n+1} for which $x_0 x_1 \ldots x_n \in R$. It is natural to require that admissible rules of choice be given by recursive sets R. Thus randomness according to Mises–Church of a sequence $x_0 x_1 \ldots$ means that:

1) the limit $p = \lim n^{-1}(x_0 + \cdots + x_{n-1})$ exists,

2) for any recursive set R of finite sequences of zeros and ones, either the subsequence of $y_0 y_1 \ldots$ obtained from the original one by the choice (preserving the order) of those terms x_{n+1} for which $x_0 x_1 \ldots x_n \in R$ is finite, or the limit $\lim k^{-1}(y_0 + \cdots + y_{k+1})$ exists and is equal to p.

4. In A. N. Kolmogorov's article that we are commenting on here, a certain more general scheme of choice is proposed. In terms of turning over cards, this generalization may be described as follows. We are allowed to turn over cards in any order (not necessarily from left to right). To be more precise, the rule of choice is given by two computable functions F and G. The first function determines in which order we shall be turning over the cards. Its arguments are finite sequences of zeros and ones and its values are natural numbers. First we turn over the card with the number $F(\Lambda)$. By Λ we denote the empty sequence. Denote the number written on it (0 or 1) by t_0. When we turn over the card with the number $F(t_0)$, denote the number written on it by t_1. The next card turned over has number $F(t_0 t_1)$ and the number is denoted by t_2. Then we turn over the card with number $F(t_0 t_1 t_2)$, etc. The second function G has finite sequences of zeros and ones for its arguments and its values are values zero and one. It decides whether or not we shall include each subsequent term in our subsequence. Specifically, the term with number $F(\Lambda)$ will be included in the subsequence if and only if $G(\Lambda) = 1$, the term with number $F(t_0)$ will be included if and only if $G(t_0) = 1$, the term with number $F(t_0 t_1)$ if and only if $G(t_0 t_1) = 1$, etc. The order of terms in the subsequence is determined by their order of choice by means of the function F. If, at some moment of time during the duration of the process, the value of the function F or the function G turn out to be undefined, then the process of choice is suspended and the chosen subsequence turns out to be finite. The same thing happens if the next value of the function F coincides with one of the previous values.

This concludes Kolmogorov's definition of an admissible rule of choice. Thus, according to the general scheme proposd by Mises, we can define randomness: a sequence α is called *random* if there exists a limit p of the frequency of ones in the sequence α and any infinite sequence obtained from α by means of an admissible (in the sense described) rule of choice also has a limit of frequency equal to p.

What we said above in this Section 4 is a detailed exposition of Remark 2 at the end of Section 2 of the commented article. The only difference consists in that we have excluded from the arguments of the functions F and G the variables ξ_i (see the beginning of Section 2 of the commented article); this does not change the class of admissible rules, since the values of ξ_i are previous values of the function F and can be recovered from its previous arguments. The same definition of an admissible rule of choice was later (but independently) proposed by Loveland [4,5]. In [5, p. 499, Note in proofs] Loveland writes: "Such a modification was proposed and used by A. N. Kolmogorov" (he has in mind the modification of the definition of randomness according to Mises-Church).

Hence admissible rules of choice in the sense described above will be called *admissible in the sense of Kolmogorov-Loveland*, while the class of random sequences that they lead to will be called the class of *random sequences in the sense of Mises-Kolmogorov-Loveland*. Obviously, any sequence that is random in the sense of Mises-Kolmogorov-Loveland is random in the sense of Mises-Church; it follows from results of Loveland that the converse is not true. For the details see [12].

5. Unfortunately, the definition of randomness according to Mises-Kolmogorov-Loveland, as well as the definition of randomness according to Mises-Church, does not correspond entirely to our intuition of randomness (about this see more details in [12, §4, 7]). Later A. N. Kolmogorov and his followers developed other approaches (different from the Mises one) to the definition of the notion of random sequence ([6, 7]). The approach from [7] may be called measure-theoretic, while the approach from [6] may be called complexity-theoretic. Both of these approaches are discussed in the present publication in the comments to the papers No. 10, 12. These approaches lead to the same class of random sequences. This class may be called the class of sequences random according to Martin-Löf (in honor of Martin-Löf, who proposed one of the definitions of this class in [7]). It is easy to show that any sequence random according to Martin-Löf is random in the sense of Mises-Kolmogorov-Loveland and are *a fortiori* random in the sense of Mises-Church. In [3, p. 6] it is stated that there exist random sequences in the sense of Mises-Kolmogorov-Loveland whose complexity of initial segments grow logarithmically. This result implies that there exist random sequences in the sense of Mises-Kolmogorov-Loveland which are non-random in the sense of Martin-Löf. The proof of this result of A. N. Kolmogorov in [3] has not been published and is not known to the author of the present commentary. On this note we conclude the discussion of the notion of random (infinite) sequence referring the reader to a more detailed discussion in [10, 12].

6. Now let us pass to our discussion of the notion of randomness (more precisely, the degree of randomness) of a finite object. (The importance of considering precisely finite objects is described by A. N. Kolmogorov in his note ("From the author") which opens the Russian version of the translation of the commented article: "The main difference from Mises's approach consists in the strictly finite character

of the entire concept and in the introduction of a numerical estimate of the stability of frequencies". The approach proposed in the commented article to the definition of randomness is naturally called "the frequency approach". A sequence is called (n, ε, p)-*random with respect to a given finite system of rules of choice*, if for any of its subsequences chosen in accordance to one of the rules of the system considered and containing no less than n elements, the mean frequency of ones differs from p by no more than ε.

Later [2,3] A. N. Kolmogorov proposed an approach to the definition of randomness based on complexity. In [2] Kolmogorov writes: "Roughly speaking, what we mean here is the following. If a finite set M containing a very large number of elements N can be defined in terms of a program of negligibly small length as compared to $\log_2 N$, then almost all the elements of M have complexity $K(x)$ close to $\log_2 N$. Elements $x \in M$ having this complexity are viewed as "random" elements of the set M". (Concerning the notion of complexity, or entropy of a finite object, see [2,3] and the commentary to these articles.) Thus what is defined here is the randomness of a given sequence x not by itself, but with respect to a certain set M containing x. If we wish to compare this approach to the approach of the commented article, we must consider in the role of M the set of all sequences of fixed length containing a fixed number of ones. In more detail, suppose $x_0 x_1 \ldots$ is a sequence of zeros and ones of length n containing k zeros. Consider the set $C(n, k)$ consisting of all sequences of length n containing k zeros. The complexity (equals entropy) of most of the elements of $C(n, k)$ is close to $\log_2 C_n^k$ and therefore to $H(k/n)n$, where $H(p) = -p \log_2 p - (1 - p) \log_2(1 - p)$ is the Shannon entropy of a random variable with two values whose probabilities are p and $1 - p$. Thus the sequence $x_0 x_1 \ldots$ must be considered random if it belongs to this majority, i.e. if its entropy is close to $H(k/n)n$. Expressing ourselves more precisely, it must be considered the less random , the further its entropy is from $H(k/n)n$. A. N. Kolmogorov established the relationship between a random sequence in this sense of complexity and its randomness in the sense of the commented paper (1982, unpublished).

REFERENCES

[1] A. N. Kolmogorov, *On tables of random numbers*, Sankhya A., **25,4**, (1963), 369-376; *On tables of random numbers*, Russian translation: 18 (1982), 3-13, Semiotika i Informatika, Moscow, VINITI.

[2] A. N. Kolmogorov, *Three approaches to the definition of the notion of "amount of information"*, Problemy Peredachy Informatsii **1,1** (1965), 3-11.

[3] A. N. Kolmogorov, *To the logical foundations of information theory and probability theory*, Problemy Peredachy Informatsii, **5,3** (1969), 3-7.

[4] D. Loveland, *A new interpretation of the von Mises concept of random sequence*, Ztschr. math. Log und Grundl. Math. **Bd. 12, H. 4** (1966), S. 279-294.

[5] D. Loveland, *The Kleene hierarchy classification of recursively random sequences*, Trans. Amer. Math. Soc., (1966, ,vol. 125, 3), 497-510.

[6] L. A. Levin, *On the notion of random sequence*, Dokl. Akad. Nauk SSSR, **212,3**, (1973), 548-550.

[7] P. Martin-Löf, *The definition of random sequence*, Inform. and Contr., **9,6**, (1966), 602-619.

[8] R. von Mises, *Grundlagen der Wahrscheinlichkeitsrechnung.*, Math. Ztschr., **bd. 5** (1919), S. 52-99.

[9] R. von Mises, *Wahrscheinlichkeit*, Statistik und Wahrheit (1928), Wien: J. Springer.

[10] V. A. Uspensky, A. L. Semenov, *What are the gains of the theory of algorithms: basic developments connected with the concept of algorithm and with its application in mathematics.*,

In: Algorithms in modern mathematics and computer science: Proc. Conf. Urgench (UzSSR), (1979).; Lect. Notes in Comput. Sci.; **122** (1981), 100–234, Berlin etc.: Springer-Verl..

[11] A. Church, *On the concept of a random sequence*, Bull. Amer. Math. Soc., **46,2** (1940), 130–135.

[12] A. Shen, *The frequency approach to the definition of the notion of random sequence*, Semiotika i Informatika, vol. 18, 1982, pp. 14–42.

REALIZATION OF NETWORKS IN 3-DIMENSIONAL SPACE
(TO ARTICLE NO.11)

(YA. M. BARZDIN)

In the original article one may read the following footnote, which was written by Andrei Nikolayevich himself:

"Several years ago A. N. Kolmogorov proved Theorem 1* for networks of bounded branching and gave the following first approximation to Theorem 2: there exists a network with $n > 1$ elements any realization of which has diameter greater than $C\sqrt{n}/\log n$, where C is a certain constant independent of n. The final versions of all these theorems (and the very idea of setting the question of the property of "almost all" networks) belongs to Ya. M. Barzdin."

In the present version (see Remark 1 in the article No.11) I slightly changed this note, also putting it in the form of a footnote. But in fact I only gave new proofs (and somewhat generalized) theorems obtained by Andrei Nikolayevich earlier, so that my achievements here are not very important. Unfortunately, I do not remember what was the occasion or event at which Andrei Nikolayevich first mentioned these results (I was not present there). I know only that the topic discussed there was the explanation of the fact that the brain (for example that of a human being) is so constituted that the most of its mass is occupied by nerve fibers (axons), while the neurons are only disposed on its surface. The construction of Theorem 1 precisely confirms the optimality (in the sense of volume) of such a disposition of the neuron network.

TO THE PAPER ON DYNAMICAL SYSTEMS
(PAPER NO.5)

A. N. KOLMOGOROV

V. A. Rokhlin noticed that Thoerem 2 of my paper no.5 is false. In my paper [1], a different definition of entropy of an automorphism is proposed. The shortest way of introducing entropy of an arbitrary automorphism was proposed by Ya. G. Sinai (see [2]). I would like also to call attention to the paper by A. G. Kushnerenko [3] who showed that the entropy of any smooth diffeomorphism or vector field of a smooth compact manifold with respect to absolute continuous invariant measure is finite.

REFERENCES

[1] A. N. Kolmogorov, *On the entropy in unit time as a metric invariant of automorhism*, Dokl. Akad. Nauk SSSR, **124,4** (1959), 754–755.

[2] Ya. G. Sinai, *On the notion of entropy of a dynamical system*, Dokl. Akad. Nauk SSSR, **124,4** (1959), 768–771.

[3] A. G. Kushnerenko, *Estimate from above of the entropy of the classical dynamical system*, Dokl. Akad. Nauk SSSR, **161,1** (1965), 37-38.

ERGODIC THEORY
(PAPER NO.5)

(YA.G. SINAI)

This paper by A. N. Kolmogorov played an outstanding role in the development of ergodic theory. First of all it contained the solution of a well-known problem which had in fact stood for more than 25 years, and the success was achieved as the result of the use, in ergodic theory, of absolutely new ideas and methods coming from information theory. Secondly, after this paper appeared, a previously totally unknown extremely fruitful direction in ergodic theory opened up, a direction rich in its deep interesting intrinsic problems and in the variety of its applications to the analysis of specific dynamical systems. It is doubtless that the appearance of the ideas of ergodic theory in physics, the progressively wider and wider use of these ideas in the analysis of systems with stochastic behavior, strange attractors, etc. has as one of its sources precisely to the paper by A. N. Kolmogorov under consideration.

The main problem of general ergodic theory is the problem of metric classification, or the problem of metric isomorphism of dynamical systems. By a dynamical system below we understand an one-parameter group of transformations of a space with probability measure that preserve this measure. Two dynamical systems are called *isomorphic* mod 0 if there exists an almost everywhere defined map of the phase space of one system on a subset of complete measure of the phase space of the other system which commutes with the group action (see [6]). The problem of metric isomorphism of dynamical system consists in finding a complete system of metric invariants, i.e., a family of invariants such that their coincidence implies metric isomorphism. The first example of a metric invariant is the spectrum of the group of unitary operators adjoint to the dynamical system.

Interest to the problem of metric isomorphism arose after the work of von Neumann [23] and von Neumann and Halmos [21], where it was shown that in the class of ergodic dynamical systems with purely point spectrum, a complete system of metric invariants is provided by the spectrum itself. Gelfand noticed that the von Neumann result may be obtained as a simple corollary of the triviality of the second cohomology group of the spectrum, which is always a countable abelian group with coefficients in S^1. It seemed that for systems with continuous spectrum, one must understand in what sense the spectrum constitutes a group, introduce its cohomology groups with coefficients in the group of unitary operators, and after that the isomorphism problem will reduce to computing the corresponding second cohomology group.

In a certain sense, this ideology turned out to be useful only for systems with continuous simple or finite-to-one spectrum, where it is related to certain problems of the theory of Banach algebras (see [6]). However, it was understood fairly long ago that the fundamental class of dynamical systems with continuous spectrum is the class of Lebesgue spectrum with infinite multiplicity. Systems with Lebesgue spectrum of finite multiplicity have not been constructed, as far as we know, even to this day. As to systems with Lebesgue spectrum of countable multiplicity, at the moment when A. N. Kolmogorov's papers appeared, they were quite well-known: these are Bernoulli automorphisms and related examples from probability theory (see [18]), ergodic automorphisms of commutative compact groups (see [10]), geodesic flows on the surfaces of constant negative curvatures (see [5]) and several other. For such systems attempts to find some generalization of von Neumann's argument turned out to be unsuccessful.

In the paper No.5, a new metric invariant of dynamical systems – metric entropy – was introduced. Entropy is a numerical invariant which assumes any non-negative value, including ∞. Its independence of the spectrum means that in the class of dynamical systems with Lebesgue spectrum of countable multiplicity it can assume any admissible value. In other words, dynamical systems with Lebesgue spectrum of countable multiplicity with different values of entropy are possible and therefore are non-isomorphic.

After A. N. Kolmogorov's paper, the problem of metric isomorphism for systems with Lebesgue spectrum of countable multiplicity acquired a totally different character than for systems with discrete spectrum. As the result of the discovery of entropy, it became clear that it should be viewed as a problem of coding, in the style of information theory, and the first examples of such an isomorphism were constructed precisely as "codes" mapping one space of sequences into another (see [7]).

No less important than entropy was the notion of quasiregular dynamical system or K-systems as they are now called, introduced in the same paper No.5. This terminology reflects both the original term and the name of the author*. The notion of K-system was inspired by A. N. Kolmogorov's general considerations on the regularity of random processes, i.e. on what happened and what characterizes the weakening of statistical relationships between values of a process and intervals of time far away from each other. From this point of view, K-systems correspond to random processes with the weakest possible regularity property.

All K-systems possess a Lebesgue spectrum of countable multiplicity (for automorphisms, this was pointed out in the paper No.5 itself, for flows it was established in [11]) and have positive entropy. No metric invariants distinguishing K-systems, other than entropy, were known, and the problem of metric isomorphism in the class of K-systems acquired the following form: is it true that K- systems with identical values of entropy are metrically isomorphic?

A great breakthrough in the solution of this problem was achieved by the American mathematician Ornstein. At first he proved that Bernoulli automorphisms with equal entropy are metrically isomorphic and then carried over this result to a much wider class of dynamical systems (the so-called very weakly Bernoulli systems)

---

*In Russian the letters K and Q are transliterated as the same letter K. (*Translator's note*).

that include Markov automorphisms and many classical dynamical systems such as ergodic automorphisms of commutative compact groups, geodesic flows on compact manifolds of negative curvature, certain billiard systems. On the other hand, Ornstein constructed an example of a K-system which is non-isomorphic to any Bernoulli automorphism. Recently a beautiful and fairly simple example of such a system was proposed (see [22]). Thus it turned out that in the class of K-systems, entropy does not constitute a complete system of metric invariants. Ornstein and Shields showed that the number of non-isomorphic types of K-systems with identical entropy is non-countable. Ornstein's theory is developed in his monograph [9].

In §4 of the paper No.5 an example of a flow, i.e. a dynamical system in continuous time, with finite entropy is constructed. Note that these examples from the point of view of traditional probability theory are fairly unnatural. In them typical realizations of such processes assume a finite or countable number of values and therefore are piecewise constant on segments of random length and change their states in accordance to a certain probability distribution. It turned out that precisely processes of the type of Kolmogorov's examples appear in the analysis of smooth dynamical systems with strong statistical properties, for example, in geodesic flows on compact manifolds of negative curvature (see [2, 12]) and certain billiard systems (see [13, 19]). There are sufficiently profound reasons for this. In finite-dimensional dynamical systems whose dynamics are determined in a strictly deterministic equation of motion, the entire randomness is contained only in the choice of initial conditions. The stationary random processes which then appear are necessarily processes with "discrete appearances of randomness". Nevertheless, for a whole series of important classes of dynamical systems to which such random processes correspond, it turns out to be possible to establish certain strong statistical properties such as the central limit theorem for time fluctuations, rapid decay of time correlation functions, etc. (see for example, [3, 26, 14, 20]). The main dynamical mechanism for this is the internal non-stability of motion (see [15, 16]). The property of unstability is possessed by Anosov systems (see [1, 8]), by the A-systems of Smale (see [17]), by the Lorentz model with strange attractor (see [15, 4]), by disseminating billiards (see [13]).

A few years after the appearance of the paper No.5, an closeknit relationship was found between the theory of unstable dynamical systems and Euclidean statistical mechanics of classical one-dimensional lattice models (see [14, 3, 26]). From the purely formal point of view this relationship arises if the large parameter T (time) of ergodic theory is viewed as a large parameter of the number of degrees of freedom in statistical mechanics. Here the possibility to carry over the notion and methods of statistical mechanics to a new situation arises. The theory which then appears is called "topological thermodynamics". Using it, it has been possible to obtain a series of new deep results.

References

[1] D. V. Anosov, *Geodesic flows on closed Riemann manifolds of negative curvature*, Trudy MIAN SSSR, **90** (1967), 3–209.

[2] D. V. Anosov, Ya. G. Sinai, *Certain smooth dynamical systems*, Uspekhi Mat. Nauk, **22,5** (1967), 107–172.

[3] R. Bowen, *Methods of symbolic dynamics*, Collected articles (in Russian translation), "Mir", Moscow.

[4] L. A. Bunimovich, Ya. G. Sinai, *Lorentz stochastic attractors*, Non-linear waves, "Nauka", Moscow, 1979.

[5] I. M. Gelfand, S. V. Fomin, *Geodesic flows on manifolds of constant negative curvature*, Uspekhi mat. nauk, **47,1** (1952,), 118–137.

[6] I. P. Cornfeld, Ya. G. Sinai, S. V. Fomin, *Ergodic theory*, Springer Verlag, Berlin, 1982.

[7] L. D. Meshalkin, *A case of isomorphic Bernoulli schemes*, Dokl. Akad. Nauk SSSR **128,1** (1959), 41–44.

[8] Z. Nitecky, *Differential dynamics*, MIT Press, 1971.

[9] O. Ornstein, *Ergodic theory, randomness and dynamical systems*, Yale Univ. Press.

[10] V. A. Rokhlin, *On the endomorphism of commutative compact groups*, Izvestia Akad Nauk SSSR, **13** (1949,), 323–340.

[11] Ya. G. Sinai, *Dynamical systems with Lebesgue spectrum of countable multiplicities*, Izvestia Akad Nauk SSSR, **25,6** (1961), 899–924.

[12] Ya. G. Sinai, *Classical dynamical systems with Lebesgue spectrum of countable multiplicity*, Izvestia Akad. Nauk SSSR, ser. mat., **30,1** (1967), 15-68.

[13] Ya. G. Sinai, *Dynamical systems with elastic reflection. Ergodic properties of scattering billirds*, Uspekhi Mat. Nauk, **25,2** (1970), 141–192.

[14] Ya. G. Sinai, *Gibbsian measures in ergodic theory*, Uspekhi Mat. Nauk, **27,4** (1972), 21–64.

[15] Ya. G. Sinai, *Stochasticity of dynamical systems*, Non-linear waves, Nauka, Moscow, 1979, pp. 192–211.

[16] Ya. G. Sinai, *The randomness of the non-random*, Priroda **3** (1981), 72–80.

[17] S. Smale, *Differentiable dynamical systems*, Bull. Amer. Math. Soc. **73** (1967), 747–817.

[18] P. Halmos, *Lectures on ergodic theory*, Math. Soc. of Japan, 1959.

[19] L. A. Bunimovich, *On the ergodic properties of nowhere dispersing billiards*, Communs Math. Phys., **65,3** (1979), 295–312.

[20] L. A. Bunimovich, Ya. G. Sinai, *Statistical properties of Lorentz gas with periodic configuration of scatterers*, Communs Math. Phys., **78,4** (1981), 479–497.

[21] P. Halmos, J. von Neumann, *Operator methods in classical mechanics. II*, Ann. Math., **43** (1942), 332–338.

[22] S. A. Kalikow, T, T^{-1} *transformation is not loosely Bernoulli*, Ann. Math. Ser. 2, **115,2** (1982), 393–409.

[23] J. von Neumann, *Zur Operatoren methode in der klassichen Mechanik*, Ann. Math., **33** (1932), 587–642.

KOLMOGOROV'S ALGORITHMS OR MACHINES
(TO THE PAPERS 1, 6)

(V. A. USPENSKI, A. L. SEMYONOV)

I. The general approach to the notion of algorithm: aims and ideas.

At the beginning of the year of 1951, the 5th year student of the mechanics and mathematics department of Moscow State University V. A. Uspensky was given a page of typewritten text with the description of a certain abstract machine by his scientific advisor A. N. Kolmogorov. The states of this machine were one-dimensional topological complexes or, more precisely, complexes with certain additional information. This additional information consisted in the vertices of the complexes being supplied with functions with values in a preliminarily chosen finite set so that the complex turned out to be labeled (by values of this function). The formulation proposed by A. N. Kolmogorov had the following two goals:

1) to give a mathematical definition of the notion of algorithm that would be as general as possible (so general that it would directly include all algorithms);

2) to verify that even this more general definition does not lead to an extension of the class of computable functions which was already clearly understood by this time on the basis of previous narrower definitions (due to Turing, Post, Markov, and others) and which, if one limits oneself to functions, coincides with the class of partially recursive functions.

The elaboration of A. N. Kolmogorov's formulation was the topic of a master's thesis of V. A. Uspenski called "The general definition of algorithmic computability and algorithmic reducibility" written in the first half of 1952.

Later A. N. Kolmogorov gave an exposition of his ideas in the report, "On the notion of algorithm" on March 17, 1953 at a session of the Moscow Mathematical Society; the publication [3] (No.1 of the present edition) is a resumé of this report.

Among the ideas put forward in this report, the following three are most fundamental:

1) the indication of the general structure of an arbitrary algorithmic process; see §§1, 2 and 3 in [3];

2) a specification of the notion of constructive object – the state of an algorithmic process;

3) an indication of the "local character" of an algorithmic process; see §4 in [3].

The general outline of the structure of an arbitrary algorithm proposed by A. N. Kolmogorov was the basis of understanding the seven parameters of an algorithm as explained in the entry "Algorithm" in the "Great Soviet Encyclopaedia" (3rd edition) and the "Mathematical Encyclopaedias" [8, 9]. As to the so-called constructive objects, they are precisely the objects with which an algorithm works. Such are, for example, finite chains of distinguishable symbols, matrices with integer entries, etc., as differing from, say, real numbers or infinite sets, which are not constructive objects. Any general approach to the notion of algorithm necessarily must include a general approach to the notion of constructive object. The point of view presented by A. N. Kolmogorov was that a constructive object (A. N. Kolmogorov spoke of "states" of an algorithmic process) is a certain finite number of elements related to each other by certain connections; these elements and connections may be of several different types and the number of types of elements as well as the number of types of connections is bounded from the outset (each specific bound of this type leads to its own type of constructive object).

Finally, the "local character" means that the actual transformation, i.e. the transformation which is carried out at each separate step, is local; this presupposes the choice in each constructive object of a bounded active part fixed in advance (the expression "in advance" means that the volume of this part depends only on the algorithm under consideration and not on the state which arises in the process of its functioning).

The results of a more detailed analysis of the situation were developed in an article by A. N. Kolmogorov and V. A. Uspenski [5]*.

II. Relationship with recursive functions.

The second of the goals mentioned in the beginning of the present commentary is discussed in the paper [5] in §3.

Let us describe the approach used there. First of all we must make the formulation of the problem more precise. Indeed, since the arguments and the results of Kolmogorov algorithms are not necessarily numbers and may be entirely different constructive objects, it is at first unclear how one may compare, say, the class of functions computable by Kolmogorov algorithms and the class of partially recursive functions whose arguments and values can only be natural numbers by definition. Of course this difficulty arises not only in connection with Kolmogorov algorithms, but also in the introduction of other computational models. Let us consider for example the family of all Turing machines whose input and output alphabet is $\{a, b, c, d\}$ and try to explain what is meant when we say that the class of functions computable on these machines is no wider than the class of partially recursive functions. To do this let us fix some natural (and therefore certainly computable in both directions) one-to-one correspondence between the set of all words in the alphabet $\{a, b, c, d\}$ and a certain set of natural numbers. This correspondence allows to assign to each function which transforms words in the alphabet $\{a, b, c, d\}$ into words of the same alphabet a certain numerical function in one variable. Suppose it turns out (and in fact it does) that under this correspondence to each function which is computable by Turing machine corresponds a partially recursive function.

---

*A review of this article was published in the review journal Matematika, 1959, No.7, p. 9 (ref. No. 6527). The present edition incorporates remarks by the referee (A. V. Gladky); in particular, it includes corrections to the misprints indicated in the review. A review of this article is also published in the journal Math. Reviews, 1959, vol. 20, No. 9, p. 948 (ref. 5735). An English translation of this article appears in as A. N. Kolmogorov, V. A. Uspensky. *On the definition of an algorithm.* American Mathematical Society Translations, Ser. 2, 1963, vol. 29, p. 217-245. (The translation published in this volume is new.)

Then we can reduce the problem of computing an arbitrary function on a Turing machine to the problem of computing the appropraite partially recursive function. And this means precisely that the class of function computable on Turing machines is no wider than the class of partially recursive functions. The same operation with the replacement of a Turing machine by the Kolmogorov algorithm is carried out in §3 in [5] (of course with the suitable modifications).

Let us note that in order to achieve the second goal under discussion, it is possible to work in a different way. Let us first illustrate this other approach on the example (considered above) of a Turing machine with input and output alphabet $\{a, b, c, d\}$. Let us define the notion of "numerical function of q arguments computable on the given Turing machine" in the following way. Let us fix some natural method that can be used to write down an arbitrary "input" numerical string $\langle m_1, \ldots, m_q \rangle$ on the input tape of some Turing machine and an arbitrary "output" number n on the output tape of this same machine. (Here we do not assume that the input and output tapes are necessarily different from each other and from the working tape, these are simply those tapes on which the input data is written down and eventually the result is formed). Each Turing machine computes the following numerical function f in q arguments: the string $\langle m_1, \ldots, m_q \rangle$ is written on the input tape and the machine is put into action; if the work of the machine ends and the output tape turns out to contain the number n, then this number n is declared to be the value of $f(m_1, \ldots, m_q)$; in all other cases the function f is assumed undefined for the arguments $m_1, \ldots, m_q$. (Without loss of generality we can assume that if the machine stops, then the output tape necessarily contains a certain number). As is well known, all numerical functions which are computable by Turing machines are partially recursive. A similar approach is applied in [5]. Namely, in §2 a method for expressing input numerical strings in the form of "initial states" of Kolmogorov algorithms are indicated, while the "output numbers" are presented in the form of "concluding states"; thus the notion of numerical function computable by Kolmogorov algorithms arises; it is indicated that each partially recursive function is computable by a certain Kolmogorov algorithm; in §3 it is proved that every numerical function computable by some Kolmogorov algorithm is partially recursive.

III. General approach to the notion of constructive object.

In the sequel we shall consider the first of the three goals mentioned at the beginning of this commentary: to find the most general mathematical definition of the notion of algorithm.

To understand Kolmogorov's approach, it is important to agree that each algorithm has to do with constructive objects of a certain fixed type only, for example, with words consisting of letters of a given rigidly fixed alphabet (and not in words in all alphabets). All possible constructive objects of the given type constitute the so-called *ensemble of constructive objects*. An important, although very simple observation due to A. N. Kolmogorov, is that the connections (between the elements constituting the constructive object) may be viewed as elements of new types and thus the entire construction may be reduced to a single type of binary connection and still to a finite (although increased) number of element types. Then constructive objects are naturally represented by graphs, one-dimensional topological complexes whose vertices are labeled by symbols of a certain finite alphabet B while their edges are not labeled. All such complexes with a given B constitute a single ensemble.

The labeling of edges, i.e., writing a certain symbol on each edge, does not lead to an essential extension of the class of objects under consideration (this follows from A. N. Kolmogorov's observation mentioned above). Indeed, a symbol on an edge joining the vertices A and C may be viewed as a vertex of a new type (i.e. a vertex with a special label) joined to the vertices A and C by a simple (unlabeled) edges.

In the general consideration of connections between elements of a constructive object, the arguments of each such connection were not ordered in [5]; however, when we pass to complexes (when the previous elements and connections become vertices and one more new connection – adjacency – is considered), the extremities of edges are not ordered (it is not indicated which of the ends of the edge is the initial one and which is the terminal one), i.e. non-oriented complexes are considered: as discovered in [5], no necessity in orientation actually arises. Nevertheless, it is possible to consider oriented complexes in which the beginning and the end of each edge are indicated as well. Then each vertex has two degrees, the incoming and the outgoing degree (which are respectively the number of incoming and outgoing edges from this vertex). If all these degrees are bounded by a certain number k, the family of oriented Kolmogorov complexes which then arises can naturally be included in an appropriate ensemble of non-oriented complexes. The situation changes if we assume bounded only the outgoing degree, while the incoming degree can be as large as we wish. Then a new type of constructive objects arises. In objects of this type one and the same element may participate, in the role of the extremity of an edge, in an arbitrarily large number of connections. Can we consider that for a fixed labeling alphabet and fixed bounds on the outgoing degree all the objects of the new type will belong to the same ensemble? Any answer to this quesiton is debatable; apparently our intuition of a constructive object and an ensemble of constructive objects is not sufficiently clear. If we accept a positive answer to the given question and assume that oriented Kolmogorov complexes with fixed alphabet and fixed bounds for outgoing degrees (but without bound on the incoming degrees) constitute an ensemble, then we can consider Kolmogorov algorithms working with elements of this ensemble. Such algorithms are called "oriented Kolmogorov algorithms" (or machines) to distinguish them from "non-oriented Kolmogorov algorithms" (or machines), which work with non-oriented complexes.

IV. Kolmogorov complexes.

We shall now present in parallel the definition of two types of constructive objects – *oriented* and *non-oriented Kolmogorov complexes*. And then in the next section V we shall define the corresponding algorithms. These definitions are taken from the reviews [13] and [14].

1. A Kolmogorov complex is an *oriented* or *non-oriented graph*; such a complex is called *oriented* or *non-oriented* respectively.

2. A Kolmogorov complex is an *initialized* graph, i.e., precisely one vertex (the *initial vertex*) is distinguished in it.

3. A Kolmogorov complex is a *connected* graph, i.e., it is possible to move from

one vertex to another along an (oriented for oriented complexes) path from the initial vertex.

4. A Kolmogorov complex is a *labeled graph*, i.e., each of its vertices is labeled by a letter from a fixed alphabet.

5. For each vertex a all the vertices b such that $\langle a, b \rangle$ is an (oriented for oriented complexes) edge are labeled by different letters.

The requirement that each complex be connected is the main difference between the definition just formulated above and that from the paper [5] which also admits non-connected complexes both in the role of input or output data as in the role of "intermediate results" or states of the algorithmic process. However, in [5] the end result, by the construction of the algorithm itself, will turn out to be a connected complex only.

Let us explain the origin of the fifth condition (corresponding to property a from item 3 of the definition of an algorithm in §2 of the paper [5]). It is natural to assume that in the processing of a constructive object, we only look at a bounded part of it located close enough to the initial vertex. At the same time we would like to have a method to reach any given vertex of the complex. This is possible only if an arbitrary vertex is *uniquely* determined by the sequence of labels on some path leading to it from the initial vertex. If we move precisely along this path, we can reach this vertex. Let us note that unlike [5] we do not require that the label of the initial vertex differ from labels of other vertices of the complex and only assume that the initial vertex is distinguished in some other way.

Thus, the objects being processed or "states" (in particular, the input data and the results) of Kolmogorov algorithms from the paper [5] and the reviews [13, 14] differ. Let us consider, however, only functions for which the arguments and the values are connected non-oriented complexes labeled in accordance to the conditions on labeling of initial vertices of input data and results as described in the paper [5]. It can be shown that for such functions the notion of computability by Kolmogorov algorithms from [5] and by non-oriented Kolmogorov machines from [13] coincides.

If the vertices of a Kolmogorov complex are labeled by letters of a finite alphabet B, we call this complex a *Kolmogorov complex over B* or *B-complex*. An oriented (respectively non-oriented) B-complex is said to be a *(B, k)-complex* if all the incoming degrees (respectively all the degrees) of its vertices are bounded by the number k.

The ensemble of oriented (respectively *non-oriented*) *(B, k)-complexes* is the set of all oriented (respectively non-oriented) (B, k)-complexes for the given finite alphabet B and the natural number k.

In connection with the general definition of a constructive object given above, the problem of realizing Kolmogorov complexes in real physical space naturally arises. This problem was considered in [4] in the following formulation. Let us fix a certain ensemble of non-oriented Kolmogorov complexes. Assume now that the vertices of the complexes are physically realized as balls of positive diameter (the same diameter for all the balls), while the edges are flexible pipes of positive diameter (the same for all types). Then the diameter of the pipes and the balls may be chosen so that for certain positive real numbers c and d we have the following: 1) any complex with n vertices may be realized inside a sphere of radius $\sqrt{n}$; 2) the volume of any realization of almost any complex with vertices is less than $dn\sqrt{n}$

("almost any" means that the ratio of the number of complexes with this property to the number of all complexes with n vertices tends to one as n tends to infinity).

V. Oriented and non-oriented Kolmogorov machines.

The definition of oriented and non-oriented Kolmogorov machines will be given simultaneously. In the case of non-oriented machines, our definition essentially coincides with the definition of algorithm in the paper [5]. The main difference consists in that in [5] the set of complexes obtained as the result of the algorithm's work does not intersect the set of complexes given at the input end of the algorithm. Indeed, in [5] the "input" complexes have a specific label, the "input label" of the initial vertex, while the resulting complexes have a different label, the "resulting label". In order that the result of the work of one algorithm be the input end of another, it is necessary to change the label of the initial vertex. The definition proposed below makes such a change unnecessary (recall that in our case the initial vertex does not carry any special "initial" label). From considerations of simplicity of the definition, we have also slightly modified the general scheme of description of algorithms proposed in [3] (and used in [5]). Namely, we require that after the signal that says the solution is found (the termination signal) appears, exactly one more application of the operator of direct processing is made. Finally, as we have already mentioned, the algorithms in [5], unlike those defined below, may be applied to non-connected complexes; the fact that the resulting complex is actually connected is guaranteed in [5], because the connected component of the initial vertex is extracted from the concluding state appearing at the last step of the algorithm and it is this component which is declared to be the "solution" (resulting output) of the algorithm. The definition given below involves distinguishing the connected component of the initial vertex at each step of the algorithm's work (so that all the intermediate states are Kolmogorov complexes). Let us note in this connection that in [5] the immediate processing was concerned at each step only with the connected component of the initial vertex (because it is precisely in this component that the active part of the state is contained).

A *Kolmogorov machine* (oriented as well as non-oriented) may now be described in the following way. (The meaning of the terms "state" and "active part" are the same as in [3]). The work of the machine Γ consists of a sequence of steps during which the *operator of immediate processing* Ω_Γ is applied to the object processed at each step (i.e. to the state). At the initial moment of time the processed object is the input data. Before each step is carried out, it is checked whether the termination signal ("solution signal") has appeared or not. As soon as such a signal appears, the operator Ω_Γ is applied once more, after that the work of the machine is considered ended and the state that has appeared is declared to be the result of the machine's work ("the solution").

Now let us pass to the definition of what we call the state of the algorithmic process, to the description of the form of the operator of immediate processing Ω_Γ and of the way in which the termination signal is formed.

The *state* of the algorithmic process is a Kolmogorov complex from a certain ensemble of (B, k)-complexes.

The *active part* of the state is a subcomplex whose vertices are accessible from the initial vertex by means of (oriented) paths of length no greater than a fixed number given for each machine.

Denote the set of all vertices of the complex G by $v(G)$. The operator Ω_Γ is defined by a finite number of commands of the form $U \to \langle W, \gamma, \delta \rangle$, where U, W are complexes; γ is a map from $v(U)$ to $v(W)$ preserving the labels; δ is a one-to-one map from $v(U)$ into $v(W)$, where for each $x \in v(U)$ the set of labels of terminal vertices of edges from $\delta(x)$ is included in the set of labels of terminal vertices of edges from x. In the case of non-oriented complexes, γ must coincide with δ. All the commands have different left-hand sides. In order to obtain a new state $S^* = \Omega_\Gamma(S)$, we must:

1) find the command, say $U \to \langle W, \gamma, \delta \rangle$, whose left-hand side coincides with the active part of the state S; after that

2) remove this active part U and

3) replace it by the complex W; the maps γ, δ are used in order to connect the vertices $v(S) \setminus v(U)$ with the vertices $v(W)$ in the following way: for all $a \in v(U), v(S) \setminus v(U)$, if $\langle a, b \rangle$ is an edge of S and $\gamma(a)$ is defined, then $\langle b, \gamma(a) \rangle$ will be an edge of S^*; if $\langle a, b \rangle$ is an edge of S and $\delta(a)$ is defined, then $\langle \delta(a), b \rangle$ will be an edge of S^*; after these connections have been established, we must

4) remove all the vertices from $v(S)$ that are not accessible from the initial vertex, as well as all the edges that are incident to these vertices. Thus the new state S^* will also be a connected graph.

The termination signal appears if the active part turns out to belong to a given finite set of complexes. In this case, as indicated above, the operator Ω_Γ is applied only once more. In more detail, if the active part of the state S belongs to the distinguished finite set, the state $S^* = \Omega_\Gamma(S)$ is declared to be the result of the work of the algorithm Γ and as soon as S^* appears, the algorithmic process terminates.

An exposition of the definition of non-oriented Kolmogorov algorithms appears in §3 of Chapter 1 of A. M. Glushkov's monograph [1]. On p. 34 the author writes: "Careful analysis of the description of the Kolmogorov-Uspensky algorithmic scheme shows that in its form this scheme is to a great extent similar to the work actually carried out by a human, when he processes information obtained from outside in accordance to certain rules of an algorithm that he has memorized. The authors of this scheme have undertaken special measures in order not to lose the generality in the character of the transformations being performed."

VI. Schönhage machines (also known as Kolmogorov - Schönhage algorithms).

On 1970 the West-German mathematician A. Schönhage proposed a computational model that he called a "storage modification machine" (abbreviated SMM) (see [11]). In the introduction to the article [12] Schönhage writes: "As it has been pointed out to me by many colleagues, Kolmogorov and Uspensky have introduced a machine model very close to storage modification machines considerably earlier (in 1958, see [5])". Schönhage machines may be viewed as a special kind of oriented Kolmogorov machine and therefore in the review article [7] these machines are called Kolmogorov-Schönhage algorithms (see Chapter 3 §1), while non-oriented Kolmogorov machines in [7] go to under the name of Kolmogorov-Uspensky algorithms. Possibly it would be more correct to describe oriented Kolmogorov machines (and therefore Kolmogorov-Schönhage algorithms) as machines and algorithms with semi-local information processing: indeed, in carrying out one step of its work, the algorithm may produce a new vertex that can be the extremity of a number of edges, this number being as large as we wish.

The definition of a storage modification machine given in [12] (a Schönhage machine, or in the terminology of the review [7], a Kolmogorov-Schönhage algorithm) is reproduced on pages 130-132 of the description [4] as an appendix to §2 of part I. Briefly, the main difference between this computational model and oriented Kolmogorov machines consists in the following: 1) the edges rather than the vertices are labeled (the unimportance of this circumstance has been pointed out by us at the beginning of section III); 2) commands are a special type of oriented Kolmogorov machine commands; 3) a certain controlling structure involving transfer operators is given on commands; 4) there are two tapes (the input and the output tapes) in which words in the alphabet $\{0, 1\}$ are written, and immediately after the input word, a special symbol is written indicating its end.

The main results related to Kolmogorov-Schönhage algorithms are described in detail in [12].

VII. Modelling in real time.

We noted above that the definition of algorithm proposed by A. N. Kolmogorov was devised to include all possible types of algorithms. And indeed, Kolmogorov machines directly model the work of all other types of algorithms with local information processes known to us. For algorithms with non-local information processing, e.g. such as the normal algorithms of A. A. Markov or machines with arbitrary access to memory [10], one requires the preliminary decomposition of their steps into local steps. The thesis claimimg that direct modeling of algorithms with local transformation of information by Kolmogorov machines is possible may at first glance raise certain objections. Indeed, the existence of a Kolmogorov machine directly modeling a Turing machine with plane memory (a plane tape) is not self-evident and requires certain delicate contructions. Such is the case, for example, in the construction of Kolmogorov machines that solve the problem of "finding self-intersections of plane trajectory" (see [6]); the existence of Turing machines with plane tape solving this problem is obvious. The thing here, however, is that actually the input of the Turing machine under consideration with plane tape is not only the argument, but the required part of the working memory, i.e. part of the plane split up into cells in the standard way. This splitting, and the neighbour-hood relation that arises as its result, may be used by the machine in its work. If the input of a Kolmogorov machine involves, besides the argument, the same kind of part of the plane divided into cells, then the modeling of a Turing machine by a Kolmogorov machine becomes obvious. Nevertheless, even without inputting a part of the plane into the Kolmogorov algorithm, it turns out to be possible, in a certain sense specified below, to make the Kolmogorov machine capable of "adding" certain necessary parts to its the memory while the computations are being carried out; the details of how this can be done appear in [12].

Before stating the corresponding theorem from [12], let us give some additional definitions. Together with the Schönhage machine, we shall consider the Turing machine with multidimensional memory, to be described below. We assume that each such machine has an input and output tape and that the motion of the reading element of the machine along those tapes is in one direction, the input tape only being used for reading, and the output tape only for writing. The working memory may contain arbitrary but fixed number of arrays of fixed dimension (tapes, planes, etc.) and the reading element can move along these arrays by one step into any cell in any direction. For any computation by such a Turing machine, as well as for any computation by a Schönhage machine, we distinguish the moments of time when the reading element moves along the input tape and the moments of printing (of the next symbol on the output tape). Now suppose $\mathfrak{A}$ and $\mathfrak{B}$ are two machines (with input and output alphabet $\{0,1\}$) each of which can be a Turing machine with multidimensional memory or a Schönhage machine. In accordance to [12], we shall say that the machine $\mathfrak{A}$ models the machine $\mathfrak{B}$ in real time if the following conditions are satisfied. For any input data x on which $\mathfrak{B}$ ends its work, suppose $t_0 < t_1 < \cdots < t_p$ is a sequence of moments of time such that $t_0 = 0$, t_p is the moment when the machine stops working while $t_1, \ldots, t_{p-1}$ are all moments of time at which the machine $\mathfrak{B}$ carries out a motion of the input tape or writes something on the output tape. Then there must exist moments of time $\tau_0 < \tau_1 < \cdots < \tau_p$ such that $\tau_0 = 0$, τ_p is the moment when the machine $\mathfrak{A}$ stops working for the argument x for any $i = 1, \ldots, p$;

1) in the computation with input data x, the part of this x read by the machine $\mathfrak{A}$ at the moment τ_i coincides with the part of x, read by the machine $\mathfrak{B}$ at the moment t_i;

2) in the computation with input data x, part of the result printed by the machine $\mathfrak{A}$ at the moment τ_i coincides with part of the result printed by the machine $\mathfrak{B}$ at the moment t_i;

3) for some c depending on $\mathfrak{A}$ and $\mathfrak{B}$, but not on x, we have

$$\tau_i - \tau_{i-1} \leqslant c(t_i - t_{i-1}).$$

Under these conditions it is obvious that the machine $\mathfrak{A}$ computes the same function as the machine $\mathfrak{B}$ or a certain extension of this function; and the computation time for the machine $\mathfrak{A}$ is no greater than the computation time for the machine $\mathfrak{B}$ multiplied by the constant c.

In the theorem from [12] mentioned above, the possibility of modeling any Turing machine with multidimensional memory by an appropriate Schönhage machine in real time has been established. As pointed out in [7], a similar result about the modeling of a Turing machine with multidimensional memory by Kolmogorov-Uspensky algorithms may be obtained . The results of the paper [2] imply the impossibility of converse modeling.

REFERENCES

[1] V. M. Glushkov, *Introduction to cybernetics*, Academy of Sciences of the Ukrainian SSR publications, Kiev, 1964, pp. 324.

[2] D. Yu. Grigoryev, *Kolmogorov algorithms are stronger than Turing machines*, Zap. Nauch. Seminarov LOMI, **60** (1976), 29–37.

[3] A. N. Kolmogorov, *On the notion of algorithm*, Uspekhi Mat Nauk 8,4 (1953), 175–176.

[4] A. N. Kolmogorov, Ya. M. Barzdin, *On the realization of nets in three-dimensional space*, Problemy Kibernetiki, **19** (1967), 261–268.

[5] A. N. Kolmogorov, V. A. Uspensky, *To the definition of algorithm*, Uspekhi Mat. Nauk, **13,4**, (1958), 3–28.

[6] M. V. Kubinets, *Recognizing self-intersections of a plane trajectory by Kolmogorov algorithms*, Zap. Nauch. Seminarov LOMI, **32** (1972), 35–44.

[7] A. O. Slisenko, *Complexity of problems in the theory of computations*, Uspekhi Mat. Nauk, **36,6**, (1981), 21–103.

[8] V. A. Uspensky, *Algorithm*, Great Soviet Encyclopaedia, 3rd Edition,, vol. 1, 1970, pp. 400–401.

[9] V. A. Uspensky, *Algorithm*, Mathematical Encyclopaedia, vol. 1, 1977, pp. 202–206.

[10] Aho A. V., Hopcroft J. E., Ullman J. D., *The design and analysis of computer algorithms*, Reading (Mass.) etc.: Addison-Wesley Publ. Co. (1974), X + 470.

[11] A. Schönhage, *Universelle Turing Speicherung*, In: Automatentheorie und formale Sprachen/ Eds J. Dorr, G. Hotz. Mannheim, (1970), 369–383.

[12] A. Schönhage, *Storage modification machines*, Soc. Industr. and Appl. Math. J. Comput., **9,3** (1980), 490–508.

[13] V. A. Uspensky, A. L. Semenov, *What are the gains of the theory of algorithms: basic developments connected with the concept of algorithm and with its application in mathematics*, In: Algorithms in modern mathematics and computer science: Proc. Conf. Urgench (UzSSR). Berlin etc.: Springer-Verl., (1981), 100–234 (Lect. Notes in Comput. Sci.; Vol. 122).

[14] V. A. Uspensky, A. L. Semenov, *Theory of algorithms: its main discoveries and application*, Algorithms in modern mathematics and its applications, vol. 1, Computational Center of the Siberian Section of the Academy of Sciences of the USSR,, Novosibirsk, 1982, pp. 99–342.

FROM A.N. KOLMOGOROV'S RECOLLECTIONS

Courses in the theory of analytic functions that year were read by two lecturers: Boleslav Kornelievich Mlodzievsky and Nikolai Nikolayevich Luzin.

Nikolai Nikolayevich's lectures were attended in principle by all members of Luzitania, including students from the upper courses and even associate professors. My first achievement, after which a certain attention was paid to my person, is related to Luzin's course.

Nikolai Nikolayevich liked to improvise during his lectures. During the lecture devoted to the proof of Cauchy's theorem, it came to his mind to make use of the following lemma. Suppose a square is divided into a finite number of smaller squares. Then for any constant C there exists a number C' such that for any curve of length greater than C the sum of perimeters of the smaller squares that the curve touches is no greater than C'. Two weeks later I submitted a small manuscript[1] to the president of the student mathematics circle Semyon Samsonovich Kovner; this manuscript showed that Luzin's lemma was incorrect.

APPENDIX 1

REPORT TO THE MATHEMATICAL CIRCLE ABOUT SQUARE PAVINGS

The question I shall talk about arose in the theory of functions of a complex variable but is essentially purely geometrical and I shall treat it as such.

1.

At one of his lectures N.N. Luzin proved the following theorem. A continuous rectifiable curve is given in the plane. No matter how we divide the plane into equal squares, the sum of the perimeters of all the squares occupied by (i.e. intersected or touched by) the curve will be less than a certain constant value K as long as the sides of the squares are less than a certain $\mathcal{E}$.

Proof. Suppose ε is the side of the subdividing squares, s is the length of the curve. Let us divide s into equal parts less than ε; to do this it suffices to take the number of parts $P > s/\varepsilon$ and it is always possible to have $P \leqslant s/\varepsilon + 1$. Each segment of the curve will occupy no more than four squares since if it occupied five, then among them there would be two non-adjacent ones and therefore the number of squares

---

[1]This manuscript has not been lost (see Appendix 1). It is dated April 4, 1921. We present it to the attention of the reader. *Editor's note.*

would be $N \leqslant 4P \leqslant 4(s/\varepsilon + 1)$. The perimeter of each square is 4ε and therefore the sum of the perimeters is $u = 4\varepsilon N \leqslant 4\varepsilon 4(s/\varepsilon + 1) = 16(s + \varepsilon)$ so that, since $\varepsilon < \mathcal{E}$, we have $u < 16(s + \varepsilon) = K$.

2.

To prove the opposite statement, I shall need the following theorem. We are given a continuous rectifiable or non-rectifiable curve $x = \varphi(t)$, $y = \psi(t)$ where t varies from a to l.

No matter how we subdivide the curve into a finite number of parts, the distance between each two points of non-successive parts will always be more than a certain constant α.

Proof. Assume the converse. Then to each positive number z corresponds a set T_z of values of τ such that the corresponding points lie at a distance less than z from points of non-adjacent parts. Obviously, if $z_1 > z_2$, the set T_{z_2} is contained in the set T_{z_1}. Subdivide the interval (a, b) in half and take as the first half the one that contains elements of all the sets T; such a half exists because if one of the halves does not contain elements of T_α, while the other does not contain elements of T_β, then both would not have any elements with indices less than α and β. Divide this half again into equal parts and choose one of them which contains elements of all the sets. Continuing this process, we obtain a sequence of nested infinitely decreasing segments. All of them contain and have a common unique point τ_ω. From the point on the curve corresponding to τ_ω, there are points as near as we like, which themselves are as near as we like, to points of non-adjacent parts, and therefore, at a distance as small as we wish from that point, there are points of non-adjacent parts or at least, if it is the common boundary of two parts, points from the opposite extremities of these two parts. Now consider the set of points T_z' located at a distance less than z from the point corresponding to τ_ω and lying on non-adjacent parts or at opposite extremities of such parts. Since for $z_1 > z_2$ the set T_{z_2}' is contained in T_{z_1}', we can find a value τ_ω' such that the corresponding point lies at a distance as small as we wish from elements of all the T'. This point cannot lie at a finite distance from the point corresponding to τ_ω and therefore coincides with this point; this contradicts the condition of the absence of double points.

3.

For a non-rectifiable curve without multiple points the sum of perimeters of squares occupied by the curve increases unboundedly when the sides of the squares decrease.

Let us show that it can be made larger than any given value K. Let us find, which is always possible, a polygonal line inscribed in the curve of length greater than K. Let us choose from it all the rectilinear segments, only even or odd, whose sum is greater than $K/2$. Subdivide the plane into squares so small that their diagonals are less than the minimal distance between the points of the parts of the curves corresponding to non-adjacent segments of the polygonal line that always exists according to the previous theorem. The sum of perimeters of the squares given by that part of the curve will be greater than the doubled length of the corresponding segment. The distinguished non-adjacent parts cannot pass through one and the same square and therefore the sum of perimeters of the squares given by all of them is greater than the total doubled length or greater than K.

4.

For the theory of functions of a complex variable the following extension of the first theorem would be important. We are given a continuous rectifiable curve. The plane is subdivided first into equal squares and then some of these squares are subdivided further into four equal square parts. After each subdivision the sum of the perimeters of the squares touched by the curve will be less than a constant K, as long as the original subdivision into equal squares was carried out sufficiently far. Without any harm for its application, we can change this formulation, taking into account only those squares that actually intersect the curve and and not those that merely touch it.

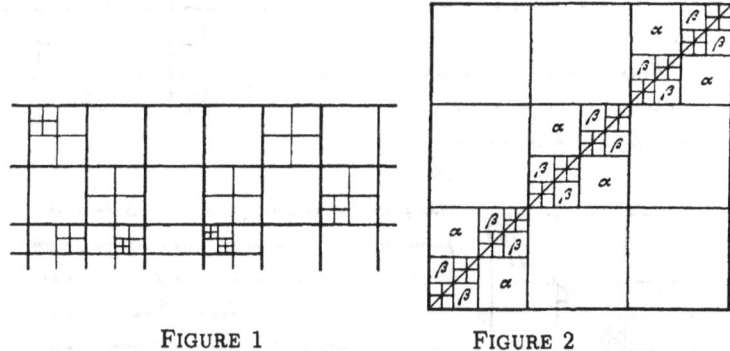

FIGURE 1   FIGURE 2

Considering two examples, I tried to show that this statement is false both in the original as well as in the modified version.

A. Instead of the curve, let us take the diagonal of the square with side M. Subdivide into smaller equal squares. No matter how small these squares are, we can subdivide them in such a way that the sum of the perimeters of the squares occupied by the diagonal will be as large as we wish. To do that, subdivide the diagonal squares. The squares α will be left and changed while the sum of their perimeters is equal to $4M$. With the new diagonal squares we repeat the same procedure. The squares β are left unchanged. The sum of their perimeters is also $4M$. Continuing this subdivision far enough, we obtain a common sum of perimeters which is as large as we wish.

B. To show the incorrectness of the modified statement, let us take the curve obtained from the straight line of the previous example by replacing the segment about the diagonal 1/10 of its length by a half circle; segments about points located at 1/4 and 3/4 of its length by length 1/100, about 1/8, 5/8, 7/8 by length 1/1000 etc., since they do not overlap. The total length of the replaced segments is

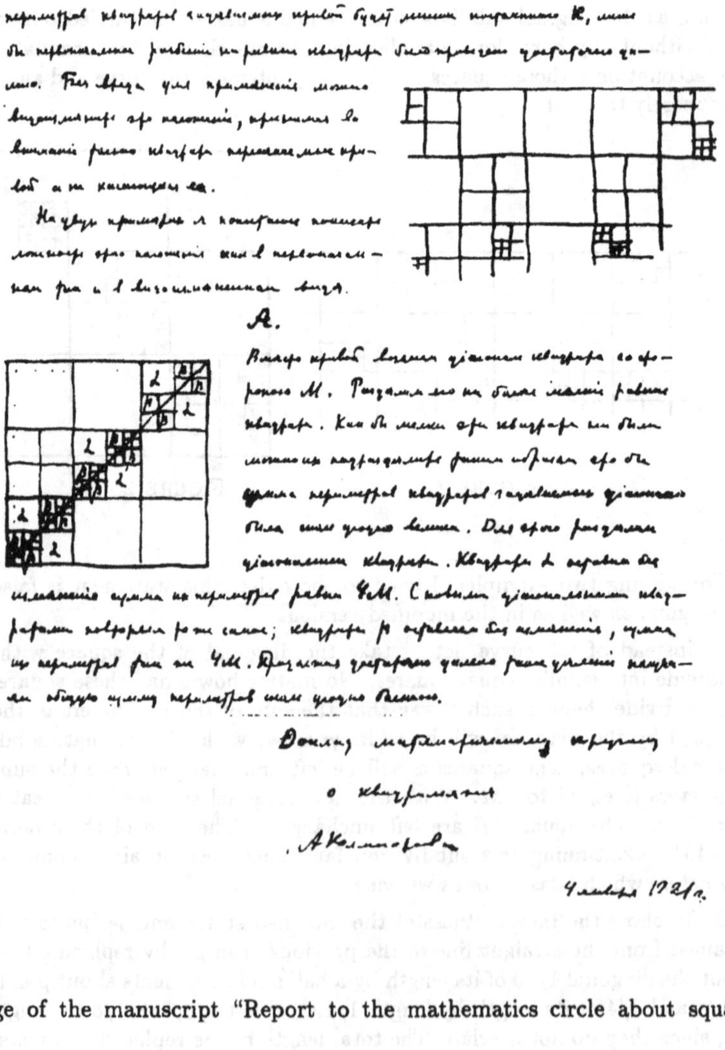

[handwritten manuscript text in Russian cursive, largely illegible]

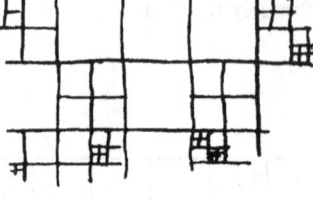

А.

[handwritten manuscript text in Russian cursive, largely illegible]

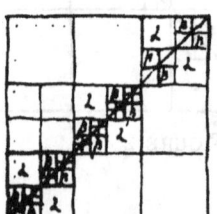

[handwritten manuscript text in Russian cursive, largely illegible]

[signature] А Колмогоров

4 января 1921 г.

A page of the manuscript "Report to the mathematics circle about square pavings"

$$\frac{1}{10} + \frac{2}{100} + \frac{4}{1000} + \cdots = \frac{1/10}{1 - 2/10} = \frac{1}{8}.$$

Let us perform the subdivision of the squares in the same way, except that in the first subdivision into equal squares each side of the larger square is subdivided into a number of parts which is a power of two. Half of the squares α, β, etc. intersect the curve, since they are located opposite arcs of radius greater than their diagonals, but there are no more than 1/8 of these in each row.

January 4, 1921

ON OPERATIONS ON SETS. II

From the author

The first part of my paper on operations on sets, written in 1921-1922, was published in 1928 in volume 35 of "Mathem. Sbornik" (see the paper No.13 in the first book of my collected works "Mathematics and Mechanics". Moscow; Nauka, 1985). The second part, which remained in manuscript form, was accessible to several researchers working in the descriptive theory of sets and developing the theory of R-operations contained in this second part. Further we publish, with small editorial modifications, the second part on the basis of the original manuscript, which was found when the first book of my collected works had already been published.

As pointed out in the text of the article No.13 mentioned above (on page 86), the origins of the theory were independently developed by F. Hausdorff and an exposition of them appears in the second German edition of his "Mengenlehre" (1927). In the Russian edition (1937) of Hausdorff's book, I actually wrote the corresponding paragraph. In view of the absence of a general definition of the operation of taking complements, Hausdorff's theory is less complete. The prototype of my general definition of the complement operation was the definition of a Γ-operation given by P. S. Alexandrov.

Further we publish the text of the second part of my paper on operations on sets[1]. Since the first part of this work ended with section V, the second begins with section VI.

VI

Let us recall that all the sets considered in the sequel are contained in the closed interval[2] $[0,1]$ of the real line. The basic class of elementary sets is the family of all sets closed with respect to the interval and contains the "empty" set.

For an operation X on a sequence of sets given by its own system of numerical chains, we define the *operation RX over the system of sets with all possible strings* of the form $(n_1, n_2, \ldots, n_k)$, where $n_1, n_2, \ldots, n_k$ are natural numbers:

$$E = RX\{E_{n_1,n_2,\ldots,n_k}\}_{n_1,n_2,\ldots,n_k}.$$

---

[1] The editors express their gratitude to L. B. Shapiro for preparing this part of A. N. Kolmogorov's work for publication. Editor's note.

[2] In this part of the paper the open interval $(0,1)$ has been replaced by the closed interval $[0,1]$. *Editor's note.*

A chain of strings U^{RX} is any family of strings possessing the following properties:

1. Strings of rank[3] one (λ) belonging to U^{RX} constitute the chain U^X.

2. If $(n_1 n_2 \ldots n_k)$ is contained in U^{RX}, then the indices λ of the strings of the form $(n_1 n_2 \ldots n_k \lambda)$ contained in U^{RX} constitute the chain U^X.

3. If $(n_1 n_2 \ldots n_{k-1} n_k)$ is contained in U^{RX}, then $(n_1 n_2 \ldots n_{k-1})$ is also contained in U^{RX}.

Obviously in the case when X is the sum operation Σ, RX is an A-operation: $R\Sigma = A$.

It is easy to see that in the case $E_{n_1 n_2 \ldots n_k} = E_{n_1}$, we have

$$RX\{E_{n_1 n_2 \ldots n_k}\}_{n_1 n_2 \ldots n_k} = X\{E_n\}_n,$$

i.e. the X-operation is a particular case of the RX-operation.

Theorem 1. *The operation over sets always yields a set.*

This statement is a particular case of Theorem II stated below but it follows more easily from the following formulas whose proof involves no difficulties: if $E = X\{E_n\}_n$, where

$$E_n = RX\{E^n_{m_1 m_2 \ldots m_k}\}_{m_1 m_2 \ldots m_k},$$

then setting

$$\mathfrak{C}_{p_1 p_2 \ldots p_k} = E^{p_1}_{p_1 p_2 \ldots p_k} \text{ and } \mathfrak{C}^*_{p_1} = [0, 1],$$

we obtain

$$E = RX\{\mathfrak{C}^*_{p_1 p_2 \ldots p_k}\}_{p_1 p_2 \ldots p_k}.$$

If we are given a finite (say X and $\bar{X}$) or countable set of operations, then constructing the operation Z according to the method used in IV for the proof of the theorem on the nonemptiness of classes (subsection 2) and then the operation RZ, we obtain an operation which transforms segments into all sets of the domain invariant with respect to the given operations and containing all segments (according to Theorem I).

Theorem II. *Whatever be the operation X, the operation RX is normal.*

Suppose the functions $t = \varphi(k)$, $j = \psi(k)$ and $k = f(i, j)$ which determine a one-to-one correspondence between the natural numbers $k = f(i, j) = f(\varphi(k), \psi(k))$ and pairs of natural numbers (i, j) satisfy the conditions

$$f(i, j) < f(i, (j + 1)), \tag{1}$$

$$f(i, j) < f((i + 1), j), \tag{2}$$

$$f(i, j) > f(1, (i - 1)). \tag{3}$$

---

[3]The number of elements constituting this string is called its rank. The empty string has rank zero and is part of any chain of strings. *Editor's note.*

Such a correspondence can easily be constructed by putting, say $f(i,j) = 2^{i-1}(2j-1)$. Then (1) and (3) imply that

$$f(1,1) = 1, \tag{4}$$

$$f(i,j) \geqslant i, \tag{5}$$

$$f(i,j) \geqslant j. \tag{6}$$

Assume that we are given

$$E = RX\{E_{n_1 n_2 \ldots n_k}\}, \quad E_{n_1 n_2 \ldots n_k} = RX\{E_{n_1 n_2 \ldots n_k}{}^{m_1 m_2 \ldots m_l}\}_{m_1 m_2 \ldots m_l}.$$

Suppose $\varphi(k) = i$. For $i > 1$ put

$$\mathfrak{E}^*_{p_1 p_2 \ldots p_k} = E^{p_f(i,1) p_f(i,2) \ldots p_f(i,j)}_{p_f(1,1) p_f(1,2) \ldots p_f(1,i-1)}.$$

For $i = 1$ put $\mathfrak{E}^*_{p_1 p_2 \ldots p_k} = [0,1]$. Denote by

$$\mathfrak{E}^* = RX\{\mathfrak{E}^*_{p_1 p_2 \ldots p_k}\}_{p_1 p_2 \ldots p_k}.$$

It is sufficient to prove that $\mathfrak{E}^* = E$.

1. Proof of the inclusion $E \subset \mathfrak{E}^*$. Suppose x is contained in E. This means that there exists a chain $\mathfrak{U}^{RX}$ of sets $E_{n_1 n_2 \ldots n_k}$ containing x such that for each of them the chain $\mathfrak{U}^{RX}_{n_1 n_2 \ldots n_k}$ of sets $E^{m_1 m_2 \ldots m_l}_{n_1 n_2 \ldots n_k}$ contains x.

Let us choose all the $\mathfrak{E}^*_{p_1 p_2 \ldots p_k}$ which satisfy the following conditions:

a) if $\varphi(q) = 1$, then the set

$$E_{p_f(1,1) p_f(1,2) \ldots p_f(1,\psi(q))} = E_{p_f(1,1) p_f(1,2) \ldots p_q}$$

is contained in the chain $\mathfrak{U}^{RX}$;

b) if $\varphi(q) > 1$, then the set

$$E^{p_f(\varphi(q),1) p_f(\varphi(q),2) \ldots p_f(\varphi(q),\psi(q))}_{p_f(1,1) p_f(1,2) \ldots p_f(1,\psi(q)-1)} = E^{p_f(\varphi(q),1) p_f(\varphi(q),2) \ldots p_q}_{p_f(1,1) p_f(1,2) \ldots p_f(1,\psi(q)-1)}$$

is contained in the chain $\mathfrak{U}^{RX}_{p_f(1,1) p_f(1,2) \ldots p_f(1,\psi(q)-1)}$;

c) the conditions a) and b) also hold for all $\mathfrak{E}^*_{p_1 p_2 \ldots p_r} < q$.

I claim that the family $\mathfrak{E}$ of all $\mathfrak{E}^*_{p_1 p_2 \ldots p_k}$ satisfying these conditions constitutes the chain $\mathfrak{U}^{RX}_1$. Indeed:

1. By c) if $\mathfrak{E}^*_{p_1 p_2 \ldots p_q}$ is contained in $\mathfrak{U}^{RX}_1$, then $\mathfrak{U}^{RX}_1$ contains $\mathfrak{E}^*_{p_1 p_2 \ldots p_r}$ also if $r < q$.

2. Among the elements of $\mathfrak{E}^*_{p_1}$, $\mathfrak{E}$ contains those for which E_{p_1} is contained in $\mathfrak{U}^{RX}$ (since $\varphi(1) = 1$), i.e. those whose indices constitute the chain U^X.

3. If $\mathfrak{E}^*_{p_1 p_2 \ldots p_q}$ is contained in $\mathfrak{E}$, then among the sets of $\mathfrak{E}^*_{p_1 p_2 \ldots p_{r+1}}$ the family $\mathfrak{E}$ contains those and only those which satisfy conditions a) and b). Here the following cases are possible.

(i) $\varphi(q) = 1$. In this case $\mathfrak{E}^*_{p_1 p_2 \ldots p_f(1,\psi(q+1)-1)}$ is contained in $\mathfrak{E}$ by condition c) hence $E_{p_f(1,1) p_f(1,2) \ldots p_f(1,\psi(q+1)-1)}$ by condition a) is contained in $\mathfrak{U}^{RX}$ since

$$\varphi(f(1,\psi(q+1)-1)) = 1.$$

Among the sets of the form

$$E_{p_{f(1,1)}p_{f(1,2)}\cdots p_{f(1,\psi(q+1)-1)}p_{f(1,\psi(q+1))}} = E_{p_{f(1,1)}p_{f(1,2)}\cdots p_{q+1}}$$

the chain $\mathfrak{U}^{RX}$ will contain only those whose last indices constitute the chain U^X. Hence by condition a) the chain $\mathfrak{U}_1^{RX}$ contains all those $\mathfrak{C}^*_{p_1p_2\ldots p_q p_{q+1}}$ such that their indices p_{q+1} constitute the chain U^X.

(ii) $\varphi(q+1) > 1$, $\psi(q+1) = 1$. In this case $\mathfrak{C}^*_{p_1p_2\ldots p_{f(1,\varphi(q+1)-1)}}$ is contained in $\mathfrak{U}_1^{RX}$ since $f(1, \varphi(q+1) - 1 \leqslant q$. Therefore, $E_{p_{f(1,1)}p_{f(1,2)}\cdots p_{f(1,\varphi(q+1)-1)}}$ is contained in $\mathfrak{U}^{RX}$. But the set $\mathfrak{C}^*_{p_1p_2\ldots p_{q+1}}$ contained in $\mathfrak{U}_1^{RX}$ are by definition those for which the sets

$$E^{p_{f(\varphi(q+1),1)}}_{p_{f(1,1)}p_{f(1,2)}\cdots p_{f(1,\varphi(q+1)-1)}} = E^{p_{q+1}}_{p_{f(1,1)}p_{f(1,2)}\cdots p_{f(1,\varphi(q+1)-1)}}$$

are contained in the chain $\mathfrak{C}^{RX}_{p_{f(1,1)}p_{f(1,2)}\cdots p_{f(1,\varphi(q+1)-1)}}$. Their indices p_{q+1} therefore constitute the chain U^X.

(iii) $\varphi(q+1) > 1$, $\psi(q+1) > 1$. In this case the set

$$\mathfrak{C}^*_{p_1p_2\ldots p_{f(1,\varphi(q+1),\psi(q+1)-1)}} = E^{p_{f(\varphi(q+1),1)}p_{f(\varphi(q+1),2)}\cdots p_{f(\varphi(q+1),\psi(q+1)-1)}}_{p_{f(1,1)}p_{f(1,2)}\cdots p_{f(1,\varphi(q+1)-1)}}$$

is contained in $\mathfrak{U}^{RX}$ since $f(\varphi(q+1), \psi(q+1) - 1) \leqslant q$. Therefore the second set is contained in $\mathfrak{U}^{RX}_{p_{f(1,1)}p_{f(1,2)}\cdots p_{f(1,\varphi(q+1)-1)}}$. The sets $\mathfrak{C}^*_{p_1p_2\ldots p_q,p_{(q+1)}}$ are contained in $\mathfrak{U}_1^{RX}$ if the sets

$$E^{p_{f(\varphi(q+1),1)}p_{f(\varphi(q+1),2)}\cdots p_{f(\varphi(q+1),\psi(q+1)-1)}p_{f(\varphi(q+1),\psi(q+1))}}_{p_{f(1,1)}p_{f(1,2)}\cdots p_{f(1,\varphi(q+1)-1)}} =$$
$$E^{p_{f(\varphi(q+1),1)}p_{f(\varphi(q+1),2)}\cdots p_{f(\varphi(q+1),\psi(q+1)-1)}p_{q+1}}_{p_{f(1,1)}p_{f(1,2)}\cdots p_{f(1,\varphi(q+1)-1)}}$$

are contained in $\mathfrak{U}^{RX}_{p_{f(1,1)}p_{f(1,2)}\cdots p_{f(1,\varphi(q+1)-1)}}$, and this holds for those p_{q+1} which constitute the chain U^X.

Thus $\mathfrak{U}_1^{RX}$ is indeed a chain. All its elements contain x, since they are either equal to the closed interval $[0,1]$ or to a set $E^{m_1m_2\ldots m_l}_{n_1n_2\ldots n_k}$ containing x. Therefore x is contained in $\mathfrak{C}^*$.

2. Proof of the inclusion $\mathfrak{C}^* \subset E$. Now suppose x is contained in $\mathfrak{C}^*$. Then there is a chain $\mathfrak{U}_1^{RX}$ of sets $\mathfrak{C}^*_{p_1p_2\ldots p_k}$ containing x.

A. For each $\mathfrak{C}^*_{p_1p_2\ldots p_k}$ let us choose $\mathfrak{C}^*_{p_1p_2\ldots p_{\cdot}\bar{p}_{k+1}}$, the set with the smallest index p_{k+1} equal to $\bar{p}_{k+1}$ contained in $\mathfrak{U}_1^{RX}$; obviously for each $\mathfrak{C}^*_{p_1p_2\ldots p_k}$, which itself is contained in $\mathfrak{U}_1^{RX}$, such a set can be found.

B. Between sets with the same $\varphi(k) = i$ let us establish the following order relation (B): the set $\mathfrak{C}^*_{p_1p_2\ldots p_{f(i,j)}}$ directly precedes (B) sets of the form

$$\mathfrak{C}^*_{p_1p_2\ldots p_{f(i,j)}\bar{p}_{f(i,j)+1}\bar{p}_{f(i,j)+2}\ldots\bar{p}_{f(i,j+l)-1}p_{f(i,j+l)}},$$

where $p_{f(i,j+l)}$ assumes a chain of values U^X under which the set obtained is contained in the chain $\mathfrak{U}_1^{RX}$, while the $\bar{p}_{f(i,j)+n}$ are well-defined by condition A.

C. For $i = 1$ let us choose all the $\mathfrak{C}^*_{p_1}$ of rank one contained in $\mathfrak{U}^{RX}_1$ and all the sets that precede them (B). The sets $E_{p_{f(1,1)}p_{f(1,2)}\cdots p_{f(1,j)}}$ corresponding to those chosen, as can be easily verified, constitute the chain $\mathfrak{U}^{RX}$.

D. For each distinguished (C) set

$$E_{p_{f(1,1)}p_{f(1,2)}\cdots p_{f(1,j)}}$$

let us choose a set of the form

$$\mathfrak{C}^*_{p_1 p_2\cdots p_{f(1,j)}\bar{p}_{f(1,j)+1}\bar{p}_{f(1,j)+2}\cdots\bar{p}_{f(j+1,1)-1}p_{f(j+1,1)}},$$

where $p_{f(j+1,1)}$ assumes a chain U^X of values for which the set obtained is contained in $\mathfrak{U}^{RX}_1$. Let us choose further all the sets that this last set precedes (B). It is easy to see that they are identical with sets of the chain $\mathfrak{U}^{RX}_{p_{f(1,1)}p_{f(1,2)}\cdots p_{f(1,j)}}$ of the system determining $E_{p_{f(1,1)}p_{f(1,2)}\cdots p_{f(1,j)}}$, which therefore contains x.

Hence E contains x.

VII

If the X-operation is Γ, we shall call the $R\Gamma$-operation an H-operation (in this case we consider the Γ-operation defined by separating chains, see p. 88 from the paper No.13).

In accordance to Theorem I from VI, the Γ-operation may be replaced by an H-operation.

Theorem III. *Any A-operation may be replaced by an H-operation.*

Indeed, the series of numbers $1, 2, 3, \ldots$ can be split into a countable set of non-intersecting Γ-chains, and we shall assume that the k-th chain

$$U^\Gamma = \{(a^k_1 a^k_2 a^k_3 \ldots)\}$$

consists of the numbers a^k_i of strings of the k-th rank. If $E = A\{E_{p_1 p_2 \ldots p_k}\}$, then

$$\mathfrak{C}^*_{n_1 n_2 \ldots n_k} = E_{p_1 p_2 \ldots p_k}$$

for $n_1 = a^{p_1}_{q_1}, n_2 = a^{p_2}_{q_2}, \ldots, n_k = a^{p_k}_{q_k}$ with any $q_1, q_2, \ldots, q_k$. We obtain $E = H\{\mathfrak{C}^*_{n_1 n_2 \ldots n_k}\}$.

Indeed, it is easy to see that the $\mathfrak{C}^*_{n_1 n_2 \ldots n_k}$ equal to $E_{p_1 p_2 \ldots p_k}$ transform each A-chain into the chain $\mathfrak{U}^H$. Conversely, from any chain $\mathfrak{U}^H$ one can choose a set whose elements are equal to the terms of a certain A-chain of the set $E_{p_1 p_2 \ldots p_k}$.

Since the H-operation is normal, this implies that the domain of the H sets is invariant with respect to the A- and Γ-operations and therefore contains the C-domain.

If we consider the Γ-operation not as an operation over a sequence of sets, but as an operation over a system numbered by strings, then the H-operation can be defined as an operation over a system of sets for all possible double strings

$$E = H\left\{ E \begin{vmatrix} n^1_1 & n^2_1 & \ldots & n^1_{k_1} \\ n^2_1 & n^2_2 & \ldots & n^2_{k_2} \\ \ldots & & & \\ n^\nu_1 & n^\nu_2 & \ldots & n^\nu_{k_\nu} \end{vmatrix} \right\},$$

where ν, k_i and n_j^i independently assume all possible values.

The definition of a chain of double strings is similar to the one given in VI if the rank is assumed equal to the number of rows, while the strings directly preceded by the given one are obtained from it by adding one more row to it from below.

Using this notation, we can obtain a convenient formula for the H-operation, but at present I shall preserve the simplified notation, considering Γ- operations over sequences.

VIII

The defining system of an H-set

$$E = H\{\delta^1_{n_1 n_2 \ldots n_k}\}_{n_1 n_2 \ldots n_k}, \quad \varphi = \{\delta^1_{n_1 n_2 \ldots n_k}\}_{n_1 n_2 \ldots n_k},$$

where the δ^1 are segments, as in the case of A-sets, may be replaced by an equivalent "proper" determining system in which $\delta_{n_1 n_2 \ldots n_k} \supset \delta_{n_1 n_2 \ldots n_k n_{k+1}}$.

To do this, it suffices to put

$$\delta_{n_1 n_2 \ldots n_k} = \delta^1_{n_1} \delta^1_{n_1 n_2} \cdots \delta^1_{n_1 n_2 \ldots n_k}.$$

Obviously, $E = H\{\delta^1_{n_1 n_2 \ldots n_k}\}$. Further I shall always assume that the system is proper.

1. Put

$$E_{n_1 n_2 \ldots n_k} = H\{\delta^1_{n_1 n_2 \ldots n_k p_1 p_2 \ldots p_l}\}_{p_1 p_2 \ldots p_l}$$

It is obvious that

$$E = \Gamma\{E_{n_1}\}_{n_1}, \quad E_{n_1 n_2 \ldots n_k} = \Gamma\{E_{n_1 n_2 \ldots n_k n_{k+1}}\}_{n_{k+1}}.$$

Also define

$$\varphi_{n_1 n_2 \ldots n_k} = \{\delta_{n_1 n_2 \ldots n_k p_1 p_2 \ldots p_l}\}_{p_1 p_2 \ldots p_l}.$$

2. Let us define the *index* of the point x with respect to a system of segments determining a H-set the number $\mathrm{Ind}_x\{\delta_{n_1 n_2 \ldots n_k}\}_{n_1 n_2 \ldots n_k} = \mathrm{Ind}_x\varphi$ by the following inductive process:

(i) $\mathrm{Ind}_x\varphi_{n_1 n_2 \ldots n_k} = 0$ if $\bar{\delta}_{n_1 n_2 \ldots n_k}$ contains x.

(ii) If all the systems $\varphi_{n_1 n_2 \ldots n_k}$ with indices less than the given natural number m is defined, then we assume $\mathrm{Ind}_x\varphi_{n_1 n_2 \ldots n_k} = m$, if it is not defined yet and if in any chain U^Γ there is a number λ such that $\mathrm{Ind}_x\varphi_{n_1 n_2 \ldots n_k \lambda} < m$.

In other words: if there is a chain U^A consisting only of such λ.

(iii) Generally if all the systems with indices less than a certain number of the second class[4] α are defined, then we assume $\mathrm{Ind}_x\varphi_{n_1 n_2 \ldots n_k} = \alpha$ if it is not defined yet and if each chain U^Γ contains the number λ such that $\mathrm{Ind}_x\varphi_{n_1 n_2 \ldots n_k \lambda} = \beta < \alpha$.

(iv) To all the points which have not obtained an index of the second class from the previous conditions we agree to assign the index ω_1. Since the system φ is numbered by the empty string, it follows that $\mathrm{Ind}_x\varphi$ is well-defined[5].

---

[4] α is a transfinite number of the second class if $\omega_0 \leqslant \alpha < \omega_1$. *Editor's note.*

[5] By the inductive definition, $\mathrm{Ind}_x\varphi = \alpha < \omega_1$ if and only if for each chain U^Γ there is a number λ such that $\mathrm{Ind}_x\varphi_\lambda < \alpha$. *Editor's note.*

3. Theorem IV. *$\text{Ind}_x \varphi = \omega_1$ if and only if x is contained in $H_\varphi = H\{\delta_{n_1 n_2 \ldots n_k}\}$.*

Proof. Indeed, if

$$\text{Ind}_x \varphi_{n_1 n_2 \ldots n_k} = \omega_1,$$

then there is a chain U^Γ such that $\lambda \in U^\Gamma$ implies $\text{Ind}_x \varphi_{n_1 n_2 \ldots n_k \lambda} = \omega_1$. If this is not so, then in each chain U^Γ there is an element n_{k+1} such that $\text{Ind}_x \varphi_{n_1 n_2 \ldots n_k n_{k+1}} < \omega_1$.

But then by item (iii) of the definition of the index, we should have

$$\text{Ind}_x \varphi_{n_1 n_2 \ldots n_k} < \omega_1.$$

Hence the family of $\delta_{n_1 n_2 \ldots n_k}$ such that $\text{Ind}_x \varphi_{n_1 n_2 \ldots n_k} = \omega_1$ satisfies conditions 1 and 2 (see VI) of the definition of the chain. Therefore, it contains the chain $\mathfrak{U}^H$ whose kernel contains the point x, i.e. $x \in H_\varphi$.

Conversely, let us show that $\text{Ind}_x \varphi < \omega_1$ implies that x is not contained in H_φ.

a) The statement is obvious in the case $\text{Ind}_x \varphi = 1$ (see the definition of the index).

b) Suppose the statement has been proved for all $\beta < \alpha$ and $\text{Ind}_x \varphi = \alpha$. In each chain $\mathfrak{U}^\Gamma$ of sets E_n there is an E_λ such that $\text{Ind}_x \varphi_\lambda = \beta < \alpha$, therefore, x is not contained in E_λ. Hence by the formulas of section VIII, subsection 1, x is not contained in E.

Thus, the complement to an H-set consists of $\mathfrak{N}_1$ sets

$$P_\alpha = \{x|\ \text{Ind}_x \varphi = \alpha\}.$$

IX

Theorem V. *Complements to H-sets, in particular all C-sets, are $\mathfrak{N}_1$-sums of increasing A-sets.*

We have

$$\bar{E} = P_1 + P_2 + \cdots + P_{\omega_0} + \cdots + P_\alpha + \cdots,$$
$$\bar{E}_n = P_1^n + P_2^n + \cdots + P_{\omega_0}^n + \cdots + P_\alpha^n + \cdots$$

Further let us denote

$$Q_\alpha = P_1 + P_2 + \cdots + P_{\omega_0} + \cdots + P_\alpha + \cdots,$$
$$Q_\alpha^n = P_1^n + P_2^n + \cdots + P_{\omega_0}^n + \cdots + P_\alpha^n + \cdots$$

If x is contained in Q_α, then for any chain $\mathfrak{U}^\Gamma$ of sets E_n we can find an E_λ such that $x \in Q_\beta^\lambda$ for some $\beta < \alpha$. Conversely, if this condition is satisfied, then x is contained in Q_α. This implies the formula

$$Q_\alpha = A\{\Sigma\{q_\beta^n \mid \beta < \alpha\}\}_n$$

(see the definition of the complementary operation in section II of the first part, Volume 1 of the Collected Works, No.13, p. 88).

The sets Q_α are A-sets. Indeed:

1. This is true for $\alpha = 1$, since $Q_1 = P_1 = \Gamma\{\bar{\delta}_n\}_n = A\{\bar{\delta}_n\}_n$.

2. If this is true for $\beta < \alpha$, then the Q_β^n are A-sets as well as their countable sums, and by the normality of the A-operation Q_α is also an A-set.

$$\bar{E} = \Sigma\{Q_\alpha | \alpha < \omega_1\}, \quad Q_\alpha \supset Q_\beta \text{ if } \alpha > \beta.$$

The H-sets themselves, as well as the C-sets, are $\mathfrak{N}_1$-products of Γ-sets:

$$E = \Pi\{\bar{Q}_\alpha | \alpha < \omega_1\}.$$

Remark 1. The question of the existence of H-sets whose complements are also be H-sets but are not C-sets remains open. Methods applied to A-sets in this case do not work.

Remark 2. The proofs of Theorems IV and V can easily be generalized to RX- and $R\bar{X}$-sets. The definition of indices then remains unchanged, but instead of Theorem V we obtain

Theorem Y*. *If an X-operation is normal (and therefore so is $\bar{X}$) and contains as a particular case the multiplication operation (therefore $\bar{X}$ contains addition), then $R\bar{X}$-sets are $\mathfrak{N}_1$-sums of $\bar{X}$-sets, while RX-sets are $\mathfrak{N}_1$-products of X-sets.*

Theorem VI. *If an X-operation on measurable sets always yields a measurable set, then the same is true for the RX-operation.*

The proof is fairly long but essentially is simply a generalization of the proof of the measurability of sets obtained by A-operation from measurable sets[6]

Theorem VII. *If to each point of the closed interval $[0,1]$ an operation X_τ is assigned, then there is an operation T replacing (see section IV, subsection 2 of the first part of this article, volume 1 of the Collected Works, No.13, p. 91) all the operations X_τ.*

A. Let us enumerate, using odd numbers, all the rational segments (with rational extremities). Let us call a *chain of numbers* U^T any infinite sequence of numbers $n_1, n_2, \ldots, n_k, \ldots$ satisfying the following conditions:

(i) the sequence of segments corresponding to odd elements of U^T converges to a certain point τ;

(ii) odd elements of U^T divided by 2 constitute a chain from the system of chains $\{U^{X_\tau}\}$ determining the operation X_τ.

B. Suppose $E + X_\tau\{E_n\}_n$. Put:

a) $\mathfrak{E}_{2n} = E_n$,

b) for the numbers $2n+1$ of a certain fixed sequence of rational segments (converging to τ) $\mathfrak{E}_{2n+1} = [0,1]$,

c) the other $\mathfrak{E}_{2n+1}$ are equal to the zero segment.

It is easy to see that under these conditions $E + T\{\mathfrak{E}_n\}_n$.

---

[6]See: N. Lusin, W. Sierpinski. *Sur quelques propriétés des ensembles* (A). - Bull. Acad. Sci. Cracovie, 1918, vol. 4, p. 35-48.

Theorem VIII. *If to each transfinite number of the second class there corresponds an operation, it is possible to find an operation which replaces them all.*

To do this it suffices to subdivide the segment $[0, 1]$ into $\mathfrak{N}_1$ sets P_α and assign to each point P_α the operation corresponding to α; then construct the operation P.

Concluding remark. If we are given a set, it is easy to present an operation which constructs it from segments; to do this it suffices to enumerate rational segments and call chain of the operation the family of numbers corresponding to sequences of segments converging to point of the set. By Theorem VII, one can effectively find, from a continual class of sets, an operation which yields all these sets from segments. But since the cardinality of the class of all operations is equal to the cardinality of the class of all point sets, one can think that from the point of view of the study of arbitrary point sets we gained little by reducing their study to the study of classes of sets generated by various operations of the type considered in this paper. The method proposed here is interesting when applied to special classes of sets generated by sufficiently simple operations.

February 1922.

(A. N. Kolmogorov)

AFTERWORD

The three books of my "Collected Works" proposed to the reader include basically all my work in mathematics, classical mechanics, the theory of turbulent motion, probability theory, mathematical logic and information theory. They do not include works related to teaching and to the history of mathematics, poetics and articles of a general character.

In certain directions what I have done seems to me sufficiently self-contained and conclusive, so that at my age of 82 I leave these results to my followers with a certain satisfaction.

In other directions the situation is different and what I have published appears to me as only being fragments of future works, about which I can only hope that they will be done by others. What has already been done in these directions is described in many cases in the commentary written by a group of my pupils, to whom I express my heartfelt gratitude.

January 25, 1985.

A.N. Kolmogorov

Printed in the United States
By Bookmasters